HEALING
THE
PLANET

PAUL R. EHRLICH

AND

ANNE H. EHRLICH

A Publication of the
Center for
Conservation Biology,
Stanford University

HEALING
THE
PLANET

Strategies for
Resolving the
Environmental
Crisis

A Robert Ornstein Book

Addison-Wesley Publishing Company, Inc.

Reading, Massachusetts   Menlo Park, California   New York

Don Mills, Ontario   Wokingham, England   Amsterdam   Bonn   Sydney   Singapore

Tokyo   Madrid   San Juan   Paris   Seoul   Milan   Mexico City   Taipei

GF
75
E4
1991

Many of the designations used by manufacturers and sellers to distinguish their products are claimed as trademarks. Where those designations appear in this book and Addison-Wesley was aware of a trademark claim, the designations have been printed in initial capital letters (e.g., Life Savers).

Library of Congress Cataloging-in-Publication Data

Ehrlich, Paul R.
    Healing the planet : strategies for resolving the environmental
crisis / Paul R. Ehrlich and Anne H. Ehrlich.
        p.   cm.
    "A Publication of the Center for Conservation Biology, Stanford University."
    Includes bibliographical references and index.
    ISBN 0-201-55046-6
    ISBN 0-201-63224-1 (paperback)
    1. Man—Influence on nature.   2. Environmental protection.
3. Environmental policy.   I. Ehrlich, Anne H.   II. Title.
GF75.E4   1991
363.7—dc20   91-2734
                                                                      CIP

Jacket design by Julie Metz
Text Design by Jennie Bush, Designworks, Inc.
Set in 12-point Bodoni Book by NK Graphics

1 2 3 4 5 6 7 8 9 -MA-95949392
First paperback printing, September 1992

TO

PETER

AND

HELEN

BING

WITH DEEP APPRECIATION

Contents

Mother Earth cannot heal herself alone. She needs our help. We two-leggeds must all come together and form a commonality of realization, a realization of potentially fatal calamities. Most of our remedies will be to cease, or drastically curtail, what we have been doing. Rising temperatures, vanishing rain forests, overpopulation, pollution of waters, and acid rain can be, and will have to be, addressed by abrupt remedies.

—ED McGAA, Eagle Man
Mother Earth Spirituality, 1990

PREFACE

I n January 1991, the United States, allied with several other countries under the banner of the United Nations, launched an attack on Iraq, which had illegally occupied and pillaged the small, oil-rich nation of Kuwait. America's interest in the affair was clear and boiled down to a three-letter word: *oil*. With its huge population (now over a quarter-billion), unprecedented affluence, and profligate use of environmentally damaging technologies, the United States is the world's most overpopulated nation in terms of its impact on Earth's fragile environment and its "need" for limited resources, although other developed nations also bear a major share of responsibility.

That overpopulation and overconsumption in rich nations are threatening global security was dramatically highlighted by the military action in the Persian Gulf. The war, undertaken to protect the access of the United States and other nations to Gulf oil, might never have occurred if American population growth had stopped in 1943 when the United States had 135 million citizens. No one has ever suggested a sane reason for having even that number of Americans, except a possible need for military manpower. And 135 million was clearly sufficient in that respect, since in 1943 America was winning the largest land war in history.[1]

Suppose the United States population had stopped growing when it reached 135 million people, and that Americans had gone on to become as wasteful of energy per person as they are now. Everything else being equal, the smaller population would not need to burn one drop of imported oil or one ounce of coal (the premier source of the greenhouse gas carbon dioxide).

Even small advances in energy efficiency in the United States would have sharply reduced the nation's interest in Middle Eastern military adventures. If Americans had continued the energy conservation programs started in the late 1970s, no imported oil from the Persian Gulf would have been needed in the 1990s. If the American private automobile fleet's average fuel efficiency had just been increased from 19 to 22 miles per gallon, no oil would be needed from Iraq or Kuwait. If average fuel efficiency were increased to 31 miles per gallon (well below half that of peppy, nimble, safer automobiles already designed, and just a little better than the 28 miles per gallon of today's new cars), no oil from Saudi Arabia or any other Persian Gulf country would be required.[2]

The connections of the Gulf situation to the environment are indirect but excellent examples of the complex webwork of the human predicament. Unwise environmental policies were central to creating the Gulf crisis, and it in turn caused enormous environmental destruction through the greatest oil spill of all time in the Gulf and the setting afire of more than 500 oil wells in Kuwait. Smoke from the fires was causing severe respiratory illness among Kuwaitis soon after the fighting stopped, as well as acid precipitation and sooty fallout hundreds of miles downwind. The fires and spill cleanup will pose problems for years.

Additional indirect damage was caused by the temporary jacking up of energy prices. Throughout the developing world, increases in the costs of liquid fuels, since the oil shocks of the 1970s, have been a major factor in increasing the unsustainable use of fuelwood. Two decades ago when oil was cheap, kerosene began replacing wood in the stoves of poor families.[3] Then, as the price of kerosene rose, people turned again to local forests and woodlots for fuel. Sadly, the impact of the Gulf confrontation on the plight of the poor in other regions was largely ignored, as has been the environmental Hobson's choice embodied in either using more oil or destroying more forests.

Yet the Gulf confrontation and war were just recent dramatic symptoms of a much deeper problem. Although too few people realize it, the gigantic and still-expanding scale of human activities has already set the stage for much vaster environmental disasters. A substantial portion of the life that shares Earth with us is now doomed to go extinct. Partly as a result, a billion or more people could starve in the first few decades of the next century, hundreds of millions of environmental refugees could be created, the health and happiness of virtually every human being could be compromised, and social breakdown and conflict could destroy civilization as we know it.

But, unlike the stunning rock-and-roll that rivets everyone's attention at the onset of a major earthquake or the ominous deluge that presages a flood, the approach of general environmental (and social) collapse is slow and insidious. Human beings have not evolved either biologically or culturally to interpret the gradual trends—stretching over decades—that are harbingers of planetwide catastrophe. They have a hard time coming to grips with what has become known as "the human predicament."

But the signs are there for those who can read them. In our everyday lives they intrude as traffic jams, air pollution alerts, increasing numbers of homeless people, wars first over water (1967) and then over petroleum (1991) in the Middle East, and so on. The signs can be seen even more clearly in statistics on the size and reproductive patterns of human populations, losses of soil to erosion, depletion of groundwater supplies, and deforestation. They can be found in readings taken by instruments measuring concentrations of carbon dioxide, methane, chlorofluorocarbons, and other trace gases in the atmosphere. They are manifest in the appearance of an ozone "hole" high above Antarctica, and hinted at by unusual weather events such as Hurricane Hugo and the drought of 1988.

The signs are also visible in trends in the distribution of income, rising health and insurance costs, and in countless speeches of politicians and economists about maintaining economic growth. They are implicit in society's preoccupation with the "health" of the economic system and people's failure to realize that the latter is utterly dependent on the health of the natural ecosystems—composed of plants, animals, and microorganisms interacting with their physical environments—that support the economy.

For those who learn to read the signs, the message is that the human enterprise is rapidly outstripping the capacity of Earth to nurture civilization. The carrying capacity of the planet for human beings is threatened virtually everywhere by degradation and depletion of such vital resources as topsoil, underground water, and the diversity of plants, animals, and microorganisms. Human alteration of the temperature and chemical composition of atmosphere, soil, and oceans stresses the ability of crucial sectors of society's life-support apparatus to supply food, water, firewood, and other necessary amenities to human populations. So does the outright destruction of natural systems to make way for burgeoning numbers of people.

Yet the signs are also telling us that our society's leaders, with a few outstanding exceptions, do not get the message. For most presidents, prime ministers, and religious leaders, and those who advise them, business as usual is the goal. They consider the environment just one more problem area kept on the political agenda by special-interest groups. Environmental deterioration is viewed as a long-term problem, and foresight is not a prime characteristic of governments or most other social institutions. No agency in the United States is charged with overall assessment or planning for the next decade, let alone the next century.

Society's difficulty in evaluating its environmental peril has long been recognized. Aldo Leopold wrote almost a half-century ago:

> One of the penalties of an ecological education
> is that one lives alone in a world of wounds. . . . An
> ecologist must either harden his shell and make believe
> that the consequences of science are none of his busi-
> ness, or he must be the doctor who sees the marks of
> death in a community that believes itself well and does
> not want to be told otherwise.[4]

But even at this late date, there remains a chance that civilization can alter its headlong rush toward the brink, that Earth can be healed, and that our descendants could live lives of plenty, dignity, and peace. How humanity might seize that opportunity is what this book is all about.

Healing the Planet is designed to help you read the danger signs. It is a companion volume to *The Population Explosion*, which focused on the population factor in our environmental dilemma. *Healing the Planet* was written to fill in the rest of the picture, to explain how overconsumption and the use of faulty technologies contribute to the deterioration of the environment, while not losing sight of the equally critical factor of overpopulation. Both books were written from the standpoint that the slowly building environmental crisis should be a unitary and overriding concern for all human beings.

Healing the Planet is not intended to be comprehensive, but rather to highlight what we consider to be the most critical environmental problems and suggest ways in which they can be alleviated. We do not delve into details of environmental economics or possible mechanisms of social and political change. Our treatment of issues puts emphasis on the threats to Earth's life-support systems we judge to be most critical at the moment. And our perspective is largely that of the developed world, especially of the United States. The environmental plight of developing nations and our interdependence with them are touched upon but not examined exhaustively. We believe these are crucial topics, and have tried to indicate how they relate to the material covered, but their detailed analysis is for other books and authors.

The goal of *Healing the Planet* is to give you a starting framework for considering what needs to be done and which personal actions you might take to help restore the integrity of our planet's environmental systems and thus secure the human future. We want to convince you to tithe to society; to dedicate at least 10 percent of your time to solving the human predicament before it's too late.

Are We Saving the Planet?

I n 1990, twenty years after the first Earth Day,[1] America celebrated another big one. Rock bands and celebrities entertained people in many American cities; hundreds of thousands turned out on the Mall in our nation's capital. Should it have been a celebration? Had humanity spent two decades rolling back the threat of environmental deterioration? Were we more secure than we had been in 1970? Had sufficient progress been made in raising ecological consciousness and turning that consciousness to good use? The answer is complicated, but we're afraid that, in sum, the Earth Day 1990 news was not good. The human predicament was much more serious on the twentieth Earth Day than when the first wave of environmental consciousness crested, and much less time remained to take corrective action. Today the situation is even worse.

That is not to say that the first Earth Day was a waste of time and effort. Far from it; the environmental movement of the 1960s and 1970s created a valuable legacy. It immensely raised the environmental consciousness of the nation and the world, and led to an institutionalization of environmental concern in business and governments. Before the movement, the United States had no Environmental Protection Agency, no Council on Environmental Quality, no National Environmental Policy Act, no Endangered Species Act, no strong national clean air or water acts, and no Superfund for toxic wastes. Most of the few environmental laws at state and local levels were weak and uncoordinated in standards and requirements.

The heightened public awareness and governmental

1

action after 1970 undoubtedly slowed environmental deterioration somewhat—and could have accomplished much more but for the Reagan administration's success in the 1980s in promoting environmental destruction for immediate gain. In spite of the Reagan years, though, air pollution is less severe than it would have been without the Clean Air Act. Pesticides are less carelessly used, and some of the most dangerous persistent ones have been banned in the United States (but not everywhere). Recycling is widely accepted in principle and increasingly practiced. The heedless destruction of biotic diversity caused by government projects has been diminished. Many people are refusing to buy furs, and the importation of endangered species or products made from them is prohibited in the United States and many other nations. Even politicians like George Bush at least give lip service to the environment. Considering how slowly social changes usually occur, catapulting the environment onto the national agenda—and keeping it there— was a monumental achievement for the movement symbolized by Earth Day.

Why, then, are we so negative in our evaluation of the situation following Earth Day 1990? The basic reason is that the environmental movement and the response to it has focused too strongly on the *symptoms* of environmental deterioration and too little on the underlying *causes*. The result is like administering cough syrup to a tuberculosis patient; the coughing may abate, but the untended disease may yet kill.

The symptoms of our environmental illness include smog, beaches awash with sewage, overflowing garbage dumps, acid rain–killed lakes, pesticides on apples, toxic substances in drinking water, and virtually continuous famines along the southern fringe of the Sahara in Africa, each of which has individually attracted much attention since Earth Day 1970.

This focus on symptoms was epitomized by Earth Day 1990, an occasion that capped a year-long effort during which too little was heard about the basic disease: the deterioration of the life-support systems of the planet and the consequent lowering of its "carrying capacity," Earth's long-term ability to support people. Too little also was heard about the deep and pervasive causes of this loss of carrying capacity: overpopulation and continuing population growth, rising per-capita consumption, economic growthism, governmental bungling and inaction, the use of inappropriate technologies and the inequitable distribution of wealth and resources.

The misplaced emphasis was encapsulated in the most popular book published for Earth Day, *50 Simple Things You Can Do to Save the Earth*.[2] All the suggestions in the book were useful, from ways to use less water on lawns and stop junk mail to supporting recycling programs and conserving energy in homes. But almost all dealt with symptoms. If those suggestions constituted the environmental agenda followed by every Earthling, Earth would still go down the drain. The single simplest and most effective thing anyone could do to save Earth was not even mentioned. That, of course, is to have no more than two children.[3] And, as we will see, there are many ways in addition to limiting the size of your family that you can do to help attack the basic causes of environmental deterioration.

Simple Things exemplified a basic reality of our society. Today, even many of the people dedicated to saving the planet (or more precisely, saving its ability to support *us*) have no clear understanding of the basic nature of the human predicament. They do not recognize that the human economy is supported by an array of services supplied free by natural ecosystems, without which we cannot survive but which we are heedlessly dismantling. So it is hardly surprising that, wishing to create the best possible world for themselves and their children, they focus their energies on the sorts of actions recommended by *Simple Things*. But to create a world that will support civilization far into the future, they should shift their focus to the more general problem of assuring the continuance of ecosystem services.

These services, which occupy our attention throughout much of this book, include: the maintenance of a benign mix of gases in the atmosphere; moderation of the weather; regulation of the hydrologic cycle that provides fresh water in a manner that minimizes the occurrence of floods and droughts; the generation and preservation of the fertile soils that are essential for agriculture and forestry; the disposal of wastes and cycling of nutrients; control of the vast majority of agricultural pests and organisms that can cause disease; pollination of many crops; provision of forest products and food from the sea; and maintenance of nature's vast "genetic library," from which humanity has already drawn the very basis of civilization. The disruption of vital ecosystem services by the ever-growing human enterprise should be our number one environmental concern, not whether pesticides, smog, or nuclear wastes are going to give people cancer or birth defects.

The proliferation of carcinogens in the environment is, of course, a legitimate concern: cancer, especially when it strikes the young, causes great personal tragedies. Similarly, other, more subtle forms of human poisoning should certainly concern us, especially when the effects are passed to future generations. Low-level lead-poisoning effects on the very young, leading to permanent disabilities, are as worrying as the lasting effects of childhood malnutrition. The challenges that lie ahead in the next generation or two will demand much of humanity, and a burden of handicaps, even minor ones, won't help.

Still, although there is ample reason to be concerned about the possibility of direct and severe damage to human beings, the greatest threat to ourselves and future generations is the indirect one posed by the dismantling of natural ecosystems. Curing all cancers, after all, would extend the average life expectancy by only a little over three years,[4] and the subtle poisonings may lower fertility or the quality of life for some individuals. But impairment of Earth's life-support systems could subtract a *decade or two* from the average person's life— and much of the quality to boot.

The continued destruction and degradation of natural ecosystems will affect most human beings primarily through declines in food security and conflicts over scarce resources, and secondarily through the enhanced vulnerability of huge hungry populations to diseases like AIDS.[5] Famine, war, and plague are much more to be feared than cancer.

A root cause of environmental deterioration is human numbers. When *The Population Bomb* was published in 1968, there were 3.5 billion people on Earth and the population was growing by some 70 million per year.[6] Some people claimed that our fears that the population explosion would degrade the human environment, lead to massive starvation, and generally diminish the quality of life—concerns shared by many of our colleagues—were unfounded.[7] Technological advances, such as nuclear agro-industrial complexes,[8] would easily allow 5 to 10 billion people to be given adequate diets, housing, medical care, social security, employment opportunities, and so on. Their basic message was comforting: keep on having lots of kids and encourage each one to consume more and more; Science (with a capital *S*) will save us from any untoward consequences of that behavior.

Our response to this technological optimism was always the same: Why not stop population growth as soon as possible and see whether humanity can properly care for the number of people present when growth ends? Once everyone has some version of the "good life," then the utility of further increasing the human population could be discussed. Could we have a bigger population without declines in per-capita standards of living? Could human numbers keep growing without degrading the future carrying capacity of the planet? What would be gained or lost if the population were to increase further or to shrink?

For the last 23 years, humanity has followed the easy course. In 1991 the world had 5.4 billion people to support, not 3.5 billion; and the population increased by roughly 95 million people, not 70 million. At least 200 million people have died of hunger and hunger-related diseases since the *Bomb* was written, and more than a billion and a half people lack the most elementary prerequisites for a healthy life—clean water and adequate sanitary facilities—at least half a billion more than were so handicapped in 1968.[9] According to UNICEF (the United Nations Children's Fund), 14 million children die annually from causes related to environmental degradation—to what Mustafa Tolba, Director of the United Nations Environment Programme (UNEP), calls "the shambles of global environmental destruction."[10] *Fewer than 1.5 billion people, less than half the number alive in 1968, have yet achieved a standard of living that most Americans (or citizens of other rich nations) would find acceptable.* In short, the test has been run. Most of the technological (and sociopolitical) miracles imagined by the optimists have not materialized. Furthermore, none seems in prospect.

What has materialized, however, is a series of nasty environmental surprises. In 1968, ecologists had no inkling of how swiftly tropical moist forests would be chopped down; the role of freons (chlorofluorocarbons—CFCs) in destroying the ozone layer had not been discovered; acid rain was essentially unknown; and global warming, with its potential for massive disruption of agriculture, was considered a possible problem for late in the twenty-first century. The theoretical threat of novel virus diseases invading an ever-larger population containing more and more hungry (and thus immune-compromised) people in an era of rapid intercontinental transportation had been recognized, but the reality of the AIDS pandemic was still more than a decade in the future.

Indeed, the only major surprise that was not entirely nasty was the success of the "green revolution" in increasing food production in poor nations such as India. That, however, was a mixed blessing, since higher yields often have been obtained at the cost of depleting irreplaceable soils, ancient groundwater, and the genetic variability essential for producing new strains to meet changed conditons. The worldwide expansion of food production now supports *2 billion* more people, but the gains achieved at the expense of nonrenewable supplies of soil and underground water may well turn out to be temporary in many areas, including India. As anyone doing a household budget knows, you can always increase consumption today by eating into your capital. Such behavior, however, has direct consequences for consumption tomorrow.

One might expect that the optimists' school would have developed some humility from the experience of the 1970s and 1980s, but far from it. In fact, it still has a few members who generate the same sorts of claims as were heard in the late 1960s. Miracles of science or social organization will permit humanity to take good care of the 10 to 14 billion people now (optimistically) projected as the population size when growth ceases. Roman Catholic bishops have even assured everyone that 40 billion people could be fed,[11] and assorted "experts" (notably a journalist and a specialist in mail-order marketing) have written books and filled the pages of the business and right-wing press on the basic theme that there are great advantages to unlimited population growth.[12]

Proponents of such views are fortunately becoming rarer, but that does not mean that the general public or decision makers are well informed on the dimensions of the population-resource-environment crisis. Even after Earth Day 1990, most people still don't recognize that the entire globe is overpopulated by a simple standard: *the human population can no longer be supported on income*—primarily the continuous daily flow of energy from the sun.

Instead, humanity is living largely on capital, on a one-time inheritance of deep agricultural soils, Ice Age groundwater, and biodiversity—the microorganisms, plants, and animals that are working parts of those vital ecosystems. Fossil fuels and high-grade mineral ores are also part of our inheritance, but their exhaustion is neither so immediate nor so potentially disastrous. Rather, the environmental costs of using fossil fuels and mining extremely low-grade ores will bring those activities to a halt long before the supplies come close to running out. The truly crucial issues of resource availability are the declines of water supplies, soils, and biodiversity.

Moreover, not only is *Homo sapiens* not supporting itself on income, but its population is so large that it is now thoroughly fouling its only home. Thus many resources that should be available have been rendered unusable.

The role of overpopulation and population growth in causing environmental deterioration is summarized in the equation: $I = P \times A \times T$.[13] The impact (I) of any group or nation on the environment can be viewed as the product of its population size (P) multiplied by per-capita affluence (A) as measured by consumption, in turn multiplied by a measure of the damage done by the technologies (T) employed in supplying each unit of that consumption. The $I = PAT$ equation shows immediately that, all else being equal, doubling a population's size will double its impact on the environment. And if, through great effort, individual impact $(A \times T)$ is halved while the population doubles, the total impact will remain the same.

All else is not equal, however. The $I = PAT$ equation is a simplification, because P, A, and T are not independent of one another. For instance, as population (P) increases, the damage done per person by the technological systems that support consumption (T) tends to increase as well. To see why, consider the natural tendency for people to use easily accessible resources first. Resources to be used by each additional person must on average be mined deeper (or extracted from poorer ores or forests with smaller trees), transported from more distant sources, grown on land requiring more mechanical cultivation, irrigation, more synthetic fertilizers or pesticides, and so on. Each person added to the population thus ordinarily has a disproportionately heavy impact on the environment compared to that of those who came earlier. More technological effort must be expended to supply each unit of their consumption—in other words, T increases with P.

By 1993, 2 billion more human beings will have been added to the world population since the first Earth Day—an addition roughly equal to the *total* human population of the early 1930s. Moreover, demographic projections make still another doubling to 11 billion or more appear inevitable *if* (a big if) death rates can be kept low. The United States population will have grown by more than 20 percent in the same period—the 50 million additional people being about the same number as the total U.S. population in 1880.

Demographic statistics are easy to find, but direct measures of affluence and the environmental damage done by particular technologies are not assessed by governments. Fortunately, however, national statistics are kept on per-capita use of energy, which can be a useful (if rough) index for A × T (AT).[14] After all, obtaining resources—iron, copper, aluminum, coal, petroleum, wheat, corn, timber—both requires the use of energy and causes environmental damage. The processes of mobilizing and using those resources contribute to pollution of air, water, and land, and to a general assault on ecosystems.

The roles of affluence and the technology that supports it in causing environmental deterioration can be seen when per-capita energy consumption is equated with AT. By this measure, an average American has roughly 50 times the impact of an average Bangladeshi.[15] Few Bangladeshi families own a couple of cars, a refrigerator, an air-conditioned and heated home, a TV set, and a VCR. They don't fly in jet planes, receive piles of junk mail, eat vegetables grown with an enormous fossil-fuel subsidy, consume fast-food hamburgers made of beef grown on pastures carved from the tropical rain forests of Central America, or in winter munch fruits grown in the Southern Hemisphere. They don't destroy natural ecosystems to build shopping malls, freeways, or golf courses (which are soused with inorganic fertilizers and pesticides). And they don't mine and transport huge amounts of ores or peteroleum. All these activities consume vast amounts of energy, and all contribute substantially to environmental destruction.

Measured by commercial energy use, each American, on average, causes some 70 times as much environmental damage as a Ugandan or Laotian, 20 times that of an Indian, 10 times that of a Chinese, [16] and roughly twice that of citizens of Japan, the United Kingdom, France, Sweden, or Australia.

Americans use about 50 percent more commercial energy than Soviet citizens, (who nonetheless have caused even more havoc by using it with minimum efficiency and virtually no effort to prevent environmental damage). In terms of per-capita energy use, only Canada, Luxembourg, and a few oil producers are really in our league. Since the United States also has a large population, our *total* impact is gigantic—about 100 times that of Bangladesh.

These numbers starkly illuminate the fallacy of the notion (common even among environmentalists) that population problems are restricted to poor nations. While rapid population growth in developing countries is a major factor in keeping them poor and does contribute to environmental destruction, the still-increasing overpopulation in rich nations poses the truly overwhelming threat to Earth's life-support systems.

Viewed in this light, the United States is the world's most overpopulated nation. It is the world's fourth largest nation in population, now numbering more than a quarter-billion people, and the average American consumes more of Earth's riches than an average citizen of any of the other "big ten" nations with more than 100 million people: China, India, the Soviet Union, Indonesia, Brazil, Japan, Nigeria, Bangladesh, and Pakistan. Furthermore, the technologies used by Americans to support their opulent consumption are environmentally damaging and needlessly wasteful of energy. Because of this combination of a huge population, great affluence, and damaging technologies, the United States has the largest impact of any nation on Earth's fragile environment and limited resources.

The I = PAT equation also underlines the perils of continually pressing for more of today's sort of economic growth, especially in rich nations. If the P and A factors in the equation both continue to expand, there is no hope of reducing I. Many people understand that the population cannot increase indefinitely but fail to realize that the same applies to consumption (affluence). Economic growth, both total and per capita, has limits as well. Some people are under the impression that the technology (T) factor can be made to shrink toward zero, allowing both population (P) and affluence (A) to grow indefinitely. Unfortunately, there are limits to how much T can be shrunk— as will be explained in the chapters that follow.[17]

Certainly, T can be shrunk significantly by cleverly

redesigning technologies, and growth can be shifted to sectors of the economy where its impact on the environment is reduced: from industrial production to services, for instance. In some cases, growth can occur by substituting information for materials, as when a computer is used to control the fuel/air mixture, make an automobile engine more efficient, and thus conserve gasoline. But the opportunities for such shifts are limited, even for a nongrowing population. People in service jobs must have tools and equipment to work with. They need (or will want) offices and shops, as well as personal possessions such as cars, appliances, and homes. Society cannot function with everyone simply providing services to one another. There are physical limits to the efficiency of machines, and even computers need to be manufactured. If the economy as a whole is expanding, the material sectors are likely to expand also.

While it is clear that further material growth in rich nations would cause greatly increased environmental impacts, it is also obvious that substantial material economic growth is vitally needed in poor nations. Yet even modest growth in developing countries would be magnified into enormous total impacts by the huge P factors in their $I = PAT$ equations.

Consider only that most crucial of potential problems: a worldwide rapid change in climate that may result from the human-caused buildup of greenhouse gases in the atmosphere. The most important of these gases is carbon dioxide, but CFCs, methane, nitrous oxide, and many others of lesser importance also contribute to the problem.[18] Besides other serious consequences, rapid climate change poses critical threats to world agriculture. Within a decade or so, largely unpredictable changes in weather and climates may appear worldwide and become essentially continuous. Since keeping global agricultural production ahead of population growth is already becoming difficult, the prospect of climatic instability is not reassuring.[19]

Most of the greenhouse-gas buildup so far can be laid at the doorstep of developed nations, especially the carbon dioxide produced by burning fossil fuels. But, as poor nations develop, they inevitably will use much more energy per person, mostly by burning fossil fuels, unless reasonable alternatives are made available. Magnified by their large and fast-growing populations, their share of responsibility for global warming accordingly will rise sharply, regardless of policies undertaken in rich nations.

To illustrate this, let us look briefly at a situation that is described in more detail in Chapter 2. Suppose both China and India were quite successful in controlling their population sizes, and planned their development so that per-capita use of commercial energy doubled, rising only to small fractions of the level in the United States today. If India and China used their abundant coal deposits to power that development, *each* nation, when development was completed, would release more additonal carbon dioxide into the atmosphere than the United States could withhold by giving up all coal use and not substituting any other carbon-based fuel. Since coal now supplies nearly a quarter of United States energy consumption, giving it up would cause enormous economic dislocations. Nevertheless, even that sacrifice could not compensate for the atmospheric effects of limited coal-based development in either China *or* India—and certainly not in both![20]

Clearly, efforts must be made to help the poor nations modernize without vastly escalating their fossil-fuel use (or other environmentally destructive activities). New approaches to the task of "development" will certainly be required. For instance, small local fermenters to turn agricultural wastes into burnable gas and small-scale solar technologies will in many cases prove superior sources of energy in the less developed world to large dams and coal-fired power plants. Labor-intensive agroforestry to yield food for local consumption usually has much more potential to aid the poor than attempts to convert tropical forest lands to the production of food for export through grazing or industrial agriculture. Too often in recent decades, shifting from a traditional way of life has meant moving from a previously sustainable system (rendered unsustainable when rapidly expanding populations outstripped the available local resources) to one that would support many more people—temporarily.

By the same token, the rich countries must wean themselves from their addiction to fossil fuels. The United States especially should restore and strengthen the alternative energy research and development programs begun so promisingly in the 1970s and then abandoned. With help from the West, the newly liberated Eastern European nations and the Soviet Union also need to modernize their energy sectors, both to reduce consumption and to curb their horrendous pollution. Above all, the economic gap between rich and poor, within or between nations, must be narrowed if there is to be any hope of achieving the degree of international cooperation that will be necessary to solve the massive problems humanity now faces.

To preserve the environment that supports our civilization, much more than catalytic converters, bricks in toilet tanks, recycling of newspapers and aluminum cans, increasingly efficient air conditioners, and spasmodic attempts to save pandas and rhinos will be required. While the symptoms must be treated, it is most important to attack the disease. The emphasis now needs to be put on reducing the scale of the whole human enterprise by shifting to sustainable development in *both* rich and poor nations—including moving toward population *shrinkage* in both. Growth of the physical economies in rich nations— growth that in the past greatly improved the lives of their citizens—must now be recognized as the disease, not the cure. People must also recognize that the *quality* of life in those nations can be increased even if the standard economic indicator, gross national product (GNP), declines. Achieving environmental security should be recognized as a top priority for all societies. These unpopular (and to the average economist, politician, and Sunday TV show pundit, utterly unacceptable) notions need to be converted into common wisdom. Then everyone will truly be an "environmentalist."

That will be a monumental task. Many people still expect miraculous technological advances to save their children's future. Our view remains unchanged since the post–*Population Bomb* debates: Let's do everything possible to halt population growth before disaster overwhelms us and see if everyone can be properly cared for. If everyone can, humanity could then determine whether further population growth could be sustained or whether *shrinkage* would be a better alternative. We believe abundant evidence already points to the latter.

In fact, we fear that time is very short; the fate of Earth as an abode for a large-scale technological civilization is likely to be determined by the turn of the century, even though many of the bills for failure will come due later. By 2000, people everywhere must understand the perils inherent in promoting rapid climatic change and allowing acid deposition, ozone depletion, soil erosion, overdrafts of groundwater, and the loss of biodiversity to go unchecked. It is also imperative that they grasp the limits to agricultural expansion and humanity's growing vulnerability to epidemic disease, and see the interrelationships of these phenomena with the decay of critical ecosystem services, the declining long-term carrying capacity of the planet, population growth, economic growth, and a welter of social problems.

By the year 2000, Earth's leaders should be pressing for the kinds of global cooperation that will be required to avoid catastrophic social and economic decline—cooperation that is unlikely to be achieved unless the plagues of racism, sexism, religious prejudice, xenophobia, and hideous poverty can simultaneously be reduced. The environmental movement itself should emphasize that these too are environmental problems and that building a sustainable world means tackling them also.

Humanity now faces both unprecedented catastrophe and a last opportunity to bequeath to all our descendants a decent planet to live on. With the end of the cold war, the financial and human capital required to accomplish that goal could be made available. We know how to put our capital to use, since more than enough is understood scientifically about the environment to know the directions in which civilization should be moving.

America now needs a leader with the courage to institute change: someone who at least senses the deepening human predicament and understands that business as usual won't get us out of it. The decade of the 1990s must be a decade of environmental *politics*—for if American, indeed global, politics cannot soon be "greened," humanity will have blown its last good chance.

The first step a concerned citizen should take is to become informed; we hope this book will help. The next steps, after becoming informed, are to take personal actions to lighten your own impact on Earth's life-support systems—steps outlined in *Simple Things* and the rest of the abundant Earth Day literature. But, beyond taking measures at the personal level, which are important in giving a feeling of participation and achievement, it is crucial that concerned citizens become politically involved as well. Remember, all your personal efforts will come to nothing if society fails to limit population size, curb runaway consumption, deploy environmentally benign technologies, protect biodiversity, and so on. Citizens all have responsibilities in seeing to it that our society's total impact is reduced. Finally, each of us can encourage our government to cooperate with other nations in designing and working for a better future for all peoples.

We believe firmly that our society can lighten its pressure on life-support systems while *increasing* the quality of life for Americans, and we believe it can be done while preserving a market-based economic system (with some important improvements over today's, however). It won't be easy, and there will be substantial costs, but the benefits to be reaped are immeasurable. Some benefits we'll enjoy in the near future; others may be reaped by our children and grandchildren. But their future is the point of it all, isn't it?

Our Life-Support Systems

Every country can be said to have three forms of wealth: material, cultural and biological. The first two we understand very well, because they are the substance of our everyday lives. Biological wealth is taken much less seriously. This is a serious strategic error, one that will be increasingly regretted as time passes.

E.O. Wilson[1]

The sun is our principal source of income. If you understand the impact of the sun's energy on the biosphere, the thin layer at Earth's surface that supports life, you will begin to grasp the most important features of the human environment. Doing that, in turn, is the first step toward seeing how human activities threaten ecosystem services on a global scale, and how weakening them will degrade the quality of human life.

First, of course, the sun warms Earth to a range in which the chemical constituents of living beings can perform their functions. Indeed, lizards, insects, and other "cold-blooded" animals often depend directly on the radiant energy of the sun to warm them and allow them to function. In other words, the sun creates physical conditions that make life possible. Second, through the process of photosynthesis, the sun provides the energy that powers the lives of almost all organisms, including you.

The sun keeps Earth's surface warm with the help of various substances in the atmosphere, especially water vapor and clouds, carbon dioxide, and methane. These and a variety of other "greenhouse" gases have the characteristic of being largely transparent to the sun's incoming rays, solar radiation

15

of short wavelengths. About a quarter of that radiation is reflected back to space by clouds and dust in the atmosphere,
which also absorb another quarter. Another 3 percent is reflected
by Earth's surface. But almost half is absorbed by the surface
and then reradiated toward space as long-wavelength (infrared)
radiation.[2]

If that were all that happened, the average surface temperature would be about $-18°$ Celsius ($0°$ Fahrenheit). Earth
would be a very cold place indeed. Instead, its surface temperature is some $15°$ C ($59°$ F)[3] because clouds and water vapor,
carbon dioxide, and several other gases absorb the outgoing
infrared radiation and reradiate it—some toward outer space,
and some back toward Earth's surface. This downward reradia-
tion is what keeps Earth warm enough for life. It is known
colloquially as "the greenhouse effect."[4]

The sun's energy does much more than keep our planet
warm; it also drives the weather. Equatorial regions get more
solar radiation than the poles, and the temperature difference
between the equator and the poles is the basic driving force.
Hot air rises; as it does so it is cooled and moisture is squeezed
out of it, producing the very moist climates in which tropical
rain forests thrive. After air heated near the equator rises and
loses much of its water content, it travels poleward at high
altitudes. Then it sinks, dry and warmed by its descent, to
create deserts at about $30°$ north and south latitudes. Beyond
the desert belt, the poleward transfer of heat depends mainly
on large cyclonic systems, which bring storms to the mid-
latitudes. These are two key elements in global circulation patterns that create what we call "climate": the average weather
in a given location.

The climatic machinery is complex, made so by Earth's
rotation (Coriolis deflection) and irregularities such as mountains
that change the course of winds and oceanic circulation, which
can cool or warm adjacent land. The climatic system is so
intricate that predicting its future course with precision is not
possible even with supercomputers.[5]

Another vital function of the sun's energy is as "fuel"
for photosynthesis. That is the process by which green plants
and some microorganisms bind solar energy into chemical bonds
of carbohydrate molecules (sugar, starches, cellulose). Plants,
animals, and microbes can then use that chemical energy to

drive their life processes, mostly by combining it with oxygen in a slow burning known as cellular respiration, or just plain respiration for short. The vast majority of non-photosynthesizers—animals, fungi, and most microorganisms—must obtain their energy from photosynthesizers, in most cases by eating them. Only a small fraction of organisms gain their energy from chemical reactions other than photosynthesis and so can exist in the absence of sunlight and without feeding on other life-forms.

With minor exceptions, therefore, biotic communities are both kept within appropriate operating temperature ranges and fueled by sunlight. A biotic community is simply the collection of plants, animals, and microbes that live together in an area. But they are more than a random collection; these varied organisms interact and depend on each other directly and indirectly. The nature of any community is determined by three things: the local climate (including underground water flows and characteristics of the water in aquatic communities), the area's soils, and its history—including the sequence in which different kinds of organisms colonized it. These factors are not independent of each other. The nature of the climate helps determine what sort of soil develops from the underlying parent rock; both climate and soil also influence which organisms can colonize where and when.

Members of biological communities interact continuously with their nonliving surroundings, and the interacting complexes are what biologists call ecosystems. Every kind of organism exchanges gases with its physical environment. The rose bush in your garden takes in carbon dioxide (CO_2) and gives off oxygen when the sun shines (and the reverse at night). Indeed, all photosynthesizing plants remove carbon dioxide from the atmosphere and water from the soil, and use the carbon from the carbon dioxide and the hydrogen from the water to build carbohydrates. The excess oxygen is released to the atmosphere. In contrast, you and other animals take in oxygen and carbohydrates (as well as other molecules necessary for life), and give off carbon dioxide, water, and heat. The latter are the exhaust products of respiration; plants also produce CO_2 as they respire and use it in their photosynthesis.

Rooted plants remove a steady stream of water from the soil and release it into the atmosphere as water vapor. The volume of this water flow, which holds plants without woody stems upright and prevents wilting of the leaves of trees and shrubs, is little appreciated. A single corn plant with a dry weight of a pound at maturity transfers some sixty gallons of water from soil to atmosphere during its lifetime of a few months. The amount of water that a single rainforest tree returns to the atmosphere in its lifetime of 100 years or more is truly prodigious—on the order of 2.5 million gallons.

Plants also help to break apart rocks and form soil, and change patterns of low-level winds (as anyone who has moved from an open meadow into a woodland on a windy day can attest). Various organisms, especially bacteria, help run vast chemical cycles in which elements such as carbon, nitrogen, sulfur, and phosphorus circulate on a global scale.

The interdependence of the biological and physical worlds can be seen in the story of how our distant ancestors migrated ashore from the sea. Until perhaps 450 million years ago (about one tenth of Earth's age), life was confined to the oceans. Then in what, geologically speaking, was a relatively short period—perhaps 40 million years—plants, arthropods (insects and their relatives), and amphibians (ancestors of frogs and salamanders) colonized the land. That sudden emergence from the deep was made possible by the activities of photosynthesizers in the oceans.

The first photosynthetic bacteria appeared in the sea 3 billion years or more before the land was occupied. Oxygen, remember, is a byproduct of photosynthesis, and all the oxygen in Earth's oceans and atmosphere was put there by that process. It gives one pause to consider that the life-giving gas most of us think of as the most important constituent of the atmosphere, one that makes up a full fifth of it, was all generated by living beings over billions of years.

Ozone is a special type of oxygen molecule formed of three (rather than two) oxygen atoms. It is formed in the stratosphere when ultraviolet radiation from the sun splits a normal O_2 molecule and one of the resultant atoms latches on to another O_2 molecule to form a molecule of ozone, O_3. Ozone absorbs solar radiation in a portion of the ultraviolet part of the spectrum known as UV-B.[6] That is lucky for life on land, since UV-B is extremely damaging to life and no other atmospheric molecule effectively blocks it out (UV-B does not penetrate water beyond 15 to 60 feet, depending on the clarity of the water).[7]

It took photosynthesizers billions of years to enrich the oceans and then the atmosphere with enough oxygen (and thus ozone) to create an ozone shield in the stratosphere protecting Earth's surface from most of the incoming UV-B. So early organisms in the oceans modified the physical world in a critical way—by giving Earth an oxygen-rich atmosphere. This enabled living things, including our distant ancestors (those amphibians), to leave the sea. Finally, ozone, a product of chemical processes in both living and nonliving systems, is itself an important greenhouse gas in the troposphere (lower atmosphere), influencing Earth's surface temperature and climate.

Because of the crucial importance of interactions between them, the living and nonliving portions of the biosphere can be viewed as two components of a single worldwide ecosystem. Ecologists consider the entire biosphere to be an ecosystem, and they view local biotic communities and the physical environments with which the organisms in the communities interact as ecosystems as well.

Two kinds of ecosystems are crucial to the functioning of human society today. The first kind is agricultural ecosystems. Their importance to society is obvious; we deal with those systems later. For the moment, just remember that they are simplified versions of natural ecosystems, artificially maintained by humanity to increase the production of commodities people need and desire. The importance of natural ecosystems is much less widely appreciated, but society depends upon them every bit as much as it depends on agricultural ecosystems. That is true in large part because agricultural ecosystems are embedded in natural ones and depend on the natural components for their productivity.

▬ Ecosystem Services: Climate and Water

Natural ecosystems provide civilization with a wide array of essential services delivered free and, in most cases, on a scale so large that humanity would find it impossible to substitute for them.[8] The preservation of Earth's millions of other life-forms is crucial to *Homo sapiens* because of their intimate involvement in the delivery of these services.

An essential ecosystem service is maintaining the gaseous composition of the atmosphere. That composition is always changing, as illustrated by the planetwide buildup of oxygen in oceans and atmosphere over billions of years of Earth's history. Ecosystems prevent changes in the mix of gases and particulate matter from being too rapid.

The slow accumulation of oxygen shifted the dominant life-forms of Earth from "anaerobic" microorganisms that obtained their energy without using oxygen to "aerobic" forms like us and all other animals and all plants. Organisms that use oxygen to "slow-burn" the energy-rich carbohydrate molecules formed in photosynthesis have long since become the dominant form. Those fascinated by the "Gaia" hypothesis that Earth is a single, organismlike, self-adjusting system whose life-forms are always improving the physical conditions in which they exist, should contemplate the fate of the anaerobes. No Gaia was keeping herself a happy home for them—they started out at the top of the heap, but many were poisoned out by accumulating oxygen. Today their descendants survive in obscure, oxygen-poor environments such as swamps, hot sulfur springs, and termite guts! That life has played a major role in shaping conditions near Earth's surface is indisputable; that Earth itself is "alive"[9] is indisputably wrong.[10]

But evolving natural ecosystems have generally kept climatic changes sufficiently gradual that life-forms could adapt to them.[11] The exceptions during Earth's long history were five events that caused sudden mass extinctions, such as the one that exterminated the dinosaurs and many other kinds of organisms 65 million years ago. All were probably due to catastrophic climatic changes. They may have resulted from huge meteor strikes or volcanic eruptions that suddenly changed the atmosphere's composition by injecting vast amounts of dust into the stratosphere. By blocking the sun's light, the atmospheric dust would have dramatically cooled the planet's surface.

The world's climate appears to have undergone other less catastrophic but relatively quick shifts from one pattern to another quite different one in response to continuous slow change in the basic forces that drive it, including small changes in Earth's orbit around the sun, pushing it over some threshold.[12] The cycling of the Ice Ages is the best-known example. Today, the emergence of *Homo sapiens* as a global force shows signs of disrupting the climate-control ecosystem service, and the prospect of rapid climatic change in response to increasing concentrations of greenhouse gases is of great concern to scientists.

The organisms in natural ecosystems influence the climate in ways other than their roles in regulating atmospheric gases. The vast rain forests of Amazonia to a large degree create the moist conditions required for their own survival. Water vapor from the Atlantic condenses into rain that falls on the eastern Amazon basin, is returned to the atmosphere by the vegetation as water vapor, and then condenses as rain again farther west. That moisture is recycled many times as it travels inland. In the western part of the basin, almost 90 percent of the rain is falling for at least the second time, and quite likely a third or fourth time.[13]

The moist climate in which the rainforest vegetation of the Amazon (and perhaps of Zaire) thrives depends to a large degree on the water-recycling function of that vegetation. This could have important consequences for the cutting and burning of the Amazon rain forest that has been so much in the news for the past few years. Many biologists believe that a critical threshold of deforestation may be reached beyond which the remaining forest will no longer maintain the climate necessary for its own persistence.[14] After that point, the loss of the entire forest (in anything like its present form) will be inevitable.

Deforestation and the subsequent drying of the climate could have serious regional effects in Brazil outside of Amazonia, conceivably reducing rainfall in important agricultural areas to the south. The degree of impact on global climate is uncertain, but substantial changes in the "albedo," or reflectivity of the region, are likely and could cause a significant change in the amount of solar energy absorbed by the entire planet. Such changes in albedo are taking place over much of Earth's land surface, and significant local or regional shifts in temperatures and rainfall patterns have been observed. Whether these yet add up to changes that have global climatic consequences is at the moment unknown.

Natural ecosystems provide another service related to their climate-control functions, that of regulating Earth's hydrological cycle. That cycle moves gigantic amounts of water continually. For instance, about 100,000 cubic *miles* of water are evaporated from the oceans every year, of which about 90 percent falls as rain back into the oceans and some 10,000 cubic miles are carried by winds over land where the water falls as rain, sleet, and snow. The same amount returns to the sea as runoff (so the continents are not soaking up water from the seas). An additional 15,000 cubic miles of water are evaporated from land and fall again on land as precipitation.

In addition to the role that plants play in the cycle by transferring to the atmosphere water that would otherwise flow back into the sea, they also affect patterns of runoff. Trees in forests break the force of falling rain and, at the same time, hold soil in place with their roots. Forest soils are thereby capable of soaking up precipitation, releasing it gradually in streams and springs, or percolating it downward into aquifers (water-bearing rock strata). When a watershed is deforested, the formerly steady flow of surface water is disrupted; rainwater runs off the surface rather than sinking in, leading to an alternation of floods and droughts downstream.

Such a change has occurred in the Parc National des Volcans in Rwanda, home of one of the few remaining populations of mountain gorillas. The park's forest covers only one percent of the nation, but is nonetheless the sponge that absorbs and meters out about 10 percent of that grossly overpopulated nation's agricultural water. The park used to be much larger, but in 1969 about two fifths of it were deforested for a pyrethrum-growing scheme (pyrethrum is a natural insecticide extracted from daisylike flowers). That reduction in forest area led to the drying up of several streams. With luck, the rest of the forest will remain intact, since the pyrethrum scheme failed, while showing the mountain gorillas to tourists has become a major source of foreign exchange.

Rwandan gorillas have already lost much of their forest habitat to the inexorable demands of a human population growing at 3.4 percent per year (a rate that, if it persisted, would double the population in 20 years). Rwanda is already almost entirely cultivated outside its few parks and cities today, and its rivers are reddish brown with topsoil washing from farms on steep slopes. By 2020, when Rwanda is projected to have a population density comparable to that of Bangladesh today (without the rich delta lands that help sustain that Asian nation), it will *really* need its agricultural water. It is ironic that the mountain gorillas may help to maintain it by providing a reason to protect the forests that provide the crucial water-metering service!

Ecosystem Services: Soils, Nutrients, and Wastes

Rwanda, however, is only a dramatic, contemporary example of a worldwide phenomenon that has been repeatedly observed since ancient times. Greece, Italy, the eastern and southern coasts of the Mediterranean, and the Middle East were well-watered, mountain-forested, richly productive regions 2,000 to 3,000 years ago. A series of brilliant civilizations rose and declined over millennia. But deforestation, followed by centuries of overcultivation, overgrazing, and soil erosion, gradually impoverished the land and helped to undermine the civilizations.[15] Today the entire region is far more arid and less productive. The exceptions are Egypt, where soils were replenished annually by the floods of the Nile, and Israel, where decades of costly and difficult restoration work have made the desert bloom.

The generation and maintenance of soils are two more services supplied by natural ecosystems. Soils are much more than ground-up rock; they are themselves complex ecosystems with a rich flora and fauna.[16] The living components of soil ecosystems are crucial to their fertility—to their ability to grow crops and forests. Charles Darwin learned that lesson more than a century ago when he intensively studied the habits of earthworms, which are extremely important because they loosen soil and allow oxygen and water to penetrate it. Other animals that help give soil its texture and fertility include insects, mites, and millipedes.

The abundance of these animals is difficult to comprehend for anyone who has not spent time sorting them from soil and studying them. In the soils under each square yard of forest in North Carolina were found an estimated 30,000 of these tiny creatures, three quarters of them mites (miniature relatives of spiders, ticks, and scorpions). But that's nothing. Under a square yard of pasture in Denmark, the soil was found to swarm with as many 40,000 small earthworms and their relatives, nearly 10 million roundworms, and over 40,000 insects and mites. And the number of animals in soils pales in comparison to the number of microorganisms. A gram (less than a twenty-fifth of an ounce) of forest soil has been found to contain over a million bacteria of one type, almost 100,000 yeast cells, and some 50,000 bits of fungus. A gram of fertile agricultural soil may contain over 2.5 *billion* bacteria, 400,000 fungi, 50,000 algae, and 30,000 protozoa.

But it isn't the numbers of soil organisms that makes them so important to us; it is the roles they play in soil eco-systems. Among other functions, microorganisms are involved in the conversion of the nutrients nitrogen, phosphorus, and sulfur into forms usable by the higher plants that we depend upon. Many green plants enter into intimate relationships with special kinds of soil fungi. The plants nourish the fungi, which in turn transfer essential nutrients into the roots of the plant. In some forests where trees appear to be the dominant organisms, the existence of the trees is utterly dependent upon the activities of these fungi. On farms, other microorganisms play similar critical roles in transferring nutrients to crops like wheat.[17]

Organisms are very much involved in the production of soils, which starts with the "weathering" (wearing away by the elements) of underlying parent rock. Plant roots can fracture rocks and thus help generate particles that are a major physical component of soils; plants and animals also contribute CO_2 and organic acids that accelerate the weathering process. More important, small organisms, especially bacteria, decompose organic matter (shed leaves, animal droppings, dead organisms, etc.), releasing carbon dioxide and water into the soil and leaving a residue of tiny organic particles, resistant to further decomposition, that make up the key soil component known as "humus." Humus particles help maintain soil texture and retain water. They play a critical role in soil chemistry, permitting the retention of nutrients essential for plant growth.

Organisms are therefore crucial to the maintenance of

soil fertility, and they also are the principal actors in soil conservation. The roots of plants help hold soil in place, slowing
erosion by water and wind. In undisturbed ecosystems, the rate
of soil loss is usually balanced by that of soil formation. (Both
rates are ordinarily measured on a time-scale of inches per
millennium.) But if plant cover is removed, as when an area is
deforested or overgrazed, soils start to disappear fast. Animals
are involved in soil preservation, too, as they disperse many
seeds and thereby often speed the revegetation of denuded areas.
To ecologists, one of the saddest sights in developing countries
is the color of many rivers, brown with silt, which indicates a
hemorrhaging of a prime constituent of any country's natural
"capital."

Soil ecosystems are the main providers on land of two
more essential ecosystem services: disposal of wastes and cycling of nutrients. When organic matter, be it dung, the fallen
branch of a tree, or a dead mouse, reaches the soil surface,
representatives of that vast category of soil flora and fauna called
"decomposers" invade and devour it. It soon disappears, gone
and forgotten. But, of course, the organic matter is not really
gone—it has just been broken down into simpler constituents
that in turn serve anew as nutrients (for example, carbon, hydrogen, oxygen, nitrogen, phosphorus, sulfur), which are essential to the growth of green plants.

Not all terrestrial decomposers are minute creatures in
the soil. Hyenas, jackals, California condors, and even grizzly
bears aid in the process of disposing of wastes by consuming
carcasses of other animals. In water, bacteria are important
decomposers whose waste-disposal capabilities are harnessed
by humanity in sewage treatment plants. But most of the action
is in the soil. In some cases, the nutrients are taken up more
or less directly by plants close to where the decomposers do
their work. In others, the products of decomposition may circulate through the global ecosystem in vast "biogeochemical
cycles" before returning to the soil and being reincorporated
into a living plant.

Among natural ecosystems, soils are one of those most
taken for granted. This neglect is both sad and potentially disastrous, because soils are among the most threatened of those
systems worldwide and perhaps the one most vitally needed for
civilization's persistence.

▄▄▄ Biogeochemical Cycles

Biogeochemical cycles are exemplified by the movement of two elements that are necessary components of all living things: carbon and nitrogen. Carbon travels through the atmosphere or oceans as CO_2 until it is taken up by a photosynthesizing plant or microorganism. Carbon can be thought of as existing in a series of "pools" through which it circulates at varying rates—the global carbon cycle.[18] One pool is terrestrial "biomass"—the bodies of all organisms living on land. That pool, roughly 600 billion tons of carbon, is about 100 times as large as the pool of carbon incorporated in the bodies of marine organisms. Recent estimates fix the size of the pool of carbon existing as dead organic matter on land as about one and a half to two times larger than that of living biomass. The pool of carbon in dead organic matter in the oceans is similar in size to that in terrestrial biomass. The amount of carbon stored in recoverable fossil fuels (all fixed by photosynthesis in the distant past) is about ten times as much. The largest pools of carbon exist in oil shale and as dispersed carbon in sediments; combined, they amount to 20,000 times the carbon in terrestrial biomass.

Flows between the pools of carbon dioxide in the atmosphere and oceans and in living and dead organic matter were for a long time approximately in balance; photosynthesis removed about as much carbon dioxide from the inorganic atmospheric and oceanic pools as plant and animal respiration and decomposition returned to them. But now the balance has been shifted because the combustion of fossil fuels and the cutting and burning of forests is adding carbon to the atmospheric pool considerably faster than natural systems can remove it. Indeed, deforestation is subtracting a major source of uptake. As a consequence, carbon dioxide is accumulating rapidly in the atmosphere. The significance of this anthropogenic perturbation of the carbon cycle will be made clear in the discussion of global warming in Chapter 3.

The nitrogen cycle is very complex. Nitrogen (in the form of nitrate or ammonia) in the soil may be recycled immediately by uptake through plant roots; or it can be released into the atmosphere as a by-product of the activities of "denitrifying" bacteria, which change nitrate into nitrous oxide and nitrogen gas. Nitrogen gas makes up some 78 percent of the atmosphere, but plants cannot use this critical nutrient in that form. Instead, it must be made available as ammonia and nitrate through a process of "biological fixation" carried out by several kinds of bacteria in soil or aquatic ecosystems. These organisms can combine atmospheric nitrogen and water to produce ammonia and oxygen. The best known nitrogen fixers are bacteria that live in nodules on the roots of legumes (plants of the pea and bean family) and fix nitrogen for their hosts in return for sugars supplied by the plants. The ammonia can then be used by plants to make amino acids, the building blocks of proteins.

The nitrogen cycle, like the carbon cycle, is being perturbed by humanity. Tens of millions of tons of fixed nitrogen are added to soils each year in the form of inorganic nitrogen fertilizers—about as much as is fixed by the bacteria associated with legumes. This nitrogen is fixed by the industrial process (the Haber process) invented by the Germans during World War I, when they needed nitrate for explosives and were cut off from conventional supplies. Overall, human activities now lead to the fixation of an amount of nitrogen comparable to that fixed by natural processes—one more indication of the degree to which humanity has become a global force.

Human interference in the global nitrogen cycle has several consequences. Since nitrate is easily leached from soils by water, inorganic nitrogen fertilization has greatly increased problems of nitrogen pollution of lakes, rivers, streams, and groundwater. It also has increased the flow of nitrous oxide, an important greenhouse gas, into the atmosphere. Fixed nitrogen from fossil-fuel combustion reaches the ground as nitric acid, an important ingredient of acid rain.

Human beings also seriously interfere in the natural cycling of phosphorus and sulfur. But the basic points about these geochemical cycles are illustrated by the carbon and nitrogen cases: first, organisms play vital roles in moving these nutrients around the biosphere; second, humanity is now a major perturbing force in these gigantic cycles.

Pest Control and Pollination

Another critical service provided by natural ecosystems is control of the overwhelming majority—an estimated 99 percent—of pests and diseases that might otherwise attack crops or domestic animals.[19] Most of the potential pests are herbivorous (plant-eating) insects, and the control is provided primarily by predaceous insects that consume them. This service has been disrupted, sometimes spectacularly, by the misuse of artificial insecticides, because insect pests are generally less susceptible to pesticides than are their predators. The populations of pests tend to be large (that's why they're considered pests) and so have a better chance of evolving resistance to the pesticides.[20] Herbivorous insects also have long been engaged in a coevolutionary race with plants. Plants have evolved many deadly compounds in attempts to poison their attackers, compounds familiar to you as the active ingredients of spices (such as pepper, cinnamon, or cloves), many drugs (marijuana, cocaine, opium), and medicines (aspirin, digitalis, quinine). In turn, the insects have evolved resistance to these poisons and thus often are preadapted to dealing with the insecticides we develop to poison them.

The bottom line is that repeated heavy application of insecticides kills off predaceous insects much more effectively than the pests. The latter quickly become resistant to the pesticides and often thrive unless dosages are continuously escalated or different insecticides substituted. Meanwhile, other herbivorous insects, previously not counted as pests because their populations were small, may be relieved of pressure from their predators, and their populations may explode. They are then "promoted" to pest status. Insecticide resistance has been documented in some 450 species of insects and mites, and is considered one of the most serious threats to both agriculture and public health, the latter because of resistance in malarial mosquitoes and disease carriers (vectors).[21]

One example of this failure of the pest-control service of natural ecosystems was the promotion of spider mites to the status of serious pests in many areas of the world when overuse of DDT and other synthetic pesticides killed off their natural insect predators.[22] Ill-advised use of pesticides against the fire ant, a nasty pest imported from South America to the southern United States, has had similar results. Despite numerous warnings from biologists since the 1950s,[23] pesticide spraying has decimated the natural enemies of the fire ant and allowed it to thrive and spread. Professor E. O. Wilson of Harvard, the preeminent authority on ants, has called the attempts to eradicate the fire ant by massive aerial spraying "the Vietnam of entomology." Mostly, those programs have simply disrupted the natural ecosystemic pest controls that would have helped keep the ant in check. The fire ant has now spread out of the South, where it has long been established, and has occupied a beachhead in Santa Barbara, California. It is expected eventually to reach Oregon and Washington.[24]

While natural ecosystems are providing crop plants with stable climates, water, soils, and nutrients, *and* protecting them from pests, they also are pollinating many of them. Although honeybees, essentially domesticated organisms, pollinate many crops, numerous others depend on pollinators from natural ecosystems. One such crop is alfalfa, which is most efficiently pollinated by wild bees.

▬ Direct Benefits

Natural ecosystems, of course, also directly provide people with food—most notably with a crucial portion of the protein in our diets in the form of fishes and other animals harvested from the seas. This service is provided by the oceans in conjuction with coastal wetland habitats, which serve as crucial nurseries for marine life that is either harvested directly or serves as a food supply for sea life that we eat.

And finally (but not exhaustively), natural ecosystems maintain a vast "genetic library" from which *Homo sapiens* has already withdrawn the very basis of civilization and which promises untold future benefits. That library of millions of different species and billions of genetically distinct populations is what biologists are referring to when they speak of biotic diversity, or biodiversity. Wheat, rice, and corn (maize) were scruffy wild grasses before they were "borrowed" from the library and developed by selective breeding into the productive crops that now form much of the feeding base of humanity. All crops, as well as domestic animals, have their origins in the library, as do about a quarter of all medicines and various industrial products, including a wide variety of timbers.

The potential of the genetic library to supply more of the same is still largely untapped.[25] Recently, scientists have found among the lowly fungi, which gave humanity penicillin and cyclosporin A (the latter used routinely by surgeons to guard against rejection of organ transplants), another medically useful compound. Gliotoxin shows promise of providing a way to make transplated organs "invisible" to the body's immune system while not compromising their other functions.[26] Its use could relieve transplant patients of the dangers of taking drugs (like cyclosporin A) that suppress the immune system and protect the transplant but also expose the patient to a serious risk of infection. Gliotoxin also has characteristics that may make it a powerful tool in designing anticancer drugs.

Substituting for Ecosystem Services

As should be apparent by now, living organisms in natural and agricultural ecosystems play enormous and critical roles in making Earth a suitable habitat for *Homo sapiens*. They have already stored enough oxygen in the atmosphere for us to breathe for thousands of years even if no more were produced;[27] they supply all our food (directly or indirectly), and they help to keep the climate equable and fresh water flowing steadily. Furthermore, these services are provided on such a grand scale that there is ususally no real possibility of substituting for them, even in cases where scientists might know how to do so.[28]

Not that people haven't tried, sometimes with a measure of success, at least initially. In developing the productive agriculture of the North American Midwest, humanity, with apparent success, has substituted corn and wheat for perennial prairie grasses (plants whose vegetative parts survive several winters and which reproduce over several summers). But the crops are annuals (they go through a complete generation each year, starting from seed), and annuals do not develop the extensive root systems of perennials. They therefore do not participate in the soil-generating service of ecosystems to the same degree as perennials; soil nutrient stores are gradually depleted, and soil itself is more readily eroded away. Inorganic fertilizers are used to replace some important nutrients, but they contribute little toward maintaining the structure of soil or its component microorganisms. Whether the depth and fertility of the prairie soils can be maintained indefinitely under cultivation remains to be seen. So far, the signs are not encouraging.[29] Meanwhile, native prairie grasses that might be essential elements in restoring more productive pastures in the Midwest are barely hanging on in places like cemeteries and railroad embankments.

The loss of ecosystem services following deforestation is especially rapid and dramatic. Ecologist F. H. Bormann explained the substitution dilemma as follows:

> We must find replacements for wood products, build erosion control works, enlarge reservoirs, upgrade air pollution control technology, install flood control works, improve water purification plants, increase air conditioning, and provide new recreational facilities. These substitutes represent an enormous tax burden, a drain on the world's supply of natural resources, and increased stress on the natural system that remains. Clearly the diminution of solar-powered natural systems and the expansion of fossil-powered human systems are currently locked in a positive feedback cycle. Increased consumption of fossil energy means increased stress on natural systems, which in turn means still more consumption of fossil energy to replace lost natural functions if the quality of life is to be maintained.[30]

The loss of the "genetic library" service is particularly severe when tropical rain forests are cleared, and crops, pastures, scrub, or other types of vegetation substituted for them, since those forests are home to somewhere between 50 and 90 percent of all of Earth's species (distinct kinds of organisms).[31]

In fact, one could conclude that virtually all human attempts at complete or large-scale substitution for ecosystem services are ultimately unsuccessful, whether it be substitutions of synthetic pesticides for natural pest control, inorganic

fertilizers for natural ones, chlorination for natural water puri-
fication, or whatever.[32] Substitutes generally require a large
energy subsidy, which adds to humanity's general impact on
the environment. And most substitutes are not completely satisfactory even in the short run.

In sum, there is little to suggest that humanity will be
able to substitute adequately for the ecosystem services that will
be lost as the epidemic of extinctions now under way escalates.
And escalate it seems bound to do. No one knows for certain
how fast genetically distinct populations and species of other
organisms are vanishing, but all biologists who deal with the
problem know the rates are far too high and are rising.

▄▄▄ The Extinction Epidemic

How do they know? It's simple. First, they're watching
them go. All field biologists have watched the flora and fauna
fading away before their very eyes. Coral reefs on which we
studied the behavior of fascinating fishes have been destroyed
by the sewage from "love boat" cruise ships. Many places where
we once studied butterflies have been converted to freeways,
parking lots, or farm fields. We've searched in vain for once-
abundant frogs in Costa Rica. In the last hundred years, ichthy-
ologists have seen 27 species of freshwater fishes go extinct in
North America.[33] Ornithologists watch in distress as populations
of many forest birds of the eastern United States decline rapidly.

That evidence, however, is anecdotal. More important
and more "scientific" evidence is what biologists know: that
organisms are highly adapted to their habitats.[34] Many eastern
warblers require extensive tracts of forest to maintain their populations; the neon tetras so prized by aquarists will breed only
in acid water (which trout cannot breed in); Bay checkerspot
butterflies require certain plants for their caterpillars to eat, and
those plants require certain kinds of soils to grow on. The list
is endless—populations of organisms are honed by evolution to
thrive in their home environments.

Thus, if a habitat is dramatically changed, most or all
the plants, animals, and microorganisms that once inhabited it

will be wiped out. Humanity today is on a rampage of changing natural habitats dramatically: cutting them down, plowing them up, overgrazing them, paving them over, damming them and diverting water, sousing them with pesticides and acid rain, pouring oil into them, changing their climates, exposing them to increased ultraviolet radiation, and on and on. The rate of destruction of tropical forests almost doubled in the 1980s.[35]

Consequently, ecologists know that Earth's biota (all living things) is being slaughtered at an accelerating pace, but it is not possible to count populations and species as they vanish.[36] For one thing, the true extent of biodiversity is unknown. Estimates of the total number of existing species range from an extremely conservative 2 million (some 1.4 million have been described and given latinized names) to well over 50 million.[37] Assuming there are 10 million species, more or less, and that on average each species consists of several hundred genetically distinct populations, one can easily postulate the existence of billions of populations.[38]

How fast is this diversity now disappearing? Although it is impossible to say with precision, the answer clearly is "frighteningly fast."[39] More than a decade ago, we estimated that mammal and bird species were going extinct 40 to 400 times as fast as they normally have since the great extinction spasm that did in the dinosaurs and many other life forms 65 million years ago.[40] In 1989, Harvard's Craoord Laureate ecologist, E.O. Wilson, conservatively estimated the annual extinction rate at 4,000 to 6,000 species, some 10,000 times the "background" rate before *Homo sapiens* started practicing agriculture. It is conceivable that the rate is actually 60,000 to 90,000 species annually—150,000 times background.[41]

Of course, biotic diversity is constantly generated by a natural process that eventually creates new species. That process of the differentiation of populations (speciation) normally operates on a time scale of from thousands to millions of years. All estimates of present-day extinction rates show them to be vastly higher than the rates at which the natural process that creates biodiversity could be expected to compensate for the losses.[42] The extinction "outputs" far exceed the speciation "inputs," and Earth is becoming biotically impoverished because of it.

To biologists, perhaps the most frightening data pointing

to the urgency of dealing with the extinction problem are those relating to the human impact on the planet's total supply of energy produced in photosynthesis—global net primary production.[43] Net primary production (NPP) is the energy fixed by photosynthesis, minus that required by the plants themselves for their life processes.[44] One can think of NPP basically as the total food supply of all animals and decomposers. Almost 40 percent of all potential NPP generated on land is now directly consumed, diverted, or forgone because of the activities of only one of millions of animal species—*Homo sapiens*. Although the human impact of NPP in oceanic ecosystems is very small (about 2 percent), that on land is so huge that we appropriate altogether about 25 percent of global NPP.

Human beings use NPP directly when they eat plants or feed them to domestic animals and when they harvest wood and other plant products. Human beings divert NPP by altering entire systems, redirecting NPP toward human ends, as when natural ecosystems are converted to cropland or pasture. And people *reduce* potential NPP by converting highly productive natural systems into less productive ones: tropical forests to pastures; savannas and grasslands to deserts; deciduous forests and prairies to farms; and farms to homes, shopping centers, and parking lots.

Since the great majority of the world's species (probably over 95 percent) now exist on land, the 40-percent human appropriation or loss of NPP there goes far to explain the extinction crisis. The amount of energy available to support the millions of other kinds of animals on Earth clearly has been drastically reduced. Plant diversity, too, is reduced because much less land, especially land with suitable soils and climates, remains to support plant growth outside human-controlled or degraded areas. One "probably conservative" estimate made on the basis of this reduction of available energy is that 3 to 9 percent of Earth's species may be extinct or endangered by 2000, an estimate in the same ballpark as the higher ones above.[45] If current accelerating trends continue, half of Earth's species might easily disappear by 2050.

The amount of terrestrial NPP available to accommodate further expansion of the human enterprise is not that great, considering that humanity has already taken over some 40 percent and the human population is projected to double in the next half-century or so. Yet expectations are for massive economic growth to meet the needs and aspirations of that exploding population. One important international study, the *Brundtland Report,* advocated a five- to tenfold increase in global economic activity in the next several decades in an effort to eliminate poverty.[46] What a substantial expansion of both the population *and* its mobilization of resources implies for the redirection and further loss of terrestrial NPP by humanity is obvious: we'll try to take over all of it and lose more in the process.

Harvard policy analyst William Clark was being extremely conservative when he wrote, "The implications of this desperately needed economic growth for the already stressed planetary environment are at least problematic and are potentially catastrophic."[47] Indeed, if anything remotely resembling the Brundtland population-economic growth scenario is played out, we can kiss goodbye to most of the world's biodiversity, and perhaps civilization along with it.

The ravaging of biodiversity, in our view, is the most serious single environmental peril facing civilization. Biodiversity is a resource for which there is absolutely no substitute; its loss is irreversible on any time scale of interest to society. But even the scientific community is not using its considerable influence to stem the tide of destruction.

We wonder how astronomers would react if they were told that, because of human action, over half of all celestial bodies would become inaccessible for investigation in the next fifty years. How would chemists feel if their ability to work with over 50 percent of all elements were to end in half a century? What if physicists by then could detect no more than half of all exisiting particles? We imagine there would be quite a fuss, even though the consequences for humanity of those events would be trivial compared to the projected loss of biodiversity. After all, those vanishing organisms won't merely be unavailable for study; they'll stop providing us with crucial ecosystem services.

Even many biologists are also complacent in the face of the destruction of biodiversity. Perhaps they have grown too accustomed to seeing their research subjects disappear, one by one; perhaps they work on only one common species; or perhaps, along with the rest of society, they have been persuaded that human "progress" and "development" are more important. Most likely, they have been trained to stick to their business and stay out of politics. Nevertheless, a substantial and growing number of biologists is taking a different view. They have decided the time has come to call a halt before the impoverishment of biodiversity swallows up all the progress. These biologists, from academics to resource managers, have banded together to found the new discipline of "conservation biology."[48]

The goal of conservation biology is to bring scientific expertise to the rescue of biodiversity—by developing the scientific tools needed for the job, by disseminating those tools to managers; and by educating scientists, decision makers, and the general public to the critical problem of saving biological resources. The field has been developing very rapidly, and involves nonbiologists (for example, anthropologists, earth scientists, engineers) who have joined in the attempt to halt the extinction crisis. An international organization, the Society for Conservation Biology, was formed in 1985 and grew rapidly to some 3,000 members in many nations—a measure of the concern for biodiversity that had developed in one segment of the global scientific community.

Conservation biologists do research on a wide variety of problems, from the best way to design nature reserves and the potential impacts of global warming on biodiversity to how to avoid genetic problems when raising endangered animals in zoos for later release and how to control invasions of exotic pests.[49] They are often concerned with finding ways to optimize the use of the very limited resources that are available for conservation. As a group, they recognize that conservation biology is, as ecologists Jared Diamond and Robert May put it, a "discipline with a time limit."[50] Conservation biologists have been very active in developing plans for global conservation of biodiversity and seeking ways to implement them.[51]

If conservation biology does not achieve significant success within the next decade or so, it will be too late. But it is comforting to realize that, for the first time in history, an entire scientific discipline has sprung up in response to a rapidly developing problem that threatens all of humanity—a discipline with the explicit goal of overcoming the problem in a comparatively short time.

We've been especially happy about this unprecedented development, because our first major conclusion about how to heal this beleaguered planet was that every effort must be made to stanch the hemorrhage of biodiversity.

    Tactics to accomplish this will be rooted in words penned by the great ecologist Aldo Leopold four decades ago: "If the biota, in the course of aeons, has built something we like but do not understand, then who but a fool would discard seemingly useless parts? To keep every cog and wheel is the first precaution of intelligent tinkering."[52] Conservation biologists are now struggling with the tactics of saving those cogs and wheels. In this volume, we focus on the essential strategy for saving biodiversity and ourselves—a strategy we will return to repeatedly: *reduce the scale of the human enterprise.*

Energy

and

the Environment

I f we consider the human assault on our planet, what sorts of images come to mind? Factory or power-plant stacks belching smoke (and invisible CO_2). Clogged freeways with myriad tailpipes belching smoke (and invisible CO_2). Bulldozers mowing down vegetation to make way for shopping malls or temporary cattle pastures. Birds on Alaskan beaches coated with oil from the *Exxon Valdez*. Giant shovels strip-mining coal in Montana. Chainsaws felling thousand-year-old trees. Plastics pouring out of factories and into landfills. Dams submerging landscapes and destroying wild rivers. Cities ablaze with lights at midnight. Crop-dusters spewing synthetic pesticides as they skim over fields.

All these images have a common element—they suggest the profligate use of energy. Energy powers the factories, cars, airplanes, electric lights, bulldozers, and chainsaws. Power plants, dams, shovels, and oil tankers use energy in the process of mobilizing more energy. Large amounts of energy are used to manufacture plastics, CFCs, pesticides, refrigerators, chainsaws, freeways, and autos, among other widely used products.

Energy is at the center of our lives—if it is not continuously made available to the cells of our bodies, we die. Energy runs the ecosystems that support society. It is also central to the life of civilization; if industrial society did not use a great deal of energy, it would collapse. Not surprisingly, then, energy use is so central to the human assault on the environment that it can serve as a surrogate in the $I = PAT$ equation. In fact, it plays such a key role in causing Earth's environmental ills that we begin our consideration of those ills and their possible cures with an examination of energy impacts and energy options.

▬▬ Energy as a Measure of Impact: The Rich and the Poor

A pivotal problem in deciding how to heal the Earth is finding some way to assess the total human impact on the planet and determining how much of that impact is generated by various nations and activities. Unfortunately, governments don't gather statistics specifically designed to measure their nations' impacts on Earth's life-support systems. Someday they might, but most government officials haven't yet realized that their environmental security is much more threatened—and much more important in safeguarding their people's well-being—than their military security.

But given the lack of regularly gathered statistics on impacts on the global ecosystem, we need some other way to estimate the Affluence times Technology factors in the Impact = Population × Affluence × Technology (I = PAT) equation. Fortunately, as we discussd in the Introduction, at least rough comparisons between countries can be made by using a surrogate statistic: total energy consumption.[1] One can simply then divide that number by P, the population size, to get per-capita energy consumption of a country to use as a surrogate for the impact of an average citizen of that nation on Earth's life-support systems. Unfortunately, though, per-capita energy use is far from an ideal proxy for per-capita impact on the environment (A × T) because not all forms of energy supply have equal environmental impacts.

A much more serious problem with using statistics on per-capita energy use as a measure of environmental damage is that those most commonly gathered record only commercial energy consumption. Thus they neglect the use of "traditional" energy sources (fuelwood, crop wastes, and dung), which comprise roughly 12 percent of energy use globally.[2] But in the poorest nations, traditional sources supply 50 to 95 percent of energy use, while in the richest they make up only about 2 percent (mainly firewood and cogeneration of heat and electricity from agricultural and forestry wastes). Very often fuelwood use is unsustainable, accompanied by deforestation and all the attendant environmental ills. So statistics on commercial energy consumption significantly understate both per-capita energy use in poor nations and its environmental impacts. How energy use in rich nations damages the environment seems evident, but the situation in poor nations deserves a closer look, which we give it later in connection with the burning of wood and other biomass.

The average American uses about 11 kilowatts (kW) of commercial energy, while citizens of poor sub-Saharan African nations average only about .03 kW (= 30 watts).[3] For comparison, an averge human beings's life processes use energy at a rate of about 100 watts, yielding roughly as much heat as a 100-watt light bulb. A kilowatt is 1,000 watts. If one assumes that the poorest nations get as much as 90 percent of their energy from traditional sources, and that half of that is on a sustainable basis and makes a negligible contribution to global problems, then the global environmental damage caused by the average person in Mali or Burkina Faso is about that done by the use of .165 kW of commercial energy. Under those assumptions, the average American does about seventy times as much damage to the global environment as a poor African. This numerical precision should not be taken too seriously, however, because of the qualitative differences in impact per unit of energy.

Another factor in poor countries that isn't adequately captured by the per-capita energy-use rule of thumb is the cutting and burning of tropical forest on a massive scale in many nations. This destruction is contributing substantially to the atmospheric buildup of greenhouse gases leading to global warming. Other severe environmental problems result from this deforestation, of course, including the loss of biodiversity and degradation of land. In this case, the environmental damage clearly has global effects, although as always the local people, mostly poor, will suffer the most direct and immediate consequences. The global effects of deforestation, however, are generated only by developing regions with significant remaining tropical moist forests, especially Brazil, Zaire, and Indonesia, as well as other parts of Central America, Central Africa, and Southeast Asia.

Even though the use of per-capita commercial energy consumption as a rough index of AT somewhat overstates the difference between rich and very poor countries, this does not invalidate the basic point that a rich person contributes much more to the wrecking of Earth's life support systems than does one living in poverty. We recognize that energy use is a very imprecise measure of the environmental impact of the average person in a society or of the differences between societies, but we do claim it provides a useful tool—the best available—for making quick approximations of the responsibility of different nations for global evironmental problems.

Even with appropriate caveats in mind, the picture painted in the Introduction of the environmental damage done by the average individual in nations at various stages of development still holds. The deleterious impact on Earth's life support systems of an average citizen of the superrich, inefficient United States is 50 or more times greater than that of an average citizen in a desperately poor nation.

▬ The Energy Trap

How can we translate all this into a picture of how the relationship between energy use and environmental destruction might change in the future? First, let's lump nations into two categories, rich and poor, and then explore some possibilities. Rich nations are defined as those with a per-capita gross national product (GNP) in 1989 of $4,000 U.S. or more.[4] The poorest of the rich nations were Greece and Taiwan, and the richest were Switzerland and Luxembourg. On average, each person in the rich nations used energy at a rate of about 7.5 kW; that is, over a year each individual was consuming about the amount of energy in 7.5 tons of coal. Some nations had much higher rates of average per-capita consumption. The average Canadian was using it at a rate of about 13 kW, and the average American at slightly more than 11 kW.

About 1.2 billion people were living in rich nations in 1990, so their total rate of energy use was 1.2 billion people times 7.5 thousand watts each, or 9,000 billion watts. A thousand billion is a trillion, and units of a trillion watts are called terawatts (TW), so the industrialized world altogether was using 9.0 TW.

The developing nations, those with per-capita GNPs of less than $4,000, ranged from Venezuela ($3,230 in 1989) to Ethiopia ($130). Although their average energy use was only 1.0 kW per person, some poor nations actually used more industrial energy per capita than some rich ones. For instance, the low-income nations Venezuela, Yugoslavia, South Africa, and North Korea all used more than 3 kW per person, while Israel, Spain, and Greece, among the rich, used slightly less.[5] A few oil-rich nations such as the United Arab Emirates have had extremely high per-capita energy use, but it is concentrated in the energy-production sector, and the rest of the economy is more typical of a poor country.

So the choice of standards to determine rich and poor is imperfect, although levels of energy use generally are highly correlated with accepted standards of wealth. There were about 4.1 billion people in developing countries in 1990,[6] and multiplying that number by their 1.0 kW average per-capita energy use yields a total for the developing world of 4.1 TW. The total energy use of both rich and poor nations around 1990 therefore was 9.0 TW + 4.1 TW = 13.1 TW. That 13.1 TW is our surrogate measure for the aggregate impact of humanity on Earth's ecosystems as we entered the 1990s. According to the I = PAT equation then, global impact [I] (total energy use of 13.1 TW) = a population of 5.3 billion [P] times the average energy use per person of 2.6 kW (Affluence × Technology) [A × T].

We know that, with the global environmental impact now being generated, the capacity of Earth's life-support systems to maintain civilization is already being severely compromised. Yet the human population is still growing by 1.8 percent annually, a rate that, if continued, would double it in less than 40 years. At the same time, it is clear that most poor nations (whose populations could double in only 33 years) must substantially increase their per-capita energy consumption if they are ever to be able to give their citizens a decent life.[7]

The combination of a population of nearly 11 billion consuming far more energy per person within a few scant decades clearly implies an enormous escalation of environmental impacts. For instance, suppose world population growth were halted at 11 billion people, while global average per-capita energy use increased to 6 kW (about 80 percent of the average for the developed world today, or half that of the United States or Canada). Under those circumstances, with everything else being equal, global environmental impact would quintuple.[8]

We might call this the "energy trap." In order to give all human beings a good life, including the 5 billion or so additional people projected to swell the population in the next half-century, it appears that vastly more energy must be used than today. Apart from the likelihood that developing and deploying energy sources on such a scale would be increasingly difficult and costly, using that much energy would pose an enormous threat to civilization's essential life-support systems.

The only way out of the trap, if indeed there is one, is to recognize that everything else doesn't have to be equal. Humanity could meet its future energy needs by seeking alternative sources and uses that are less environmentally destructive. After all, energy is basically a means to an end: a way of providing numerous comforts and conveniences, such as warm (or cool) homes, offices, and stores; illumination, appliances, transportation, and communications. The same conveniences might be provided through other, less damaging means, and society would be no poorer for it. New technologies doubtless often would cost more initially, but when the environmental costs of the old technologies are factored in, society might even make a net gain.

▬ The Holdren Scenario

Energy expert John Holdren, of the Energy and Resources Group at the University of California, Berkeley,[9] has developed an "optimistic" energy scenario for the planet, based largely on maximizing the efficiency of energy use, that might provide a way out of the trap. In his scenario, poor nations develop fast enough to increase their per-capita energy use by 2 percent per year between 1990 and 2025, doubling it from 1.0 to 2.0 kW (while their combined populations increase by two-thirds). Simultaneously, the rich nations go all out to *reduce* their per-capita use by 2 percent annually through increased efficiency, dropping their use per person from 7.5 to 3.8 kW (while maintaining or increasing benefits). Population growth in rich nations during this period will be less than 10 percent for the whole 35 years. Then, in the remainder of the next century, both rich and poor nations converge on an average per-person energy use of 3 kW. Meanwhile, the world population peak size of 10 billion people is reached around 2100, after which it begins to decline slowly. When the peak is reached, total energy use would be 10 billion × 3 kW, or 30 TW. Holdren's scenario is summarized in the following table.

The Holdren Scenario

		Population [billions]	Energy/Person [kilowatts = kW]	Total Energy Use [terawatts = TW]
1990	**Rich**	1.2	7.5	9.0
	Poor	4.1	1.0	4.1
		5.3		13.1
2025	**Rich**	1.4	3.8	5.3
	Poor	6.8	2.0	13.6
		8.2		18.9
2100		10	3	30

Holdren's optimistic scenario assumes both that population size can be limited to 10 billion and that a high standard of living can be achieved with a per-capita rate of energy use only one fourth to one third of that now seen in the United States. This assumption seems reasonable based on technologies already in hand. Indeed, depending on assumptions made about total energy use in the future—which technologies supply what fraction and for how long—efficiency itself could make available 10 to 40 TW by 2050. Holdren's scenario depends on increased efficiency "supplying" some 45 TW by 2100—the difference between the scenario's 30 TW and the 75 TW that would be required to give 10 billion people a life-style resembling that of the rich in the 1990s, fueled by 1990 technologies requiring 7.5 kW per capita.

Even with all that efficiency, of course, Holdren's scenario yields a total energy use more than twice that of 1990, a situation that would still produce catastrophic environmental impacts, unless the mix of energy technologies were substantially altered.

Fortunately, that the mix must be changed is already widely recognized.[10] The main thrusts behind this recognition are the clear limits to readily accessible supplies of petroleum and natural gas, and increasing public opposition to unacceptable environmental risks and tradeoffs (such as oil spills in fragile coastal or polar areas, or the sacrifice of prime farmland to strip-mine coal). Perhaps most important is the growing recognition of the connection between the present energy economy and the potential for global warming (discussed in detail in the next chapter).

In our view, the global warming problem alone means that the era of fossil-fuel dominance must soon be brought to an end. Coal, which releases almost twice as much carbon dioxide per unit of energy yielded as does natural gas (with petroleum falling between) should be phased out first.[11] Substantially reducing dependence on fossil fuels will require economic adjustments and considerable time to accomplish, since fossil fuels now provide nearly 80 percent of the world's energy. Coal itself drives more than a quarter of the world's energy economy.

What energy options are available, then, for playing out Holdren's optimistic scenario? First, suppose we look at some characteristics of today's patterns of energy use. Almost 90 percent of civilization's energy use is commercial; the rest is traditional (also called "biomass," since it is all from recently living organisms). Of the world's commercial energy, oil supplies 38 percent, coal 30 percent, natural gas 20 percent, hydropower 7 percent, and nuclear 5 percent. Of the biomass sources, perhaps 60 percent is fuelwood and 40 percent is crop wastes and dung.

Commercial energy use worldwide has been rising at about 1.9 percent annually in recent years, just slightly faster than the population growth rate of 1.7 to 1.8 percent. Since the late 1970's, energy use in developed nations has risen much more slowly than in previous decades, an averge of about 1 percent per year, largely because of improvements in energy efficiency in rich market economies, spurred by oil price rises. The increase in per-capita GNP in those nations (averaging about 2 to 3.5 percent annually in the 1980s)[12] at a much faster rate than energy use is an indication of the potential for maintaining or increasing prosperity while reducing energy use (and environmental impact).

We might guess that traditional energy use has been growing at slightly less than the population growth rate in developing nations; fuelwood is becoming scarce in many areas, and in the poorest regions (especially Africa), the rate of increase in agricultural production (and therefore of available crop wastes and dung) has fallen behind population growth. Overall, it seems fair to assume that per-capita impact on the environment $(A \times T)$ has remained essentially constant since about 1980, and that total impact measured in terms of energy use has grown more or less in step with population growth.

The fundamental pattern of energy use around the world is one of increasing dependence on nonrenewable rather than renewable resources or, putting it in economic terms, stock rather than flow resources. Stock resources—coal, petroleum, natural gas, and uranium—accounted for nearly 95 percent of commercial energy and almost 85 percent of all energy use by around 1990 (assuming half the fuelwood is not used sustainably and therefore must be considered a stock resource). Energy use clearly is a primary area where humanity is living on capital, not on income.

■■■ What Limits Energy Choices?

What are the constraints to increasing energy use? One common misapprehension can be cleared up immediately. There are no serious limitations on fossil-fuel supplies now or in the immediate future. At the 1990 rate of consumption of commercial energy, each of several fossil sources—petroleum liquids, conventional natural gas, and heavy oils—by itself could power civilization for more than 40 years (albeit with increasing costs and environmental risks). Coal alone could keep us going for over 400 years and oil shale for more than 2,400 years.

Supplies of nuclear fuels are also large. Uranium in standard nuclear reactors could maintain society for perhaps 250 years, and uranium in breeder reactors for at least 250,000 years.[13]

In other words, with respect to energy supplies, we have a great deal of capital, and, everything else being equal, we ought to be able to live on it for a very long time. But, as usual, everything isn't equal, While fuel supplies themselves might not be a problem, there certainly will be environmental and economic constraints on which and how much of those supplies can be used.

Today, short-run economic considerations, the failure of market prices to reflect the actual relative costs of different energy sources and the unrealistically low level of energy prices (which in general are far below social costs),[14] prevent society from fully exploring alternative energy choices—even though in the future enormous advantages would accrue from their development. In particular, the low price of oil since 1982, combined with lack of government initiatives, has substantially retarded the development and deployment of various solar technologies, despite their promise for delivering both long-term sustainability and relatively small environmental impacts.

The low oil prices, and the overproduction by OPEC nations behind it, ironically were partly a result of reductions in energy consumption in the United States from 1975 to 1985.[15] But the incentive for continuing efforts to increase the efficiency of energy use was killed by low oil prices, and momentum was lost. Meanwhile, government programs to encourage both energy efficiency and development of alternative energy sources were dropped by the Reagan administration, just as they were beginning to pay off. By mid-1990, the United States was again importing half its petroleum, domestic production was twenty years past its peak and falling, while consumption was rising rapidly. The oil import bill, already the biggest component of the trade deficit, soared even higher when oil prices rose after Iraq's invasion of Kuwait in 1990.

Interestingly, the lack of concern shown by Japan in supporting the United States and the United Nations coalition against Saddam Hussein was in part based on the much more efficient use of energy in their economy. Despite their greater long-term vulnerability because of an almost complete dependence on imported oil, the Japanese were confident of their

ability to withstand a temporary supply interruption or higher
prices.[16]

Another constraint on energy choices, one closely tied
to economics, is safety. Nuclear fission power was once supposed
to produce "electricity too cheap to meter," but it never happened. On the contrary, environmental and safety concerns—
among other factors—caused the costs to multiply severalfold.
Today, electricity from fission is roughly as expensive as that
generated with fossil fuels.[17] No new nuclear plants have been
commissioned in the United States since the mid-1970s, largely
because of the risks associated with generating power with nuclear fission. The risks include (roughly in order of increasing
seriousness) exposure of the public to routine releases of radiation, hazards associated with managing nuclear wastes for
tens of thousands of years, the danger of catastrophic accidents
(as at Chernobyl), and proliferation—the spread of nuclear
weapons.[18]

Guarding against these risks, especially of accidents, is
such an exceedingly expensive proposition that private insurance companies will not offer coverage that remotely approaches
the possible losses. As a result, the majority (about 50 to 95
percent) of insurance against large-scale nuclear accidents in
the United States today is provided by the taxpayers,[19] who also
footed the bill for most of the billions of dollars of research and
development costs of the reactors themselves. And one more
cost is about to come due: that of decommissioning old power
plants whose useful life is over. Dismantling the extremely radioactive reactors and their components is an expensive proposition that generally was not included in the original cost
projections. Once again, taxpayers will get the bill.[20]

Of course, there are important safety and health considerations associated with nonnuclear technologies for mobilizing energy, too. Coal mining is an extremely hazardous
occupation, and burning of fossil fuels in general adds considerably to humanity's burden of respiratory and circulatory diseases. Coal burning even releases small amounts of radioactive
materals into the environment. Nor are safety and health concerns limited to the processes of mobilizing energy, of course;
they also must be considered in its use. Electricity from the
cleanest possible source can be used in enterprises such as the
fabrication of nuclear weapons or the manufacture of ozone-

destroying CFCs or gas-guzzling automobiles. On the other hand, power from a nuclear plant about to melt down, or a coal-fired plant releasing huge amounts of CO_2 and toxic air pollutants, can be used to light operating theaters in hospitals or process grain to be sent to starving people.

No energy technology can be perfectly safe. Even energy conservation carries some dangers. People who live or work in very air-tight, energy-efficient buildings have suffered from severe indoor air pollution. That problem is being solved by the installation of heat exchangers that permit the air in a building to be changed completely once an hour with little loss of energy.

John Holdren has observed that human problems with energy arise not so much from "too little, too late" as from "too much, too soon,"[21] His point was that, even though many human beings lack access to enough energy, a smaller segment of the human population is using so much energy so recklessly that it is threatening the ability of Earth to support *Homo sapiens.* Holdren has long emphasized that the most important constraints on the growth of energy use are those environmental impacts—another indication that per-capita energy use is a handy surrogate for $A \times T$ in the $I = PAT$ equation.

Escaping the Energy Trap

The processes by which energy is mobilized and the things human beings do with energy assault the natural systems that support civilization. The assaults can be reduced in two ways: by minimizing the amount of energy mobilized by humanity and by changing the mix of technologies used in that mobilization to emphasize more environmentally benign ones. Neither strategy will be particularly easy; both will be absolutely necessary. Programs substituting alternative energy sources for the now-dominant fossil-fuel sources obviously must also be backed by efforts to improve efficiency. It would be folly for society to invest in new technologies and then waste the hard-won energy!

The first key, then, to escaping the energy trap and fulfilling the Holdren scenario is to increase the efficiency with which energy is used. Society must strive to extract the maximum amount of goods and services from each unit of energy provided. There is no question that efficiency is the quickest, cheapest, and most environmentally benign energy "source" available.[22] International comparisons and detailed studies of technical possibilities show that the energy system of the United States has great potential for increasing efficiency—far beyond that achieved between 1975 and 1985.

Most western European nations and Japan are much more efficient than the United States in some sectors of their economies.[23] Overall, they use about 70 to 90 percent as much energy per person as Americans, but the standards of living achieved with that energy are difficult to compare: Japan, for instance, makes much greater use of mass transportation than does the United States, but much of its energy "efficiency" can be traced to crowded housing. Unlike the United States, however, those nations are making plans to become even more efficient. Sweden intends to double the efficiency with which it uses electricity, and Germany and Denmark plan to reduce their carbon dioxide emissions dramatically, primarily by increasing their energy efficiency.[24] Eastern European nations and the Soviet Union, on the other hand, have been even more wasteful of energy than the United States, and they suffer horrendous pollution problems partly as a result.

The only chance we see for a planet on which everyone could lead a decent life depends on successful improvement of energy efficiency everywhere.[25] Achieving much higher efficiencies in most developed nations would not be terribly difficult if people were given incentives to do it, backed up by appropriate government policies. The technological means for getting more goods and services from each unit of energy used are already largely in hand. At the moment, however, there are no efficient nations—only efficient industries and practices within nations. Everyone has ample room to make improvements.[26]

The energy bills for American buildings could be halved without lowering comfort levels by such simple expedients as carefully scheduling the operation of air conditioners and keeping them properly maintained; coating windows to reduce solar heating in summer; using shade trees and reflective roofs to reduce home cooling bills in summer; storing heat in specially constructed walls, floors, and ceilings in winter; putting reflectors in light fixtures, installing high-efficiency compact electric light bulbs, and so on. In summer, energy can be saved in office buildings by turning off air conditioning after people have departed at 5:00 P.M. and letting the buildings heat gradually past the comfort level, then turning it back on before dawn when outside air is cooler (and electricity is cheaper). That precooling lowers demands on air conditioning during the day (when electricity is expensive).[27]

Efficient heating and lighting often depend on computers and temperature and light sensors to attune sources of heat (or cold) and light closely to needs. Other high-technology contributions to efficiency include devices such as "superwindows" that retain heat as well as a wall (nine times better than a single-glazed window) and special controllers that match electric motor speeds to the power demands of the devices they are running.

According to one estimate,[28] if the 1.5 billion lighting fixtures in the United States were fitted with innovative lighting installations (such as compact fluorescent lamps) that were commercially available in 1990, a fifth to a seventh of all the electricity used by the nation could be saved. Great energy savings are also potentially available through more efficient electric motor systems installed in pumps, fans, and other machinery. The potential for innovative energy-saving programs is vast. Much was gained in the 1970s and 1980s, but substantial scope remains for further progress in energy-saving consumer goods.

In the industrial sector of the American economy, large amounts of energy can also be saved. Electric arc furnaces, which can use virtually 100 percent scrap, are extremely energy-efficient. Only a little over a third of American steel is made with such furnaces, but an estimated 60 percent could be.[29] Similar opportunities exist in the aluminum industry and in such diverse activities as manufacturing chemicals, refining petroleum, and making cement and paper. Energy savings can be gained from recycling instead of using virgin materials, and from reducing the quantities of materials required to deliver the goods and services that people want.[30]

There is also great potential for increased efficiency in transportation, of course. In terms of people moved, filling buses and trains instead of commuting with one person per car is one obvious step. Most important, though, redesigning cities and suburbs so that less motorized transport is required will be an essential component of long-term solutions.[31] Revising American cities so that homes for people of varied income and employment levels are near offices, shops, schools, and even industries will be a challenging task, involving cooperation among local officials, businesses, trade unions, developers and redevelopment planners, architects, and builders. Most important, the public must support the effort.

During the lengthy transition from automobile as master to automobile as servant, though, the efficiency of automobiles themselves can be greatly increased. Progress has already been made; the average American car now burns about one-third less gas per mile than it did in 1975,[32] pushing toward 20 miles per gallon (mpg), although there was a slight decline in efficiency between 1988 and 1990.[33] There is enormous potential for further savings, perhaps moving to 60 to 70 mpg, levels already achieved by European experimental cars such as the Volkswagen Eco-Polo and Volvo's LCP 2000.[34] But simply improving automobile efficiency will not do the job of sufficiently reducing the environmental impacts of automobiles—especially since the number of motor vehicles has been growing two and a half times as fast as the number of people.[35]

In market economies, progress will depend not only on the actions of firms, but on billions of individual consumer decisions to do such things as buy those energy-efficient automobiles and appliances and install long-lasting, efficient light bulbs—bulbs that deliver each unit of illumination more cheaply and for many times longer than a conventional incandescent bulb, but cost more to buy. Similarly, energy-efficient appliances are sometimes more expensive initially, but soon make up the difference in energy savings.

A key question with all such technological fixes is whether enough consumers will be willing to pay the higher up-front costs to allow them to come into wide use. Past experience suggests that industry and business will do so if the ensuing energy savings are clear, but homeowners may need incentives or financing assistance. Those up-front costs seem to be the reason that many otherwise easy and sensible steps toward energy efficiency have not been taken. In economic terms, consumers have very high "discount rates" with respect to their consumption behavior. They would rather buy a cheap but energy-inefficient (and ultimately more costly) refrigerator than a

more expensive, efficient, and *ultimately* cheaper one—even
when they completely understand the long-run costs.[36] Obviously, there is a role for government and industry to help
spread out those up-front costs so that the socially desirable
goal of energy efficiency can be achieved.

After the "energy crises" of the 1970s, federal, state,
and local governments and many power companies in the United
States encouraged consumers and businesses to invest in energy
efficiency through subsidies, discounts, and tax incentives.
Some of these programs still exist. Taunton Municipal Lighting,
in Taunton, Massachusetts, now leases efficient compact fluorescent bulbs, costing some $15 each retail, to its customers
for 20 cents per month apiece. That amount covers the cost of
the bulbs, marketing, breakage, and so on. The average saving
to Taunton's customers is $50 during the life of each lamp.[37]
Some utilities (including ours) have even given the efficient light
bulbs to their customers and offered rebates to people who
purchased energy-efficient appliances such as refrigerators.
Such policies could and should be renewed and extended.

Simply by creating the incentive of higher energy prices,
perhaps through a higher gasoline tax, the government could
encourage considerable movement toward efficiency and directly
address problems of urban congestion and air pollution. Every
penny of tax on gasoline in the United States would now fetch
$1 billion a year toward reducing the deficit (as well as re-
straining gasoline consumption in the long run). A small step
in this direction was taken in the 1992 federal budget passed
by Congress, which increased the national gasoline tax by nine
cents a gallon, to be phased in over a few years. Some states,
including California, have also boosted gas taxes. Additional
possible tax approaches to encourage energy efficiency are discussed later.

▬ Fueling Holdren's Optimistic Scenario

Of course, the entire world can't be run on efficiency.
No matter how efficient humanity becomes in its energy use,
large amounts of energy will still need to be mobilized if people

are to have the fruits of industrial civilization. Society must turn its collective cleverness to exploring ways to mobilize that energy with an absolute minimum of environmental impact.[38] In the discussion that follows, we shall try to look at more distant future possibilities in addition to technologies now in hand or expected in the near future. But crystal balls get clouded—in 1960 we could not have foreseen powerful calculators the size of credit cards!

Humanity may be able to find energy-mobilizing technologies that are environmentally superior to those discussed here. We certainly hope so, because the 30 TW world of Holdren's optimistic scenario will probably be difficult to achieve under any circumstances, let alone make sustainable. At the moment, Holdren's is the best scenario that appears feasible with foreseeable technologies, and prudence requires our aiming in its direction. After all, even if superior technologies did emerge, there is no guarantee that they would be sufficiently more benign to change that feasibility very much.

What, then, appear to be humanity's alternatives for supplying that 30 TW? One important alternative is to make greater use of renewable energy sources.[39] Renewables have many advantages, but some disadvantages as well. We'll examine both for each technology.

Of the renewable alternative energy sources already on line, hydropower at first glance seems the most desirable. Hydropower establishments, unlike electricity generated with fossil fuels, produce little in the way of direct global impacts; no carbon dioxide or precursors of acid rain are released. They seem to pose less risk of catastrophic accident than nuclear fission, although dam failures can kill thousands of people at a time—and have done so.

Building dams usually causes extensive local ecological disruption, both by flooding the area to be occupied by the reservoir and by changing the characteristics (temperature, flow rate, seasonal cycle) of the aquatic ecosystem downstream and in the reservoir area above the dam. The recreational benefits of wild rivers are also sacrificed (although they are replaced with others associated with lakes).

Dams can set up conditions that increase diseases in local human populations: for instance, they may create water conditions that favor the larvae of malarial mosquitoes, spillways or downstream conditions that support the larval stages of the flies that transmit river blindness, or bodies of water suitable for freshwater snails that serve as intermediate hosts for the parasites (schistosomes) that cause bilharzia. Dams also may adversely influence coastal fisheries, as did the Egyptian Aswan Dam by trapping nutrients required by Mediterranean sardines. And they may promote the flooding of coastlines by trapping silt that ordinarily replaces material lost to beach erosion; some 10,000 people lost their homes to the sea on the coast of Togo because of the Volta Dam.[40]

In addition to these disadvantages, two things make hydropower no panacea for humanity's long-range energy problems. First, if all practical sites were utilized, hydropower would only generate 1 to 1.5 TW (equivalent to 3 to 5 TW of fossil fuel used for electricity generation); only a modest contribution to the 30 TW required in the Holdren scenario.[41] That does not mean, of course, that hydropower should not be part of a future mix of energy technologies, just that its potential is limited, and the human and environmental costs of the projects can be substantial.

Second, and perhaps more critical in the long run, hydropower projects generally also have a limited lifetime. Reservoirs behind the dams eventually silt up, often very rapidly when watersheds are deforested. Ultimately, the reservoir is filled in and the dam is replaced by a waterfall, from which power usually can no longer be extracted profitably—since the storage capacity of the lake is necessary to take advantage of periods of high runoff (when more water than the turbines can turn into power is available, water that needs to be stored to generate electricity in periods of low runoff).

Wind power could be an important future source of energy in many places where it is suitable. But, like hydropower, wind power can make only a modest contribution to save us from the energy trap. If the windiest 3 percent of Earth's land area were covered by efficient windmills (a gigantic undertaking), only about 1 TW of electricity might be generated (equivalent to 3 TW of fossil fuel used for electricity generation).

In the age of environmental awareness, the windmill as a generator of electricity has been making a comeback.[42] In California alone, partly as a result of the efforts of Governor Jerry Brown, "wind farms" were able to produce nearly 1,500

MW of power in 1987 when the wind was blowing at the maximum speed the windmills could handle. On average, they yielded about 15 percent of that, mainly because the wind does not blow hard enough much of the time. Nevertheless, they provided about 2 percent of the electricity for northern California in 1988. Thousands of very small wind turbines have been installed in China in the last decade, largely to run TV sets in a nation much of which has no network of high tension wires to distribute power![43]

As far as can be foreseen now, the environmental impacts of wind power are minimal. Windmills can often be erected on land also used for grazing or other agricultural activity, and the fraction of the wind's energy extracted is so small that regional or global impacts on climate should be negligible.

Both hydropower and wind power are forms of solar energy. Heating by the sun evaporates water from Earth's surface so that it can fall as rain, form rivers, and drive turbines. Solar energy also drives the winds that move the water around in the atmosphere and rotate the blades of wind turbines. Another form of solar energy is "ocean thermal." This energy source takes advantage of the temperature difference between the sun-warmed upper layer of the ocean and the cooler depths. In the tropics, ocean surface temperatures are typically about 80° F, and waters at a depth of 1,000 meters are about 40° F. In theory, ocean thermal energy has greater potential to contribute significantly to the global energy mix than either wind or hydropower; if fully exploited, it could yield a flow of several terawatts.

Unfortunately, though, the technical problems of mobilizing that potential are great, and it seems highly unlikely that more than a small fraction of the theoretically available energy will ever be used to benefit society. Other sources of energy in the oceans—tides, waves, and currents—have a very small energy potential on a global scale. Each is unlikely to yield even as much as 1 TW, so it seems certain they will either not be employed commercially or will be restricted to places with especially favorable conditions.

Much the same can be said of geothermal energy. It heats virtually all the buildings in Iceland, an island located astride the rift where the Atlantic Ocean floor is being continuously generated by an upwelling of molten rock, to the accompaniment of a great deal of volcanic activity. Geothermal power could contribute to electricity generation or space heating in many other parts of the world; indeed it already does in Hawaii, northern California, Japan, and New Zealand. With today's technologies, however, it is available only in limited areas. Some specialists believe the heat in Earth's depths, available everywhere, will eventually be economically harnessed; but unless that occurs, geothermal seems destined to remain a minor component of any global energy regime.

Humanity now obtains about 1.5 TW by using biomass fuels—fuelwood, crop wastes, and dung. These, of course, are plant materials whose production is fueled by the sun's energy converted into chemical energy by the process of photosynthesis. Even in industrial countries, biomass contributes to the energy economy. In the last decade, wood-burning power plants with capacities in excess of 45 megawatts have been constructed by four United States utilities. Wood also supplies about 10 percent of the heating in American homes and fills more than half the energy needs of the pulp and paper industry.[44]

Wood and other biomass materials are major energy sources in poor nations, where burning them may create more pollution than burning an equivalent amount of fossil fuel— especially when the biomass is burned in open fires or inefficient village stoves.[45] But the quantity used is small compared to that of fossil fuels burned in industrial nations, and the burning of traditional fuels contributes relatively little to worldwide pollution problems such as global warming. Emissions of hydrocarbons and particulate matter from burning traditional fuels mainly affect people locally (especially in homes while cooking), but account for most of the worldwide population exposure to these pollutants.

Yet a watt of power produced by burning branches from the local woodlot may on balance have less negative environmental impact than a watt obained by burning coal or oil in rich countries, if the wood harvest is *sustainable*—that is, if the trees are replanted or the branches regrown. It is important to realize that any sustainable use of biomass adds no net CO_2 to the atmosphere and thus makes no contribution to global warming. As they grow, the photosynthesizing plants take up the CO_2 released in burning of the previous generation of plants.

Carefully managed woodlots of fast-growing shrubs and

trees on marginal land could provide useful energy supplies on a sustainable basis to farms and villages in developing nations. Such plantings could also help curb soil erosion, and leguminous plants such as acacias and *Leucaena* would enrich soil with nitrogen. The leaves of the harvested plants could provide a nutritious feed supplement for livestock or be used as fertilizer for crops.

There appears to be great potential for increasing the energy yield from biomass. Suppose 10 percent of Earth's land surface were devoted to growing energy crops: wood for fuel; corn and sugar cane for conversion to ethyl alcohol (ethanol); and other plants (or plant parts) whose energy could be used to power human enterprises. Suppose further that those plant materials could use incoming sunlight with 1 percent efficiency (that is, 1/100 of the sunlight falling on the plants could actually be converted into chemical energy in the plants). Then biomass could theoretically provide 25 TW, or more than 80 percent of the energy requirements in the Holdren scenario.

We say "theoretically" because production of biomass is basically an agricultural process and is therefore fraught with the same environmental difficulties and constraints that beset agriculture (discussed in Chapter 7).[46] Of course, land used for production of biomass to fuel the human economy generally will not be available for the production of food to fuel human beings themselves.

The exceptions are grazing lands that could produce animal wastes and cropland from which residues could be gleaned to use as energy sources. But using those materials for energy amounts to mining vital nutrients from the soil and can rapidly impoverish it, since dung and crop residues both serve as fertilizers crucial for maintaining soil structure and fertility. Burning residues or dung is a desperation measure taken by destitute people lacking other fuel sources. Using these materials as fuel leads to a steady depletion of soil nutrients and fertility, causing the deterioration of farmland—a stark example of how the environmentally damaging activities of the poor contribute to their own impoverishment.

Most of the energy in biomass can be captured, however, and most nutrients retained, if the material is processed by a procedure known as "biogasification." Biogasification involves the breakdown of biomass by bacteria in the oxygen-free environment of enclosed containers called "digesters," which can be very simple designs. The fermentation process yields methane and CO_2, and the methane can be removed to yield a clean-burning gas that is suitable for cooking, heating, generating electricity, or for running farm machinery that has been modified to permit it to use methane rather than gasoline. The solid residue from biogasification makes a fine fertilizer.

Biogasification of crop and animal (including human) wastes has been successfully used at farm and village levels in several developing nations, including China and India, with admittedly mixed success. Clearly, the opportunity exists to improve the technology and gain significant additional amounts of energy. This solution is superior to simply burning dung or crop wastes, for the reasons already indicated and because the methane would not be released to the atmosphere (methane is a more potent greenhouse gas than the CO_2 produced by burning it) as it could be if the residues were left to decompose in soil as fertilizer. An analogous approach to biomass energy sources is the direct burning of urban wastes or, better yet (again from a global-warming standpoint), harvesting methane from decomposing materials in landfills to use as fuel.[47]

Overall, though, biomass proposals can be put most clearly in perspective by noting that the amount of land that would have to be devoted to producing the hoped-for 25 TW is nearly equal to that already under cultivation, and the best land nearly everywhere is already planted in crops meant to feed people (or their animals) or to provide them with cotton, oils, or other important products. Even though some biomass crops can be grown without competing with other crops for land, enough land to produce that much biomass is unlikely to be available. Still, even though the full, first-blush potential probably cannot be realized, biomass can certainly play an important role in future energy economies.

At the other end of a scale of technological sophistication from wood-burning and backyard biogasification is the possibility of fueling the Holdren scenario with nuclear fusion. Many people hope that nuclear fusion will become an energy source that will suffer from fewer constraints than most technologies. The effort to build controlled fusion reactors has been a daunting task; although fusion reactions can be obtained, so far none of the machines yields an energy profit—that is, more energy is required to run the machine than is produced by the fusion process.[48] Fusion machines, therefore, are not candidates now for commercial energy production. But when they do make an energy profit, will humanity's energy supply problems be solved?

Not necessarily. Making an energy profit is not the same as being economically viable. Because of the technical difficulties involved, fusion machines may require so much tender loving care or yield so little energy above energy invested that they will never be commercially practical. But fusion might become practical; and if it does, it should present substantially fewer direct risks to human health and world peace than do fission reactors. If nuclear fusion can be made to work, energy to power civilization for millions of years would be available, making fusion essentially a "renewable" energy source, one with gigantic potential.[49] That seems appropriate, since all the other large-scale renewables depend on that one huge fusion reactor sited 93 million miles away—the one that rises every morning and sets every evening.

Perhaps the most promising renewable energy alternative that could help meet the needs of society under the Holdren scenario uses that distant fusion reactor to generate electric power. The sun's energy can either be captured in solar cells that directly generate electricity, or it can be used to heat a fluid and drive turbines to generate electricity—a process called "solar thermal." If 1 percent of Earth's land could be covered with solar collectors, and if the sun's energy could be converted to electricity with 20 percent efficiency, then 50 TW of electricity would be produced—much more than the 30 TW needed for the entire Holdren scenario. Solar cells have an additional advantage especially important in developing countries, in that they do not need to be deployed in single huge arrays. Rooftop panels and other small deployments of cells already power lights, pumps, television sets, and other devices in Third World villages.

Furthermore, with some loss of efficiency, part of the captured solar energy could be used to dissociate water to yield hydrogen for use as a portable fuel. Of course, less hydrogen fuel would be needed in rich nations if they were reasonably successful in shifting more people into electric-powered mass transportation and in reducing the need for commuting in urban areas than if current commuting patterns prevailed. Similarly, if efficient batteries that supplied convenient, economical power to private vehicles were developed, less portable fuel would be needed, even if commuters couldn't be enticed out of their cars. Hydrogen has a vast environmental advantage over other fuels in that the only effluents generated by its use are heat, water vapor, and, in some cases, oxides of nitrogen.

On paper, a solar-hydrogen economy (one in which solar energy is used both to generate electricity and produce hydrogen fuel) appears to be the most attractive of the alternative energy schemes.[50] Reasonable estimates suggest that the energy obtainable from an ounce of silicon used in a solar cell could be roughly that potentially extractable from a pound of nuclear fuel used in a breeder reactor.

A solar hydrogen economy is not without potential problems, however. First, 1 percent of Earth's land surface covered with solar energy collectors may not seem like much (and presumably it would be concentrated in desert areas with abundant sunshine). But that amounts to about 1.5 *million* square miles—equal to a square about 1,250 miles on each side, or most of the lower 48 United States west of the Mississippi. That's a lot of solar collector!

There is a small possibility of climatic effects from changing the amount of solar energy absorbed by the planet, although at first calculation the chance seems negligible. Regional climatic effects from extensive solar collector deployment might be significant, however. In some solar power-plant designs, toxic or corrosive working fluids may be circulated, which could create some hazards from leaks. Environmental impacts also would result from the resource extraction and industrial/assembly processes necessary to cover 1.5 million square miles with apparatus.

There will, of course, be safety considerations with the use of hydrogen as the fuel of the future, as there are with all energy systems. Hydrogen molecules are small and therefore hard to contain, and hydrogen also is highly explosive. There doubtless will be occasional fires and explosions with everyone driving hydrogen-powered cars, refueled at the local service station. We are sure, however, that the system can be made acceptably safe rather easily, especially considering how seldom problems with explosive fuel arise today when everyone drives around with a gasoline "bomb" in his or her car. Experience on a large scale will be needed to know exactly how safe hydrogen can be made in comparison to gasoline. Prospectively, while it would seem to require a different set of precautions from those associated with gasoline, it appears inherently to be about equally safe.[51]

Solar power as a substitute for other central-plant-generated electricity also presents the technical problem of storing massive amounts of energy to use at night. At the moment, battery technologies are not up to the job. Another possibility is to use solar-generated electricity to pump water uphill into a reservoir during the day, then run that water through turbines at night to generate electricity again. Solar energy could also be used to heat a fluid during the day and the fluid used to heat rocks to store the energy. That stored energy could then be used to generate electricity at night. The rock "reservoir" for a 1,000 megawatt plant would be substantial—a cube about 300 yards on a side.[52]

In spite of various potential difficulties, we believe that, watt for watt, solar electric power appears to produce the fewest problems of deployment, the mildest environmental impacts, and the fewest direct health hazards of any large-scale novel energy technology that is undoubtedly technically feasible. A potential snag with all the renewable energy flows discussed here is that, with the exceptions of hydropower, some forms of biomass, and wind power, they have never been deployed commercially on a large scale. Consequently, many of their potential costs and benefits are subject to some uncertainty, which is likely to persist until large-scale operations have been undertaken for an extended period and more experience has been gained. Some government will have to decide to take the chance and help finance the enterprise—starting, in our best judgment, with some version of solar hydrogen.

Moving Away from Fossil Fuels: Nuclear Power

Humanity now finds itself in an interesting predicament; the era of cheap, abundant liquid fossil fuels is over.[53] Even if it weren't, environmental constraints dictate a rapid reduction in the use of energy sources that deposit large amounts of CO_2 in the atmosphere. The only *prudent* course for humanity is to start now on a program of sharply reducing emissions of greenhouse gases. There is no time to waste—especially since energy transitions in poor countries like China and India will require not only substantial outside help but also time measured in decades. Moreover, the time to embark on alternatives is as soon as possible, since planning for a particular energy technology and building a system based on it from the beginning is easier and less costly than changing it later. Starting more or less fresh is an option for developing nations, not for industrialized ones—one of the few advantages poor countries have!

Of the technologies now deployed on a large scale, only nuclear fission holds any promise of being able to supply enough energy to be a mainstay for Holdren's optimistic scenario. If nuclear technologies similar in their use of fuel to today's light-water fission reactors are employed, world uranium resources that probably will be available at tolerable cost would be sufficient to support the Holdren scenario for about 100 years. If "breeder" fission reactors were used, the time would increase to 100,000 years.[54]

But fission suffers from the four hazards enumerated earlier: routine release of radioactivity, disposal of long-lived wastes, risk of catastrophic accident, and proliferation of nuclear weapons. Even if costs could be held to an affordable level, the industry's record with burner reactors is such that it is doubtful that public opinion in many countries would support the worldwide deployment of thousands of reactors, which would be required if fission were to be a major support for the human energy economy.

Indeed, any attempt at deployment on that scale would be extremely risky, since the chances of an accident would rise with the number of power plants, and one more accident like Chernobyl would almost certainly make fission power globally unacceptable.[55] In fact, for fission to be a candidate as an important power source in the future, we believe several hurdles must be overcome. First, the present generation of reactors probably should be decommissioned everywhere with all the care and deliberate speed societies can muster. As long as any are still operating, some analysts believe too high a risk of catastrophic accident remains *for humanity to gamble its energy future on it.*[56]

Second, a new generation of reactors must be designed with satisfactory characteristics.[57] Our reading of the present situation is that routine releases present no serious problems (even in the present generation of reactors). It seems likely that precautions against catastrophic accident, at least in burner (as opposed to breeder) reactors, and a suitably safe system for storing nuclear wastes can be achieved without raising the costs of fission-generated electricity by more than 25 percent.[58] The environmental advantages of such a safe nuclear technology, we suspect, would more than compensate for the burden of any added cost. The problem of nuclear weapons proliferation will not be so amenable to technical fixes, however. All combinations of reactor type and fuel cycle end up producing plutonium (from which bombs can be made), except those that use fuels made of fully enriched uranium (from which bombs also can be made). Thus the creation of the key ingredients of nuclear weapons cannot be avoided, although some fuel cycles make them more difficult to get at.[59] All breeder reactors, which will be required if fission is to serve as an energy source for many centuries rather than many decades, must actually recycle plutonium— greatly increasing the risks of diversion.[60] The diversion problem probably can be solved only by bringing the entire global nuclear fission power system under careful international control.

The nuclear power question really boils down to a complex mix of risk assessment, politics, and economics. Some experts think fission power can never be made both acceptably safe and economical.[61] Others are convinced it can.[62] The history of attempts to assess the hazards of nuclear technologies would fill a small library and is loaded with extremely divergent views of safety. Estimating the probabilities of catastrophic reactor failures (as nearly occurred at Three Mile Island and did occur at Chernobyl) has proven especially difficult; there fortunately is little experience on which to base the assessments. But the possible sources of accidents are myriad, which adds to the difficulty. Not only must numerous types of failures due to errors in design and construction be considered, but also those due to operator error (so far the most common), as well as external dangers such as earthquakes and sabotage.

A new generation of reactors designed from the ground

up as power reactors and with safety a prime consideration might lessen some of this uncertainty. Today's most common reactors are in essence scaled-up submarine reactors, the basic design of which required high power density (extraction of a lot of power from a small volume), since small size makes submarines harder to detect. But reactors with high power densities are not as easy to secure against accident as those with lower power densities. Possibly the most serious mistake ever made by the nuclear industry was to be in too much of a rush to show the benefits of the "peaceful atom." The world would have been better off if the first deployment of nuclear power had featured reactors designed from scratch to run electric grids rather than nuclear submarines. It seems unlikely to us now that the arguments on safety will ever be fully resolved until and unless a new nuclear technology has been deployed on a large scale— that is to say, unless society decides to run what may be a very risky experiment.

▬▬▬ Moving Away from Fossil Fuels: Solar Power

On the other hand, solar-thermal and direct solar-generated electricity (called photovoltaics) are promising but represent relatively untested technologies. Electricity generated by them today is still comparatively expensive, although the cost has been dropping rapidly in recent years.[63] Whether solar technologies would remain expensive if deployed on a large scale is unclear and (like the safety of nuclear power) might well remain unclear until some time after such deployment. The best-informed guesses are that they will not. But perhaps the biggest barrier to the deployment of solar technologies is that they are novel. The Chinese have chosen to develop nuclear power as the mainstay for their future energy because they think they know what they will be getting; solar electricity on a large scale is a much bigger unknown, and so it seems to them a bigger gamble.

Nevertheless, India has gone forward with photovoltaics, and by 1990 solar cells had been installed in 6,000 Indian villages for electricity.[64] In this case, extending the conventional power grid was judged costlier, slower, and less reliable than going solar. Some specialists in solar power believe that photovoltaics could be competitive with conventional power sources by the end of the decade.[65] The United States alone could be producing as much as 1,000 megawatts in photovoltaic capacity by 2000 at costs competitive with peak-hour electricity from conventional sources. Solar thermal with a capacity of nearly 200 megawatts has been installed in southern California, backed up by natural gas when the sun isn't shining. With an efficiency factor of 22 percent of the sunlight's energy converted to electricity, the cost by 1990 was approaching 8 cents per kilowatt-hour, below the 1989 cost of electricity from natural gas for peak power and less than twice that of base-load power from coal-fired plants.[66] Additional systems are contemplated for Nevada and Brazil.

If costs for solar technologies continue to fall, as they undoubtedly will, they will soon be competitive with conventional power sources—especially if, as also seems likely, higher world petroleum prices prevail during the 1990s. Solar systems are safer and can be deployed more quickly than nuclear power technologies. In developing nations, they have the further advantage of being independent of conventional power grids. Villagers needn't wait for the powers-that-be to get around to extending power lines to their village; they can install solar cells on their roofs and start watching television with the rest of us!

━━━ Toward an Energy-secure Future

Human energy futures, then, look very different, depending on one's evaluation of the safety of future nuclear burner and breeder reactor technologies, the possibility that commercially viable fusion reactors can be developed, the potential for successful deployment of large-scale solar technologies, and the feasibility of using hydrogen as a portable fuel generated with the power from these technologies. Regardless of how one evaluates each of these possibilities, it seems inevitable that the necessity of abating global warming, if not many other environmental factors, will bring the era of gross overdependence on fossil fuels to an end.

Since the global mix of energy sources is bound to change dramatically in the future, the international political dimensions of energy supplies, prices, and choices must be considered. Just as all nations are not equally blessed with fossil-fuel deposits, their potential to develop flow resources—hydropower, thermal, wind, biomass, or solar sources—is also unevenly distributed. Some nations also are better candidates for increasing energy efficiency than others; for example, investments in making Poland or the Soviet Union more efficient are likely to yield much greater environmental gains than equal investments in making Japan or Germany more efficient.

Furthermore, a nation's international behavior often is tightly tied to its energy position. Soviet adventurism in the late 1970s and early 1980s was fueled by a flow of dollars generated by oil exports in the era of high oil prices.[67] One need only contemplate the connection of petroleum to the 1991 outbreak of war in the Persian Gulf region to see that future energy development will contain many political pitfalls, including the possibility of generating more international conflicts.

So the global energy economy is already deeply intertwined with international politics, and the situation will, if anything, grow more complex and potentially more tense as well. If civilization begins making a transition away from dependence on fossil fuels by first reducing the use of coal, coal miners will suffer disproportionately within nations, and coal-rich nations will be disadvantaged in the world community.

Consider the case, summarized in the Introduction, of China, where coal forms the very backbone of the energy economy.[68] For China to fulfill even a small part of its ambitious development and industrialization plans, per-capita consumption of commercial energy must greatly increase. Today it stands at about 7 percent of per-capita energy use in the United States. Suppose China scaled back its development plans so that per-capita energy use only doubled during the first half of the next century, to about the level of Algeria today. Suppose also that Chinese population growth halted now at about 1.1 billion, rather than soaring (as some demographers expect) to at least 1.6 or 1.7 billion. If China's development were fueled with its abundant supplies of coal,[69] then, even under all these *optimistic* assumptions, China would be depositing more *additional* carbon dioxide into the atmosphere than the United States is now releasing from coal burning—and coal supplies almost a quarter of America's (much higher per-capita) energy use.[70]

These optimistic assumptions are unlikely to hold. Despite huge investments in the energy sector, in 1987 about a quarter of China's total manufacturing capacity was idled for lack of electricity. The government plans to correct this situation by increasing its energy production from roughly 850 million tons of coal-equivalent in 1987 to 1.4 to 1.5 billion tons of coal-equivalent in 2000 (with coal still supplying more than three-quarters).[71] If this actually happens, China's carbon-dioxide emissions by the end of this decade will nearly equal those of the United States in 1990.

Could China reduce its CO_2 emissions substantially by shifting to other conventional energy sources? It has little natural gas but substantial oil reserves. More oil than before is being burned domestically rather than being exported, and to the degree that oil is substituted for coal, CO_2 emissions per unit of energy gained will be reduced, but by only 20 percent. The nation's hydropower potential is the world's largest; about 380,000 megawatts could economically be developed. If brought on line, that much hydropower could reduce annual coal use by 380 million tons—more than half the amount consumed annually now—but might also lead to inundation of a considerable amount of scarce prime farmland. Since China is struggling to reforest and has a marginal food situation, biomass sources are unlikely to make major contributions to its energy economy, although establishment of sustainable village woodlots for fuelwood and improvement of the biogasification projects would be of considerable benefit to the rural economy.

Nuclear power is thought by the Chinese government to be their energy source for the future, however, and two plants are now under construction. Both hydro and nuclear power plants take seven to ten years to construct and require large initial outlays of capital. Plants that burn fossil fuels, in contrast, take only about three years to construct and are cheaper (although the Chinese believe the energy will be more costly in the long run).

It is against this background that one must view any possibility that China will substantially cut back or give up coal burning to help mitigate a global problem. It seems clear that China is heavily committed to expanding the use of coal. Only massive aid from other nations to deploy alternative technologies that add less to the CO_2 buildup in the atmosphere could possibly prevent a huge increase in China's contribution to one of humanity's most serious threats to global life-support systems. The principal technologies for which outside help would be useful appear to be nuclear fission and solar-hydrogen.

The situation in India is similar to that in China. India now consumes only 3 percent as much commercial energy per capita as does the United States (as a nation, India consumes about 10 percent as much, with a population more than three times as large). Supposing India underwent minimal development, entailing an increase to the per-capita energy-use level of China today (to 7 percent of the United States level), fueled mainly by coal. That would make India's annual additional contribution to the atmospheric CO_2 load a century hence as large as that now contributed by coal burning in the United States. This increase is largely due to the projected gigantic growth in the P factor of India's $I = PAT$ equation. If a demographic miracle occurred, dropping India's completed family size from 4.2 today to 2.2 (replacement level) over the next 35 years, then the nation's population would not cease growing until nearly 2100, topping out at some *2 billion* people. Even small increases in per-person CO_2 emissions amount to whopping quantities of CO_2 when multiplied by 2,000,000,000.[72]

While we have limited consideration here to the two largest developing nations, the prospect of greatly expanded use of fossil fuels, especially coal, applies to all of them to some degree. Other developing countries include some 2 billion more people, and most have rapidly growing populations and aspirations for development. They too will require assistance from rich nations to realize those aspirations without enormously increasing their per-capita energy use and accelerating the buildup of greenhouse gases.

Given the dangers in vastly increasing coal burning in developing nations, careful evaluation of both nuclear fission and solar options will be key in shaping future energy paths. It seems clear that China would welcome aid in deploying a nuclear fission technology; it is an alternative already favorably viewed by the government. What the Chinese attitude would be toward adopting a relatively untried solar technology (with considerably less information on possible costs than surrounds the nuclear option) is much more problematic. Still, the potential for solar technologies to employ large numbers of moderately skilled people, rather than a handful of highly trained scientists and technicians, might have appeal.

The contribution that energy efficiency can make to solving the energy problems of poor nations should, of course, not be neglected. For instance, some 300 million incandescent light bulbs account for about 35 percent of India's peak power demand. If only 20 percent of them were replaced by compact fluorescents, India could forgo building 8,000 megawatts of new coal-fired generating plants.[73] The proportionate waste of energy in less-developed countries is probably greater than in industrialized nations, because the technologies are often so primitive. The developing nations, therefore, have the possibility of being two or three times better off in material well-being while using the same amount of energy as they do today. An energy strategy aimed at providing basic human needs for the huge populations of poor people in developing regions that focused on efficiency would be far more effective and environmentally benign than the one now in place, which keeps trying to enlarge the energy-intensive industrial and farm export sectors.[74]

Humanity has created a gigantic and complex system for supplying energy to produce goods and services for itself. Even though by American standards the needs of most human beings were still unmet, by the 1970s the world energy system depended on burning every year about 17 billion barrels of petroleum, 3 billion tons of coal, over a trillion cubic meters of natural gas, and something like 2 billion cubic meters of wood.[75] The amount of electricity generated by hydroelectric plants was equal to what would be generated by a third of the average runoff of water from all Earth's continents falling 100 meters. The effort to supply energy even in the 1970s required drilling for oil off sensitive coastlines, destroying entire landscapes by stripmining coal, building nuclear power plants of dangerous design, deforesting large areas for fuelwood, and doing numerous other things that damage environmental and human health.

In the 1970s, a yearly investment equivalent to about 300 billion 1990 dollars was needed to keep the energy system growing at about 3 percent annually. Holdren estimates that in the 1990s an annual investment of over $500 billion would be needed to maintain that rate of increase. In a world where investment capital is short and the needs of an exploding human population ever more pressing, the allocation of sums that large would require careful decision making, even if the fate of Earth's life-support systems were not intimately intertwined with the mobilization and use of energy. But since it is, we will return to the key dilemma of the energy trap when we discuss global warming in the next chapter. It will recur later as well, as we try to outline solutions to the human predicament that do not leave the vast majority of human beings mired in poverty.

Global Warming: The Beginning of the End?

Nowhere is the energy trap more apparent than in the prospect of global warming. And nowhere are the potential perils of that trap clearer, since warming could cause dramatic changes in regional climates. The nonhuman organisms of the planet, and the ecosystems they inhabit, are highly dependent on those climates for their very existence. We in turn depend on them—on "biodiversity"—for *our* very existence. The agricultural ecosystems on which human beings most directly depend may be among the most vulnerable to changes in weather and climate.

The factors pushing humanity deeper into the trap are clear. After all, why shouldn't the Chinese and Indians want to drive automobiles and have air-conditioned houses? Where is it written that only members of a small fraction of the human population should each benefit from having roughly 75 energy "slaves"? Using 7.5 kW of commercial energy, in a way, is like having 75 slaves working for you, each using energy to run their life processes at a rate of 100 watts. The problem, of course, is that the use of those "slaves" is responsible for much of the deterioration of Earth's life-support systems. And now, every four days there are a million more poor people who eventually will want to become large-scale holders of energy slaves.

Carbon Dioxide and Climate

Energy slaves affect our environment perhaps most seriously by adding carbon dioxide to the atmosphere.[1] Carbon dioxide (CO_2) is the most important of the greenhouse gases whose atmospheric concentration is being increased by humanity. Its role in regulating Earth's temperature has been speculated on since the early 1800s, and its effects have been calculated for almost a century. In 1896, a Swedish chemist, Svante Arrhenius, estimated that Earth's surface would warm 7° to 10° F if enough fossil fuels were burned to double the amount of CO_2 in the planet's gaseous envelope. That is more than sufficient warming to cause serious climatic change, about the same temperature difference as between the depths of the last Ice Age (when ice covered New York a mile deep) and today. Interestingly, Arrhenius's simple "back-of-the-envelope" calculation gave a result fortuitously close to those recently given by supercomputers.

Although various scientists subsequently noted the potential effects of CO_2 emissions, only after publication of a classic paper on atmospheric accumulation of CO_2 in 1957 by Roger Revelle and Hans Suess of the Scripps Institute of Oceanography was interest rekindled in the scientific community.[2] Revelle and Suess were right on target when they noted that "human beings are now carrying out a large scale geophysical experiment of a kind that could not have happened in the past nor be reproduced in the future."

Before the industrial revolution, the atmosphere contained about 275 parts per million (ppm) of CO_2; by 1991 the concentration was some 355 ppm and rising by about 1.5 ppm (0.4 percent) annually. At an observatory about 11,000 feet high on the slopes of Mauna Loa in Hawaii, the atmospheric concentration of CO_2 has been continuously recorded since 1958, thanks to the foresight of C.D. Keeling of Scripps, and of Revelle, then Scripps's director, who encouraged him to start the measurements. The observatory is built on a lava field in a beautiful setting; it seems a strange place for what might be regarded as a lookout post for the world's future.

But that's what it is. Keeling's records, when plotted,

produce an oscillating line on a graph. The graph shows CO_2 concentrations climbing from about 315 ppm in 1958 to their present levels, but not steadily. Instead, a peak each April is followed by a valley in early fall, producing a zigzag upward trend. The graph shows not only the rise in CO_2 but the pulsing metabolism of the Northern Hemisphere. In winter, ecosystems—especially forests—release more CO_2 than they take up because respiration by plants and by the microrganisms that decompose organic matter predominates over photosynthesis. In summer, the reverse is true, and photosynthesis sucks vast amounts of CO_2 from the atmosphere. So much more land and vegetation are in the Northern Hemisphere that its effects swamp those of the Southern Hemisphere. Nevertheless, the regular ups and downs do not mask the inexorable rising trend.

The main source of that trend, of course, is the burning of fossil fuels by industrial civilization. The amount of CO_2 in the atmosphere by convention is expressed in terms of its carbon component.[3] About 6 billion tons of carbon (in the form of CO_2) are added annually by fossil-fuel combustion to the atmosphere, whose total pool around 1990 was about 750 billion tons.[4] For comparison, the estimated pool in recoverable coal and oil is about 4,000 billion tons.

The worst offender in emitting CO_2 is coal, which accounts for about one-third of fossil-fuel combustion worldwide. Coal consists almost entirely of carbon and therefore puts substantially more CO_2 into the atmosphere per unit of energy obtained than oil (which accounts for 45 percent of global fossil-fuel combustion) and nearly twice as much as natural gas (only 22 percent of global combustion).

Burning fossil fuels basically helps complete an age-old cycle. The energy stored in the fuels is solar, produced by photosynthesis hundreds of millions of years ago. Fossil fuels are the converted remains of ancient plants that escaped decomposition by being buried in bogs or sediments. There oxygen was absent, and anaerobic bacteria could not complete the breakdown of the plant materials. Over eons, heat and pressure transformed those materials into carbon-rich substances whose burning for over 150 years has powered industrial civilization. That burning has returned to the atmosphere the CO_2 long ago extracted from it by photosynthesis. But the rate of return is many thousands of times faster than that at which the CO_2 was originally taken out.

The atmosphere today is receiving carbon dioxide from a second major source: deforestation. There is substantial dispute over the exact net release of CO_2 from the rapid destruction of tropical forests. Some uncertainty arises from a lack of precise information on the amount of tropical forest being cut down and burned each year, although information is increasing in accuracy and extent. Estimates of carbon (as CO_2) released from deforestation and other biomass burning are in the vicinity of 2 to 5 billion tons per year. Much of that carbon is taken up again immediately by terrestrial vegetation, however, so about 1 to 2.5 billion net tons of carbon are added to the atmosphere from these sources. Most of the addition apparently comes from permanent deforestation and forest degradation, largely in the tropics.[5] The total pool of carbon in plant biomass was perhaps 560 billion tons around 1990. If *all* the tropical forests were removed, an estimated 150 to 240 billion tons of carbon would be released to the atmosphere—and, of course, biomass uptake would be severely reduced.[6]

Much of the extra carbon dioxide put into the atmosphere from both fossil fuels and deforestation, perhaps as much as half of it, is quickly taken up by the oceans.[7] So the net annual addition of carbon to the atmosphere is about 3 to 4 billion tons, amounting to about a half-percent increase in the 750 billion ton pool.[8]

Some, but not much, attention was paid to Keeling's climbing zigzag trend of CO_2 increase in the 1960s, but enough so that environmental scientists who were not climatologists began to be concerned. Still, what might happen to the climate was unclear. In 1968, about the only conclusion we could draw in *The Population Bomb* about a greenhouse warming was that humanity was conducting a dangerous experiment in the atmosphere.[9]

▬ Other Trace Gases and Global Warming

As so often has been the case with environmental trends, the scale of the problem and the speed with which it could be upon us were underestimated. It was thought in 1968 that low-level clouds and air pollution could counterbalance global warming. Today that no longer seems likely, although they may still serve to slow it.[10] Only in the mid-1970s did climatologists such as Veerabhadran Ramanathan and James Hansen begin giving serious attention to the contributions that trace gases other than CO_2 might be making to global warming.[11] In the mid-1980s, work by Ramanathan, Ralph Cicerone, and others demonstrated quantitatively that several other greenhouse gases, present in trace amounts but also increasing, could together almost double the greenhouse impact of CO_2.[12] Suddenly, the rising concentrations of methane, nitrous oxide, and chlorofluorocarbons (CFCs) became an important part of the picture.

Methane, a simple molecule consisting of four hydrogen atoms hooked onto a single carbon atom (CH_4), comes primarily from anaerobic respiration (a form of respiration that takes place in the absence of oxygen) by anaerobic bacteria. Moist places that lack oxygen are the principal methane factories. Natural wetlands, including swamps, peat bogs, tundra, and poorly oxygenated wet soils; rice paddies; landfills; and the guts of ruminant animals (cattle, sheep) and termites are all significant sources, accounting for perhaps four-fifths of the methane emitted to the atmosphere.[13] The rest comes from leakage and venting in coal mining and oil and gas exploitation, and from incomplete combustion of organic materials, including biomass burning in the tropics.[14]

The atmospheric concentration of methane has doubled in about the last 150 years and is now rising at about 1 percent per year—more than twice as fast as CO_2. Much of the increase, 30 to 40 percent, has occurred since 1950, and the present concentration of methane may be the highest since Earth was young.[15] It is far higher than at any time in the last 160,000 years, according to analysis of air trapped as bubbles in the ancient ice of ice caps.[16] Human activities, especially the expansion of rice cultivation, cattle herds, and garbage dumps, as well as increased coal mining and extraction of natural gas, are thought to be major causes of this surge in atmospheric methane concentration.

Currently, the residence time of a methane molecule— the length of time the average molecule remains in the atmosphere—is about a dozen years, one of the shortest times among greenhouse gases, around a tenth or less of that of CO_2.[17] But, by actions such as emitting carbon monoxide from sources like automobile exhausts, humanity may be interfering with the natural processes that continually remove methane from the atmosphere. Carbon monoxide (CO) reacts with and reduces the atmospheric pool of a chemical called hydroxyl, which also reacts with and removes methane. The likely result is to lengthen methane's residence time and accelerate its buildup.[18] Hydroxyl acts as a "sink" not only for both CO and methane, but also for ozone, another common pollutant that is a greenhouse gas in the lower atmosphere. Thus the "competition" among the three gases for the hydroxyl sink depletes the hydroxyl pool and increases the residence times and warming effects of the competitors.

No one is certain why nitrous oxide, two nitrogen atoms hooked to one of oxygen (N_2O), is building up in the atmosphere, but its concentration is rising about 0.2 to 0.3 percent annually.[19] That may seem inconsequential, but it indicates a yearly net addition of about 3 to 4 million tons. Moreover, nitrous oxide is one of the longer-lived greenhouse gases, with an estimated residence time of 150 years, and is a relatively efficient absorber of infrared radiation. Possible anthropogenic sources are nitrogen fertilizers, forest clearing, and burning of nitrogen-rich fuels (especially coal) and organic materials (biomass).[20]

There is no question where chlorofluorocarbons (CFCs) come from. These compounds are hydrocarbon derivatives, made up of simple carbon backbones on which chlorine and fluorine atoms are substituted for hydrogens (for instance, CCl_2F_2, known as CFC-12). None occurs naturally; all are made by *Homo sapiens*. They are used in a wide variety of industrial applications and are better known for their role in depleting the stratospheric ozone layer (discussed in detail in the next chapter).

Releases of CFCs to the atmosphere were rising at a rate of about 4 to 5 percent a year in the 1980s, but recent international agreements to phase out their manufacture should put a stop to that—*if* the treaty signatories are vigilantly monitored and *if* the phase-outs are carried out properly. Even so, between 1990 and 2000, the CFCs released into the atmosphere will increase by 30 percent.[21] The most promising ozone-friendlier substitutes, moreover, are chemically similar and are also greenhouse gases, if somewhat less potent or long-lasting ones.[22] CFCs persist for decades to centuries in the atmosphere, so even if production were halted today, those already released will continue contributing to global warming for a long time.

The warming influence of a trace gas is primarily a function of two characteristics: its ability to absorb infrared radiation, and its residence time in the atmosphere.[23] A gas that, molecule for molecule, is a rather weak absorber of infrared but persists a long time may have a much greater effect per ton added to the atmosphere than a powerful absorber that is quickly removed. Generally, a molecule of methane has about ten times, nitrous oxide 180 times, and CFCs thousands of times the warming potential of a molecule of CO_2, if both absorption effectiveness and residence times are taken into account.[24]

But, of course, for each of the greenhouse gases, the annual amount added to the atmosphere and the rate of buildup differ considerably, so warming potential and actual warming contributions over time are very different things. By one estimate, CO_2 had contributed almost 72 percent of the warming potential through 1985, and each of the others less than 10 percent.[25] If only the *additional* heating potential added during the 1980s is considered, however, CO_2 accounts for only 57 percent of the heating potential, and all the other trace gases 43 percent. Thus the high-potential non-CO_2 trace gases are gaining in importance relative to CO_2.

So a question that twenty years ago appeared to be a potential concern for late in the twenty-first century has become an immediate and profound challenge. Concern spread to the public, fueled by the very hot summer and serious drought in much of the Northern Hemisphere in 1988.[26] The press was full of speculation that the drought was a harbinger of global warming, and climatologists conceded that it was unprovable but possible.

Among atmospheric scientists, concern had already been aroused by the announcement in 1985 of the totally unexpected appearance of a hole in the ozone layer over Antarctica (see Chapter Four). Although it was not directly related to global warming, that discovery reinforced a general conclusion among climatologists that the climatic system "operates on small differences between large numbers" and that sudden major shifts in the system were possible. Abrupt changes from one kind of climatic regime to another seem to have occurred in the prehistoric past.[27] A gradual warming of Earth might similarly lead to a sudden massive alteration in climate.

The 1988 drought, however, was not necessarily an indication that a change in climate due to increased greenhouse gases had been detected. Indeed, it is not even certain that the small increase in surface temperatures recorded in this century, about 1° F, can be attributed to the buildup of excess greenhouse gases—although that is certainly a good bet. But scientists *are* sure that, if those concentrations keep on rising, a temperature increase clearly attributable to them *will* be detected, sooner or later.

Sometimes the popular press writes about the greenhouse "theory," implying that there is great uncertainty about whether the mix of gases in the atmosphere is a major determinant of Earth's temperature. But there is *no* scientific debate about that. Atmospheric scientists have long recognized the radiation-trapping properties of water vapor and CO_2 and have concluded, for instance, that Venus is so much hotter than Earth largely because its heavy atmosphere is made up almost entirely of CO_2.

Moreover, a clear relationship has been detected between temperatures and concentrations of CO_2 and methane in the Earth's atmosphere in prehistoric times. A team of French glaciologists headed by Claude Lorius collaborated with a Russian team at Vostok base in central Antarctica, to examine air trapped in bubbles in a core sample drilled from the Antarctic ice cap.[28] The bubbles in the deepest ice were some 160,000 years old—so the core contained samples of air spanning two great glaciations ("Ice Ages") and two warm interglacial periods. Analysis of the forms of oxygen in the air samples also allowed temperature trends to be determined.[29]

The record correlates well with earlier, shorter ice-core records in periods of overlap and with those derived from other sources, indicating that the average global temperature changed over a range of about 10° F during that long period, and that concentrations of CO_2 and methane changed along with it. In warm interglacial periods, the atmosphere contained about 25 percent more CO_2 and twice as much methane as during the cold glacial periods (but never as much as today). So temperature has been correlated with greenhouse-gas concentrations for a long time.

Still, it is important to remember one of the classic cautions of science: "correlation is not causation." It is possible, for instance, that rising concentrations of greenhouse gases caused the great ice caps to melt at the end of glacial periods by warming the atmosphere. It is also possible that warming due to other causes at the end of glaciations increased the rates of respiration (yielding more CO_2) and promoted methane release, thus enhancing the concentrations of greenhouse gases. Perhaps the most likely explanation is that the major driving force of glacial cycles are cyclic changes in Earth's orbit and the tilt of its axis, but when warming occurs, the generation of more greenhouse gases creates a positive feedback, whereby warming itself leads to more warming.[30]

Detecting the Warming

Is there a similar correlation of CO_2 and methane concentrations with temperature changes over the past century, when CO_2 has increased some 25 percent and methane has doubled? The answer appears to be yes, although the question is still surrounded by some uncertainty and debate.[31] The climate is an extremely "noisy" system; everyone knows weather is very changeable. Some periods are naturally warmer or cooler than others, often for reasons that are at best only partially understood. It is simply not yet certain whether the greenhouse signal has been sorted out from all that noise. Doing so is difficult; climatologists are trying to detect with assurance a warming of only about 1° F over the past century, and then nail down its cause.

Part of the problem is simply one of measurement. Many factors must be considered and compensated for when attempting to compare temperatures of 100 years ago with those of today. These factors include the possible inaccuracy of early instruments, movements of weather stations up and down mountains and from city centers to airports, changing conditions (for example, many stations were established in population centers that have grown enormously and become urban "heat islands"), and improved techniques to measure seawater temperatures. The task is difficult, but it has been done to the satisfaction of most climatologists, who agree that there has been a warming of about 1° F (0.5° C) between 1880 and 1990.[32] Furthermore, the 1980s were the warmest decade and contained six of the seven warmest years in the century.[33] And 1990 proved to be the warmest year since people started keeping records.[34]

Warming does not necessarily mean *greenhouse* warming, however. There might have been a natural warming trend caused by something other than increased trace gases; perhaps thermometers are just showing some of the normal climatic "noise." But recent analyses suggest this is not the case, as mounting evidence points to greenhouse gases. Most climatologists now seem to lean toward the view that a greenhouse warming has been detected, but that the strength of the effect is not unambiguously clear from the "noisy" data.[35] Another decade or more will be needed to achieve a high degree of certainty and to make sound quantitative estimates of the degree of warming. Such data will come from the giant experiment we are running on the planet, not from a lab or a computer, just as Revelle and Suess predicted in 1957.

Future Warming

Regardless of whether the greenhouse effect has now been sorted out of the climatic noise, what are the prospects for global warming in the future? Of course, that mainly depends on future emissions of greenhouse gases. Those in turn will depend on circumstances discussed earlier, such as the energy sources humanity selects and the human population's size. But the speed and extent of future warming will also partly depend on the inertia of the climatic system and possible feedback mechanisms.

The planet may already have been committed to a further average warming of several degrees because of the inertia of the climatic system. When more solar energy is trapped near Earth's surface by increased greenhouse gases, the previous balance (equilibrium) between incoming and outgoing solar radiation is disrupted. Also disrupted are the previously balanced flows of energy between various elements of the climatic system: oceans, atmosphere, ice, clouds, vegetation, land, and so on.[36] This imbalance (disequilibrium) cannot be quickly restored, even if there are no further increases in greenhouse-gas concentrations. It takes a long time to warm up the huge mass of water in the oceans, which provides the main climatic inertia. Until the oceans reach a new equilibrium with the other parts of the system, they will serve as a heat sink, drawing energy from the atmosphere and slowing its warming.

As a result, even if greenhouse-gas concentrations in the atmosphere could be stabilized immediately, we would have to wait until the system returned to equilibrium before being able to measure the new stable average temperature of the atmosphere. But estimates of that temperature that climatologists have made with computer models indicate that the atmosphere at Earth's surface is already committed to a further average warming of 1° to 3.5° F beyond the roughly 1° F that has already occurred. How soon most of the additional warming will actually occur depends primarily on the length of the "ocean thermal delay," somewhere between ten and fifty years—more likely the latter, and possibly even more.

If human augmentation of atmospheric greenhouse-gas concentrations continues, the system will remain out of equilibrium, and the surface temperature rise will continue to lag because of the ocean's thermal inertia. Consequently, whenever humanity decides to do something about the greenhouse buildup, it will already be committed to perhaps 2° F of further warming when it makes the decision. And, since greenhouse-gas buildups cannot feasibly be halted instantaneously, humanity's warming "commitment" at any time before the buildup is stopped will clearly be substantially greater.

Positive feedback mechanisms that might be triggered by global warming itself are another cause for concern. Indeed, a serious concern of some scientists has been that such self-amplifying mechanisms might lead to a catastrophic runaway greenhouse effect, which would be uncontrollable by any human action. Methane is particularly involved in positive feedbacks. One important source is northern peat bogs, which may release methane more rapidly as the warming climate speeds up the metabolism of the anaerobic bacteria in the bogs. As a longer-term concern, a huge amount of methane is entrapped in frozen formations called methane clathrates, lattices of ice crystals and methane gas, in the permafrost deep beneath the arctic tundra. Warming of the tundra could melt the permafrost and liberate the methane, leading to further temperature increases. In fact, this process may already have begun, but significant releases are considered unlikely for hundreds of years.[37]

One recent study, however, indicates that bacteria in tundra soils rapidly oxidize methane to CO_2 (a much less effective greenhouse gas), releasing some in respiration and absorbing the rest. This activity could work against the positive feedback, since it would be likely to increase as warmer temperatures evaporated moisture from the soils and lowered water tables.[38] A question remains, though, whether the methane-absorbing capacity of drier tundra soils can offset the additional methane released from newly waterlogged tundra soils, defrosted by global warming.

Gigantic quantities of methane are also stored in methane clathrates deep under the oceans—indeed, the clathrates altogether may contain much more carbon than exists in the world's known coal reserves.[39] Under high pressure in the cold ocean depths, the clathrates generally remain stable. But warmed ocean waters could destabilize them, causing sudden huge releases of methane. On one occasion, Soviet scientists detected a 500-meter-long plume of methane belched from clathrates off the eastern Siberian coast.[40] That plume could have been the result of destabilization of subsea clathrates by the waning glacial period over the past 18,000 years. Many other suspicious plumes have been observed by satellites in the Soviet arctic, but their methane content was not confirmed.[41] At the moment, methane releases from clathrates appear to be a potential positive feedback factor that might affect long-term climate change. They may have played an important part in the cycling of the Ice Ages, but do not now seem likely to influence climate significantly over the next few centuries.

Carbon dioxide sources are also subject to possible positive feedback mechanisms. Respiration is very sensitive to temperature, with rates increasing in warmer conditions. A temperature rise of 2° F can boost respiration rates as much as 10 to 25 percent.[42] Vast amounts of carbon are stored in dead organic matter in soils. Bacteria and fungi can decompose that matter, leading eventually to the release of the greenhouse gases CO_2, nitrous oxide, and methane. Ecologist George Woodwell is rightly concerned that higher temperatures will raise the respiration rate of those decomposers and accelerate the flow of greenhouse gases from soils into the atmosphere. Woodwell estimates that, if there is no increased uptake of carbon through photosynthesis, the warming that has already occurred will, for a time, cause about 1 to 6 billion tons to be added annually to the atmospheric pool through increased respiration. Even the lower end of that estimate is not an insignificant amount compared to some 6 billion tons released annually by fossil-fuel burning and roughly 1 to 2.5 billion tons from burning biomass.[43]

If civilization is extremely lucky, the climatic system may contain some powerful negative feedback mechanisms that are as yet unrecognized. Negative feedbacks would be mechanisms in which the warming itself triggered mechanisms that worked to cool the planet. (The classic example of a negative feedback system is a thermostat that increases the output of a heater as a room gets cooler and reduces its output as the room warms, keeping the temperature within the comfort range.) Northward expansion of boreal forests as the climate warms, if there were no southern retreat, would be a negative feedback, since the growing trees would sequester carbon in their biomass. Another possibility is that warmer air will evaporate more water from the oceans and create more clouds. Clouds, depending on their form and altitude, can contribute either to warming or cooling. If cooling were the predominant effect (recent evidence indicates that it would be), the warming might be limited, but regional climates would change dramatically.[43A]

The possibilities of both positive and negative feedbacks are major elements in the greenhouse uncertainty. Of course, society should hope that negative feedbacks might slow the appearance of serious consequences of the enhanced greenhouse gas of the atmosphere. Nevertheless, it should prepare for the unattractive possibilities that feedbacks will balance each other out and global warming will essentially follow the projections, or even that positive feedbacks will dominate and accelerate the warming trend.

Should We Care?

So what? Why should society care if the planet gets a little warmer? If a one-degree Fahrenheit increase in average global temperature over the past century has occurred, it has scarcely been noticed. Even the 3° to 9° F of warming that models indicate are likely over the next half-century don't seem so awful. Sure, Dallas might get hotter than hell, but Minneapolis, Fairbanks, and Moscow might have livable winters.

In this vein, we heard Michael Boskin, the chief economic advisor to President Bush, pooh-pooh the whole idea of global warming. At a meeting of business executives in Virginia, he pointed out that virtually all who were there had experienced temperature changes of more than a few degrees as they traveled to the meeting and were none the worse for it!

Perhaps the best way to put in perspective the apparently small change expected in global average temperature is to compare present conditions with those of the last Ice Age. The coolest time in the last million or so years was 18,000 years ago, when the planet was only about 9° F cooler than it is today. At that time, an ice cap a mile or more thick covered most of Canada and much of the northern United States and Eurasia. The glaciers tied up vast amounts of water and lowered the sea level so far that you could have walked from England to France or from Alaska to Siberia without getting your toes wet. Later, melting the ice and rewarming the planet took about 10,000 years. Humanity may face as much *further* warming in just the next century: an equally big change, but as much as 50 to 100 times as fast.

Scientists are still debating about the amount of warming ahead of us and how fast it will develop.[44] The real concerns, of course, are how the average global temperature increase will affect climatic patterns, weather extremes, and sea level. While present computer models do a fine job of projecting the overall changes in the Northern Hemisphere as spring turns into summer, autumn, and then winter, they are not very good at predicting the precise *regional* effects of a given amount of average global warming. The models do agree that the warming generally will be greater farther from the equator. Thus, if the average global temperature rises 1° F, the tropics may have only a fraction of a degree rise, while temperate regions experience increases of 1° to 2°, and polar regions may be several degrees warmer.

Assume for a moment that the warming will be on the order of 5° F over the next century and that some of the regional effects that various climate models generally agree on actually come to pass: a somewhat pessimistic, but by no means extreme set of possibilities. What could be the results?

━━━━ Consequences for Agriculture

One of the most worrisome potential effects is a substantial decline in soil moisture in the centers of the large continents in the temperate zones, due to higher temperatures and reduced rainfall.[45] More than half of humanity's food is grown in those regions: central North America, Europe, the Soviet Union, and China. If the world's breadbaskets became even a little less productive, the consequences for humanity could be disastrous (as we shall explain in Chapter Seven).

Losses of productive capacity in some areas are unlikely to be compensated by gains in others, at least in the short term, for several reasons.[46] First, even if a loss of agricultural productivity in one area were balanced by the potential for improvement in another, there would certainly be an overall production decline while the system adjusted. Iowa farmers are not likely to pack up quickly, move, and start clearing boreal forest in northern Canada to grow corn, soybeans, or wheat. Farmers, on the basis of centuries of experience, normally think of unfavorable weather as a temporary phenomenon—something to take in stride and wait out. It will take some time for the true nature of the situation to become obvious, and still more time to decide that farming is worth trying elsewhere (assuming they are financially able to sell out, move, and buy a new farm).

Second, it would be a mistake to presume that global warming will simply redeal the cards, giving deuces to some previous winners and aces to some previous losers. Actually, there is good reason to believe that, for the next century at least, change will be rapid and *continuous*. It won't be just a matter of replacing farming operations in area A by opening up area

B; it will be a matter of matching production to a moving climatic target.

Third, if an area of the Canadian shield[47] now covered with boreal forest acquires temperatures resembling those of Iowa in 1970, it would not be an equally suitable place to grow corn or soybeans.[48] Soils play a very big role in determining which plants will grow where, and growing corn-belt crops on the often thin, nutrient-poor, acidic soils of the shield would be very difficult, to say the least.[49] Other soils farther west in the boreal forest zone, not on the shield, could prove suitable, but only after expensive drainage systems were installed.

Altered rainfall patterns are among the most likely, most unpredictable, and most deleterious effects that greenhouse-related climate change may have on agriculture.[50] Since more water will be evaporated from the oceans with higher global temperatures, average rainfall worldwide is bound to increase. But changes in circulation patterns may cause reduced precipitation in some areas or rainfall at different times. In agricultural systems, too much water, or water delivered at the wrong time, can be just as damaging as too little. Flooding and waterlogging would not help agricultural production, and might be more difficult to cope with than dry conditions (since much is already known about techniques of irrigation). Increased problems with flooding seem quite likely in monsoonal areas of the Indian subcontinent, a region that can ill afford *any* additional problems with food production.

Changed rainfall patterns resulting from warming-induced climate change are more likely than not to reduce crop yields. Most crop strains are finely attuned to the climates in which they are customarily grown, especially to the amount and timing of rainfall. Also, crops are very sensitive to day length and intensity of sunlight, which would be different farther north. At the least, new crop strains responsive to new conditions would have to be bred, and that would take still more time.

Another agricultural problem that could develop in response to greenhouse warming is that many fruit and berry crops (apples, pears, walnuts, cherries, grapes, apricots, peaches, and nectarines, for example) require chilling (vernalization) before fruit can be produced. This requirement, under rather simple warming assumptions, could cause considerable difficulty for horticultural crops in certain parts of Australia, among other areas.[51] Adjustments of tree crops to climate change would be especially slow, since the first harvest comes many years after planting. A similar problem might plague winter wheat, which normally undergoes freezing conditions in winter after sprouting (and which also requires summer water, which could be in short supply). And the quality of wheat grown under warm conditions may decline unless new, more heat-tolerant strains can be created.[52]

Since CO_2 is generally a limiting factor in photosynthesis, however, the possibility exists that the rise in CO_2 concentration itself will benefit agriculture by promoting plant growth. Supplementary injections of CO_2 into greenhouses have been shown to accelerate plant growth, and it is reasonable to assume that, in a CO_2-enriched atmosphere, plant productivity in general would rise (although no such rise has been clearly identified as a result of the 25-percent increase in CO_2 concentration already seen). Unfortunately, the plants might not allocate any additional growth to parts desired by *Homo sapiens* (although eventually this might be accomplished through genetic manipulation or conventional plant breeding).

Outside of a laboratory greenhouse, an atmosphere enriched with CO_2 might encourage pest and weed growth more than crop growth.[53] More CO_2 might also alter relationships between crops and pests in complex ways that do not necessarily benefit the crop. For instance, some pests have been shown to feed at much higher rates in an atmosphere with enhanced CO_2.[54] They apparently did so because concentrations in the leaves of the key nutrient nitrogen were low, so the insects ate more to make up for the lack.[55] Some insects, though, seem unable to compensate for the lack of nitrogen in their food and grow poorly on plants grown in CO_2-rich atmospheres.[56]

Today, one great advantage of temperate over tropical agriculture is an automatic pest-control service supplied by winter. In tropical regions that lack even a marked dry season to interrupt the benign growing conditions, insect pest populations can multiply year-round; this is one of the factors that operate against the practice of "conventional" agriculture in moist tropics. In higher latitudes where freezes occur, pests cannot reproduce, and so suffer high death rates in winter. To the extent that global warming prevents or moderates winter frosts and freezes, insect pests will gain an advantage. By favoring pest outbreaks and weed growth, global warming could lead to even more use of herbicides and insecticides, with all the environmental problems that additional use of these toxic substances would entail. The bottom line here, as far as can be determined so early in the game, is that pest problems might be worsened by global warming; at the very least, relations between plants and insects that feed on them will probably be disrupted.

The world food situation today is so marginal that grasping at the straw that increased CO_2 will enhance agricultural productivity is delusionary.[57] If any increases in yields of major crops actually materialize, they will almost surely be more than offset by reductions caused by disrupted and changed weather patterns, at least during a prolonged period of adjustment until a stable new regime can be established and adapted to. Furthermore, perhaps half of the warming ahead will be caused by trace gases other than CO_2, which have no positive effects on plant growth (and tropospheric ozone has a known negative effect). The longer it takes for civilization to come to grips with global warming and stabilize concentrations of greenhouse gases, the longer will be the period of disrupted climate and agricultural adjustment and the greater the degree of disruption.

Some complexities of the human dilemma were illustrated earlier in our discussion of the energy trap. Another, closely related trap lies hidden in the global warming phenomenon, which we might call the "food trap." This trap, quite simply, is that human attempts to keep increasing food production may create another positive feedback by releasing more greenhouse gases, thus accelerating global warming, which in turn will reduce food production.

Three greenhouse gases with increasing concentrations are linked to agriculture: carbon dioxide, methane, and nitrous oxide. Important sources of methane are rice paddies and digestion in ruminant animals, mainly cattle. The sources of the nitrous oxide increase are less well understood, but apparently include nitrogen fertilizers, without which a substantial further expansion in food production is unlikely to be realized. Clearing of tropical forests to create pastures also seems to increase the flow of nitrous oxide into the atmosphere.[58] And, of course, clearing forests to open new land for agriculture and the use of fossil fuels in the production and distribution of food all contribute CO_2 to the atmosphere. Unless considerable attention is given to finding and deploying alternative methods of cultivation and cattle feeding that produce lower emissions, while efforts are also made to limit both forest clearing and fossil-fuel use, increases in food production will almost inevitably accelerate the buildups of these greenhouse gases.[59]

Impacts on Natural Ecosystems

The same kinds of impacts that rapid climate change would inflict on agricultural ecosystems would also affect natural ecosystems. Altered climates and increased CO_2 levels would lead to complex changes in ecosystems that are now only partially understood.[60] Many forests and other natural ecosystems will be unable to migrate fast enough to "keep up" with an unprecedented pace of climatic change. Even if rates of seed dispersal and tree growth were adequate in theory, human-created barriers in most areas would prevent the compensatory migration of some plant species. On the other hand, human-caused transport, purposeful or accidental, would speed the dispersal of others.

Animals, especially migratory species and those with small populations or very strict habitat requirements, would suffer as well, especially as climate change interacts with other human insults to the environment.[61]

The rare Kirtland's warbler is a symbolic case. This beautiful species, now down to a few hundred individuals, lives in just six counties in Michigan where the habitat it requires is found: five- to six-year-old jack pine stands growing on sandy soil. Climate warming could cause the jack pines to disappear from the sandy soil, and in turn would almost certainly push Kirtland's warbler to extinction.[62]

Innumerable species that need special habitats around the world will follow if climates change very rapidly—quite possibly including virtually all arctic and alpine flora and fauna. Lakes in boreal regions may lose populations of characteristic organisms such as lake trout, a trend previewed by the response of lakes in northwestern Ontario to a 3.8° F rise in air and lake-water temperatures between 1969 and 1988.[63]

Global warming may have strong impacts on marine ecosystems as well. The incredibly rich and fascinating faunas of coral reefs might toally disappear, since the corals themselves appear to be vulnerable to even small temperature changes.[64] Extensive deaths of corals were seen throughout the Caribbean in 1987 and again in 1990, apparently associated with a slight warming of the water. There is no evidence that the warming of Caribbean waters is connected with the greenhouse warming, but it does suggest how susceptible crucial oceanic ecosystems may be to unusually rapid changes in temperature.

Even in tropical forests, where temperature rises are expected to be relatively small, considerable disruption could result. Many tropical organisms are adapted to highly stable climates, and even small changes in temperature and moisture could result in local extinctions and replacements of species.[65]

Climate change quite clearly could enormously exacerbate the current worldwide extinction episode. The decimation of species in natural ecosystems in turn will almost certainly degrade the services they supply, including the indispensable services to agricultural ecosystems. Both direct and indirect benefits will be lost. Hydrological cycles and local climates could be further disrupted by forest death and massive forest fires that could afflict many regions as trees weaken and die. Valuable fish species will disappear from warmed streams. Pest-control and pollination services could be impaired as local populations of the organisms providing them die out. Starlings, cockroaches, rats, kudzu vines, and other highly adaptable pest organisms that thrive on disturbance will doubtless become even more common as natural pest-control services are further compromised.

The relationship between the decline of biodiversity and the provision of ecosystem services (aside from the genetic library service) is not well understood now, and may never be known in detail. About all one can say with assurance is that the more rapid climatic change is, the more severe will be the pressure towards extinctions. And the more extinctions, the more likely are serious disruptions of ecosystem services. Certainly, if climate change causes biodiversity to decline in many areas over a few decades, losses of ecosystem services could be disastrous.

Foretastes of the effects of severe ecosystem disruptions

and biotic impoverishment can be seen in such diverse cases as forest destruction in eastern Europe by air pollution, the desiccation of the Aral Sea in the Soviet Union, and desertification in the Sahel region of Africa. These are admittedly extreme cases of what, in connection with the Vietnam War, has been called "ecocide"; with the additional stress of global warming, extreme cases could become commonplace.

▬ Sea Level Rise

Rapid climate change will not be the only consequence of global warming that *Homo sapiens* will face. As the planet warms, the sea level will rise, for two basic reasons. First, as water gets warmer, it expands. The oceans will creep up beaches just as mercury creeps up the sides of a gradually warmed thermometer tube. Second, glaciers will melt, depositing water and ice into the sea.[66]

There is considerable uncertainty about how fast and how much the sea level will rise. Part of the uncertainty traces to the projected increase in precipitation expected as more water evaporates from the warming oceans. Some of that precipitation will fall as snow and add to ice caps, but warmer water and air will also melt them around the edges. How the two processes will balance out is not at the moment predictable.[67] If the planet warms enough (in several centuries), sea level could rise some 15 to 30 feet, changing coastlines everywhere in ways difficult to imagine today.[68]

A reasonable estimate of the possible sea-level rise in immediate prospect is about a foot by 2050,[69] although that rise might be very unevenly distributed among the oceans.[70] Beyond the uncertainties connected with the warming itself, there are further uncertainties in the response of the ocean-precipitation-ice system. The oceans' waters will not expand uniformly; rather, expansion will depend on temperatures and salinities in different places.

A one-foot rise may not seem like much, but the effects would be far from neglible. The first would be increased flooding of many of today's coastal wetlands. Some wetlands might "migrate" inland, although whether that is possible would depend on the rate of sea-level rise. A one-foot rise over the next half-century or so is just about the rate at which marshes might "keep up" by upward accretion and inland spreading.[71] But in many places, natural topography or elevated roads, dikes, or other human-made structures would slow or prevent these processes.[72]

Coastal wetlands are highly productive biologically and are already seriously endangered ecosystems.[73] Their further diminution through global warming would have serious direct impacts on many valuable fisheries such as that of crabs in Chesapeake Bay, and would indirectly harm many commercially important deep-sea fishes that either depend on coastal wetlands to nourish their young or feed on wetland-dependent species.

Coastal erosion would also be seriously aggravated by even a modest rise in sea level, resulting in shoreline losses of between 10 and 100 feet, depending on local conditions.[74] Vulnerability to storm damage would also be increased, since storm surges would be provided the launching pad of a higher sea-level base. Storms would cause many hundreds of millions of dollars in additional damage annually in the United States with just a one-foot sea-level rise, unless extensive protective measures were taken. In Bangladesh, three times in recent decades storm surges have killed tens or hundreds of thousands of people.[75] Even a small sea-level rise would be nothing less than catastrophic there, especially since warming-induced increases in monsoonal rainfall could add 50 percent or more to the flows of the Ganges and Brahmaputra during the flood season.[76] Losses of farmland in Egypt's Nile Delta would exacerbate that nation's already severe food production problems, and a foot of rise would be extremely threatening to low-lying island nations such as the Maldives and the Bahamas.

A rising sea level would also create problems through the salination of coastal aquifers and the inland creep of salt water in estuaries. Many coastal aquifers are already contaminated by salt because of overpumping, and even a one-foot rise would threaten the water supplies of some cities that extract water from rivers near their mouths.

Of course, if a one-foot rise were all that occurred by 2050, that would probably be just a start. Without concerted efforts to control greenhouse-gas emissions, a half-century after that—within the lifetimes of our great-grandchildren—additional rises of two feet or so would be entirely possible, and still greater ones are conceivable.[77] Such rises would initiate a worldwide coastal catastrophe: millions of deaths caused by storm surges, cities abandoned, prime agricultural land inundated, many aquifers salinized, and millions of ecological refugees created.

Other Impacts

Another problem created by global warming may be an intensification of hurricanes. Hurricanes (tropical cyclones) can be considered as atmospheric devices for transferring heat from tropical to temperate regions. Some climatologists believe that, as the oceans warm, hurricanes will become much larger and more powerful.[78] Hurricane Gilbert in September 1988 was the biggest storm ever recorded in the Western Hemisphere. Hurricane Hugo, a year later, was nearly as gigantic; it caused billions of dollars in damage to the South Carolina coast and did massive damage to natural ecosystems.[79] Both represent the sort of climatic event that could become more frequent as global warming progresses. On top of a one-foot sea-level rise, such storms could easily convert "100-year record floods" into more common occurrences.

Water supplies also may become a problem in many areas. California, whose water comes largely from mountain snow melt, is projected to receive more rain and less snow. This change would lead to a heavy early runoff, followed by late spring and summer deficits just when the water is needed most.[80] Water shortages, already causing trouble in much of the United States West, could become acute in many areas.[81] Similar difficulties may beset drier parts of the Soviet Union, Asia, Africa, and South America as well.[82]

Threats to health also may arise with global warming. One of the most serious potential hazards is the extension of the ranges of disease-carrying vectors, such as mosquitoes that carry diseases like malaria, dengue fever, encephalitis, and yellow fever, into temperate regions.[83] One tropical mosquito species associated with dengue fever has already been found in the United States. Of course, heat itself will be a problem. The frequency of extremely hot days will almost certainly rise dramatically, especially in the warmer areas of the temperate zones, such as the southern and central United States. The annual number of summertime heat-related deaths (mostly among the elderly) in the United States could more than double in a world with the equivalent of twice the preindustrial concentration of atmospheric CO_2. The additional discomfort (and loss of productive work time) would also be great.

Finally, a variety of other problems caused by heat, from distortion of railroad tracks to melting of asphalt, would also plague societies and cost millions of dollars in repairs and preventive measures. Some of these problems, however, would presumably be compensated by milder winters.

In this situation, another kind of positive feedback could come into play. In the developed nations of the temperate zones, intensified summer heat would lead to increased use of air conditioning and refrigeration, which are generally more energy-intensive than heating and also entail use of CFCs. If fossil fuels were the source of power, more CO_2 and CFCs would be released as a result.[84]

Surprises

But then, maybe none of the untoward effects mentioned above will happen. Maybe some as-yet-unknown negative feedback mechanism will stabilize the climate about where it is now, or perhaps the warming will be so small and slow that civilization can easily adapt. Such an outcome cannot be completely ruled out.

But then neither can some very unpleasant possibilities. Perhaps the dynamics of the Antarctic ice cap are not properly understood, and deglaciation might occur much faster than expected. Or suppose the ocean "conveyor belt"—the global circulation of waters of different temperatures and salinities—proved sensitive to changes in the average global temperature, as recent computer modeling by Uwe Mikolajewicz and his colleagues at the Max-Planck Institute for Meterorology in Hamburg, Germany, suggests it is.[85] If so, ocean circulation patterns might change dramatically, totally revamping Earth's climate. Dramatic changes in the circulation could occur in only fifty years, and possibly could themselves greatly reinforce the greenhouse warming by releasing heat[86] or by slowing the absorption of CO_2.

If the upwelling of cold water from the depths were slowed or suppressed by the presence of a warmer surface layer, the capacity of the oceans to serve as a sink for CO_2 would be reduced. This would create a positive feedback mechanism that would enhance the warming, much to humanity's detriment. Also, upwellings bring nutrients from the ocean bottom to the surface, where they are accessible to photosynthesizing plankton, the basis of marine food chains. These upwelling regions support some of the world's richest fisheries; the loss or displacement of the nutrients could have serious consequences both for marine ecosystems and for fisheries on which societies depend for food. Such possibilities remind us that the climate works on small differences between large numbers, and the ocean-atmosphere system can still be full of surprises.

The fundamental question behind all this still is the one on global warming that we asked in *The Population Bomb* so long ago: "Do we want to keep it up and find out what will happen? What do we gain by playing 'environmental roulette'?"[87] Or, to put it another way, would the costs of slowing global warming be so high that we should simply gamble that everything will turn out all right, that we or our descendants will be able to deal with any warming-related problems that arise?

The answer, in our view, is a resounding "NO." Most of the measures that should be taken to slow global warming indeed *would be beneficial even if they had no effect on greenhouse gas concentrations and the climate were locked in the favorable mode of, say, the 1960s.*

How Should Humanity Respond?

What, then, should be done about global warming? The first goals must be to slow the pace of warming and to limit its extent as far as possible without incurring unsupportable costs. Slowing the greenhouse gas buildup and the rate of climate change will give agricultural systems time to adapt to changes, allow coastal wetlands more opportunity to build upward and/or migrate inland, provide coastal human communities more opportunities to adjust, and so on. Changes in many areas could be very substantial and many will be unpredictable; the more time societies have to plan and implement strategies to cope, the more satisfactory the outcomes are likely to be.

Because of the excess greenhouse gases already emitted, the lag times in the climatic system, and the slow response times of societies and economies, however, it will be impossible to avoid some warming and some additional greenhouse-gas buildup. International cooperation will be needed to organize responses and to deal with impacts that are unavoidable.[88] Fortunately, the first signs of such cooperation are appearing.

The first and most obvious step to limit emissions of greenhouse gases is a substantial reduction in the present human dependence on fossil fuels. The pattern of their use today is the result of a gigantic market failure, since they are generating, in the language of economists, enormous "externalities." (Externalities are social and environmental costs that are not reflected in the prices of the products whose production is creating the externalities.)[89]

To keep the level of greenhouse gases from climbing further, emissions of CO_2, N_2O, and CFCs would have to be drastically reduced as soon as feasible.[90] The distinguished Scientists Working Group of the Intergovernmental Panel on Climate Change (IPCC) warned in 1990 that "the long-lived gases would require immediate reductions in emissions from human activities of over 60% to stabilize their concentrations at today's levels. Methane would require a 15–20% reduction."[91] Moreover, this enormous task will have to be accomplished in the face of a rapidly rising population and a legitimate demand for more development in developing nations.

The sensible starting place is to begin curbing CO_2 emissions through serious programs to increase energy efficiency and switch to alternative energy sources (as described in Chapter 2). The more benefit that can be extracted from each ton of coal, gallon of petroleum, or cubic foot of natural gas that must be burned, the less CO_2 will be released into the atmosphere.

Increased efficiency would also save consumers money; lower costs and enhance profits (and competitive positions) for innovative corporations; lessen corrosive, crop-damaging, and health-threatening air pollution; and help abate acid rain. It is an ideal long-term "win-win" strategy for society. Unfortunately, it is not seen as such by most corporations or consumers today because the initial changes needed are not perceived to be in their *short-term* interests, and because there will inevitably be some associated costs that will (also inevitably) not be equally shared.

As quickly as possible, societies also should replace energy-mobilizing technologies that release abundant CO_2 (starting with those based on coal) with others that either release less CO_2 (preferably natural gas),[92] or none at all. Hard decisions will be required about which substitute technologies to deploy and for how long, and what sorts of policies should be employed to encourage the transition. Our recommendation is to put the primary effort into various solar technologies, both to make portable fuel (hydrogen) and supply electricity. Still, fission or fusion reactors should not be entirely ruled out as future possibilities, at least until or unless other, safer technologies prove adequate to meet needs.

Major decisions—those that lock in whole systems to a particular energy source, such as a city's subway net or an expanded freeway network—should be made only after careful technical and economic evaluations are made of the various options. These evaluations should based wherever possible on operational or experimental deployments (pilot projects). The most important choices must be made more or less democratically in pluralist societies, so keeping the public well informed and conducting tests fairly and openly will be essential.

Many questions should be carefully addressed before actions are taken, such as whether it would be more or less costly to replace hundreds of coal-burning power plants outright with solar or safe fission technologies, rather than converting the plants first to oil or gas, then phasing them out later as solar or nuclear replacements come on line. Test conversions and replacements might be necessary before sound judgements could be made on the best approach nationwide. Pilot projects in both paths could result in some immediate reductions in CO_2 emissions without undue economic disruption; the remaining plants could be converted or replaced according to the most economically *and* environmentally practical choice.

Of course, these are transitional measures, useful for easing the economic strains of switching away from fossil fuels. But if overall fossil-fuel combustion is not substantially reduced in the near future, no other combination of measures will do the job.

One favored mechanism for reducing CO_2 emissions in developed market economies like that of the United States is through some form of a carbon tax levied on coal, oil, and natural gas in proportion to their carbon contents.[93] Placing a tax on carbon would make the price of fossil fuels reflect at least part of the environmental costs of their burning, thereby encouraging both efficiency in their use (cars with high gas mileage, insulated buildings, energy-efficient heating, cooling, and lighting, and so on) and development of alternative energy technologies.

Such a tax is likely to be regressive, since poor people generally pay a higher proportion of their incomes than rich people do for fuels and electricity, and often live in poorly insulated homes.[94] Still, the revenue from a carbon tax could be used to lower other regressive taxes, such as the social security tax, although cutting personal income taxes in the lower brackets might be a more politically palatable alternative.

The combination would both shelter the poor and have a double-barreled positive effect. Imposition of a carbon tax would cause energy markets to operate more efficiently because energy prices better reflected the true costs of mobilizing and using energy. Making the carbon tax revenue-neutral by coupling it with lower social security taxes or personal and corporate income taxes would encourage individuals to work harder and earn more, while corporations would have an incentive to hire additional employees rather than buy energy-using machinery.

A carbon tax has another big advantage over classic "command-and-control" mechanisms such as governmental regulations to limit fossil fuel-burning. Supposing the government established standards for energy efficiency for businesses, a plant that was operating within the government standard for the industry would have no incentive to reduce energy consumption any further. But a carbon tax would provide that incentive. A carbon tax also has an advantage over gasoline taxes because it would cover most sources of CO_2.[95] One disadvantage, though, is that it might be more politically difficult to enact than efficiency standards or even increasing the gasoline tax. Overall, while it would be neither cheap nor easy to administer, a carbon tax mechanism would probably wean Americans from overuse of fossil fuels at a smaller aggregate social cost than any command-and-control mechanism.

How large a tax would be needed to hold emissions at the 1990 level in the United States, and what its possible impacts on the economy would be are somewhat uncertain, although experience with the sudden oil price hikes of the 1970s is revealing.[96] The tax might have to be several hundred dollars a ton (or about a dollar per gallon of gasoline) to reduce consumption and induce significant increases in efficiency. The Congressional Budget Office in 1990 estimated that a tax of $28 per ton of carbon content in fossil fuels would stabilize American CO_2 emissions and raise $163 billion in revenue over five years, while a tax of $113 per ton, beginning in 1991, might reduce emissions by 10 to 20 percent by 2000.[97] The big jump in fuel prices caused by the higher tax could also lead to a small increase in unemployment for a time and perhaps a percent or two less growth in GNP, at least while the economy adjusted to the change.

The studies done so far of economic impacts have focused almost exclusively on the negative impacts, leaving out the numerous subsidiary benefits that would flow from depressing the use of fossil fuels, quite apart from those of slowing global warming.[98] Among the benefits would be new alternative energy industries springing up; less deterioration of roads, bridges, and other infrastructure; and more employment of the construction industry to build alternative transport (such as rail) systems, which might be less expensive than freeways to construct and maintain; abatement of air pollution and improved health of people and crops; and reduced dependence on foreign oil, to name just a few.

Deforestation

Second only to reducing fossil-fuel use, stopping deforestation (and where possible, restoring forests) is an imperative strategy for reducing CO_2 emissions. A great deal of carbon is stored in mature tree trunks, and even more could be taken out of the atmosphere by planting billions of young trees. Again, almost everyone wins in the *long run* by taking this step (if almost everyone won in the short run, it would be happening on its own). Just stopping the destruction of the remaining closed tropical forests could withhold as much as 1.4 billion tons of carbon from the atmosphere annually (about 20 percent of the net input).[99] There would be other beneficial effects on climate as well, to say nothing of those accruing to local and regional human populations dependent on the forests.[100] Finally, conserving and regenerating forests would help preserve and maintain all the other essential ecosystem services that they supply, including their invaluable inventories of biodiversity.

But stopping deforestation is much easier said than done. In poor nations with burgeoning populations, shortages of farmland for peasants will make it extremely difficult to preserve tropical forests. In more arid developing regions, fuelwood is increasingly scarce, and the depletion of scrub forests and woodlands for this purpose contributes both to global warming and to desertification. For the well-being of rural people dependent on wood for fuel, as well as to reduce CO_2 and other greenhouse-gas emissions, woodlots and other fuelwood sources need to be restored and nurtured. But this, too, is difficult in the face of rapid population growth, poverty, and rising pressures on the land.

Even in a rich nation like the United States, stopping wholesale deforestation and replacing it with sound forest management practices will not be simple. America is burdened with a long history of legislation that distorts property rights and hence removes many incentives to practice sustainable forestry. The recent scalping of the Northwest and Alaska was a result of such a market failure, as well as a failure to internalize the externalities associated with various forestry practices (that is, to include such costs as destruction of fisheries following clearcutting of watersheds as part of the timber companies' business expenses and thus in the price of lumber). Legislation passed by Congress in 1990 corrected some of these imbalances, but the struggle to save the forests is not over. If it can be won soon enough, jobs now being lost in the timber industry could be preserved (see Chapter Six).

Reforestation will also be difficult to accomplish. In part, this is because the same pressures that now threaten virgin and older second-growth forests—cutting for fuelwood, timber, and pulp, and clearing land for agriculture—are certain to intensify as the human population continues to expand. Reforestation probably will be more and more difficult as climate change itself threatens the survival of forests. Indeed, saplings usually are more susceptible to drought stress (which might occur in many areas) than are mature trees, so selecting appropriate tree species to plant could be a problem.

Still, if it can be carried out, massive reforestation could significantly help mitigate greenhouse warming. Richard Houghton and George Woodwell estimate that reforestation of 400,000 to 800,000 square miles, about one and a half to three times the area of Texas, would take up about a billion tons of carbon annually.[101] They also note that ample amounts of land could be made available for such purposes, given appropriate adjustments in land use. But in many areas, irrigation might be required and water might be short; moreover, obtaining the necessary land for large-scale forest planting might prove more difficult than anticipated, given rising pressures on productive land worldwide.

Planting trees outside forests, including in urban areas, of course, could be an important part of the solution and would relieve the pressure on forests. President Bush has proposed planting a billion trees throughout the United States, and the proposal was included in the 1990 Farm Act. Not only would each tree make its tiny contribution by taking in CO_2, very local benefits would include shade and reduced demand for air conditioning in summer. In the countryside, trees help control soil erosion, stabilize watersheds, and provide shelter for other organisms.

Reforestation on a massive scale, not to mention protecting and managing the resultant forests, would be a gigantic undertaking to organize and finance. Moreover, replanting forests will help reduce CO_2 in the atmosphere only while the trees are growing—some 40 to 60 years. Mature forests are in rough equilibrium with the atmosphere with regard to CO_2; they give off about as much as they take in. If the forests are cleared and burned later, most of the CO_2 they sequestered will be released again. It would also be released, but more slowly, if the trees were left to decay.

In short, growing a forest amounts to a one-time removal of CO_2, and the CO_2 remains out of the atmosphere only as long as the forest (or the wood from it) is preserved. From this standpoint, construction and furniture are beneficial uses of wood;

paper and fuel are not (and here is another opportunity for tax incentives to influence economic activities in less destructive directions). Despite all the difficulties, we believe the benefits of forest restoration and widespread tree-planting would far outweigh the costs. Drawing CO_2 out of the atmosphere by planting trees now will help buy time that may be badly needed later.

Curbing Other Greenhouse Gases

Since carbon dioxide emissions are only about half of the story, steps must be taken to curb releases of other greenhouse gases, especially the long-lived ones that are more potent than CO_2 in trapping heat. In the agricultural sector, efforts must be made to limit the flows of methane and nitrous oxide into the atmosphere. Unfortunately, reducing emissions of N_2O may prove more of a problem than reducing CO_2 because of the intimate connection of N_2O to food production and because its sources and sinks are poorly understood.[102] On the other hand, methane's short residence time (about ten years) may provide an opportunity to compensate for N_2O's long one.

The first task in addressing both methane and N_2O emissions will be carrying out extensive research in natural and agricultural systems to determine approaches to fertilizing, rice cultivation, and cattle feeding that might minimize emissions without loss of productivity. Research is needed also to understand how the greenhouse gases interact with each other and compete for sinks (mechanisms by which the gases can be removed from the atmosphere, such as methane's reactions with hydroxyl). Reducing emissions of carbon monoxide, a product of incomplete combustion, may be one of the best ways to curb methane concentrations in the atmosphere, because CO competes with methane to react with hydroxyl and thus may lengthen methane's residence time.

In some cases, interactions complicate the issue; some fertilizing regimes, especially in rice cultivation, appear to reduce methane but increase nitrous oxide emissions. Such counterproductive efforts obviously must be avoided if possible. Given a choice, allowing increases in the shorter-lived methane is the better one, especially since it may be easier to manipulate its sink.

Fortunately, another set of long-lived gases may soon be eliminated altogether. An essentially complete ban on the production of CFCs and halons is absolutely necessary, not only to help slow greenhouse warming, but also because of the critical need to protect the stratospheric ozone layer. The ozone threat is now so clear that international agreements to eliminate the offending chemicals are coming into force (see Chapter Four), but care must be exercised to be sure substitutes are not potent greenhouse gases themselves.[103] These agreements and the process by which they were reached may serve as a model for the even more comprehensive and complex international actions now needed to deal with global warming.[104]

Global Greenhouse Control

The need for worldwide action to address greenhouse-gas emissions is patent. The allocation of effort among nations, however, will be complicated. If every rich nation agreed simply to reduce CO_2 emissions by 20 percent in twenty years, for instance, some nations would have to work harder than others. A few, such as Japan and Sweden, which have relatively energy-efficient industries, would need to invest proportionately much more than countries whose industries are more profligate users of energy, like Poland. Developing countries may suffer disproportionately from the consequences of global warming, including coastal flooding and effects on agriculture, but they are less able than rich nations either to help prevent the buildups or to cope with the consequences.[105] Many poor nations have very few realistic energy choices if left to their own resources, as illustrated by the case of coal in China and India.

One way around these dilemmas is to let nations make their contributions to reducing greenhouse emissions either at home or abroad. Rich countries could help poor ones find ways to develop without burning huge amounts of coal. Thus Japan's contribution might be primarily to help China deploy a solar-electric technology, whereas Poland and the United States could concentrate first on cleaning up their own messes and moving away from coal. [106] The details of making arrangements equitable could be left to treaties and protocols developed by international working groups and conferences of the sort pioneered in the effort to save the ozone shield, but regulating CO_2 alone (let alone methane or nitrous oxide) will be enormously more complex.

Some mechanism that separates the assignment of responsibility (that is, who pays) from the better opportunities to reduce emissions (who saves energy or plants forests) is needed. One proposal is a glorified carbon tax—a "world fund"—created from proceeds of fossil-fuel taxes contributed by nations. But this has the disadvantage of requiring some form of world government to administer it: collect the taxes and decide how and where they should be spent. [107]

A better idea is to develop an international market-based mechanism, thus avoiding the need for centralized international control—basically a world government. One possibility is to issue international tradable emission permits, which so far have seen only limited use. The permits are issued by a government or a regulatory agency to limit emissions of certain pollutants (usually air pollutants) and maintain a regional standard of environmental quality. Firms that can generate their products while emitting less than their allotted amount of pollutant can sell their unused permits to other firms that cannot. The costs and the regulatory effort required to achieve a given reduction in overall emissions are minimized, and an incentive is provided for industries to develop and install more effective emission-control technologies. If applied to controlling CO_2, the permits would encourage use of energy-efficient or alternative energy technologies.

Like a carbon tax, tradable permits would establish a monetary value for reducing CO_2 emissions. How well and how equitably tradable permits would translate to the international scene if applied to both industrialized and developing nations is quite unclear, although several variations have been put forward, including government-owned and traded permits. The latter also would probably require a large element of centralized government control, though perhaps exercised at the national level, which might hinder free international exchanges. [108]

Two energy analysts, Joel Swisher and Gilbert Masters of Stanford University, have proposed the use of "international carbon emission offsets" (ICEOs) as a currency for investing in carbon-saving services, such as forest planting, energy-efficiency measures, or alternative energy sources.[109] The ICEOs could be held by private firms or governments, and could be bought and sold within or between nations or even traded for debt relief. Within nations, they could be used in conjunction with any carbon-regulating regime—a carbon tax, emissions ceilings, or tradable permits—and would supplement national policies by allowing individual entities to seek the best investments. The ICEOs thus would be a means for meeting national goals that helped to distribute costs among nations more fairly.

The ICEOs would give developing nations a financial incentive to use clean energy technologies in their development, and to protect or replant forests and woodlands. In effect, carbon-saving measures would become an export commodity for poor countries, and the usual conflict between preserving natural resources and the need for development would be partly avoided.

Similarly, the wasteful energy use of developed nations (including the woefully inefficient nations of the former Eastern Bloc) could be mitigated through the easiest, least costly means first, with nations trading credits and debits to meet their national carbon-reducing goals. Rich nations also could gain credits by paying for forest planting in poor nations (a tactic pioneered by a New England power company that contracted to plant a forest in Guatemala to offset the CO_2 emissions from a new coal-burning power plant).

The ICEOs would require a minimum of bureaucratic management, which could be handled by an existing international agency such as the United Nations or the World Bank. Ideally, it would be administered within a structure such as a treaty for CO_2-emission reductions that incorporated each nation's reduction goals, but it could be initiated by a few nations even before a treaty was completed. Success with ICEOs, however, would depend on a fair and reasonable allocation of emission rights among nations, a potentially painful process that cannot reasonably be avoided. Even though ICEOs have the potential to help redistribute costs more fairly among nations, the negotiations leading to their initial distribution might be very difficult. The creation of valuable assets causes emotional reactions (think of the battles over television and radio station licenses) that may hinder or even halt progress.

▬▬ Heel-Dragging in the United States

The United States so far has not initiated any measures specifically designed to slow global warming (although the EPA has pointed out that the tougher regulations of the 1990 Clean Air Act will have a positive effect).[110] In part, this lack of action has been due to the failure of scientists to communicate effectively with each other and with decision makers. Statistician Andrew Solow testified in 1990 that he considered the probability of unprecedented climatic change in the next century to be "low." Stephen Schneider, a distinguished climatologist from the National Center for Atmospheric Research, considered it "uncomfortably high." It turned out that they agreed that the chances were at least 50 percent! Schneider understood that to decision makers a 50-percent chance of catastrophe is high; Solow was using the arbitrary (and in this case, utterly inappropriate) standard of routine statistical analysis that the result of an experiment is "significant" if the odds are at least 95 percent that it was not caused by random chance alone.[111]

This sort of miscommunication has maintained the impression that vast disagreement exists among scientists over the seriousness of the threat of global warming. The *lowest* probability we have heard any competent scientist attach to the chance of unprecedented climatic change in the next century is 25 percent. The Scientists Working Group of the IPCC, a panel of 200 atmospheric scientists whose work was reviewed by 200 others, reached a consensus in 1990 that there were at least even odds that a warming of 3° to 10° F will occur by the middle of the next century. Their "best guess" was about 5° F by 2050. They also estimated a sea-level rise of between 3 and 12 inches by 2030, and noted that continental interiors are likely to experience drier conditions in summer.[112]

Most of us, of course, would consider a 10 or 25 percent chance of such events more than high enough to take action unless the costs of so doing were absolutely prohibitive. After all, safety-belt use in automobiles is mandated to reduce the odds of an individual being killed in an accident—a chance without the belt of less than 0.01 percent per year. And you probably would skip lunch rather than eat in a restaurant where you knew there was "only" a 10 percent chance of fatal food poisoning.

British writer John Gribbin has suggested that sources of serious environmental hazards, unlike people accused of wrongdoing, should be considered guilty until proven innocent. He calls this approach the "precautionary principle."[113] Considering the severity of the probable consequences of global warming, and the relatively small cost of taking out some "insurance," waiting for final "proof" rather than acting on 50–50 odds would be the height of imprudence.

Greenhouse Economics

Unfortunately, confusion has been added to the debate by a series of one-sided estimates of the economic costs of greenhouse-gas abatement.[114] One estimate was that between 1990 and 2100, the costs of the necessary increases in efficiency and technological substitutions would be about a trillion dollars.[115] But that is less than $10 billion per year, an amount that could be saved by reducing our annual oil-import bill (which in 1990 topped $50 billion) by 20 percent. Moreover, the estimate may have been too high and evidently took no account of economic (let alone the noneconomic) benefits that would accrue from increased energy efficiency and cleaner technologies. Even if the cost could not be compensated, it would still be a small payment for insurance against economic losses that could run into tens of trillions of dollars over that period if the worst happened: American agricultural production decimated, coastal areas flooded, frequent wars over water (and perhaps food), and the economy generally disrupted.

Some economic analysts have displayed deep ignorance of the natural world and the potential costs of unabated global warming. In the guise of a sober cost-benefit analysis, Yale's William Nordhaus (who has been one of the more forward-looking of economists) focuses on dollar values rather than the relative importance of economic sectors to life support.[116] Thus he considers the vulnerability of natural systems to climate change as insignificant, in part because agriculture and forestry account for only 3 percent of the United States economy. One wonders how he thinks the other 97 percent would be kept growing if we had no food to eat![117] Nordhaus also shared with George Bush's chief economic advisor the classic misapprehension about the significance of a small change in Earth's average temperature: "The change in temperature between 8:00 and 9:00 a.m. on an April morning in Washington, D.C. is normally greater than the expected change from 1989 to 2089."[118] Nordhaus's analysis is a graphic illustration of how little understanding even many of the best minds in industrial nations (especially urban-oriented economists) have of the physical/biological underpinnings of their society.

Moreover, Nordhaus presents his estimated economic costs of reducing the long-lived greenhouse gases by 60 percent[119] as if no money would otherwise be spent in the relevant sectors. That is, he did not subtract from his cost estimates the expected "normal" expenditures on energy, transport, and other relevant activities that would be made regardless of global warming. He assumed no new cars would be manufactured, no other transport facilities constructed or renovated, no new or renovated factories, offices, stores, or housing, no investment in agricultural improvement or coastal development, and certainly no new energy technologies. A great deal of the cost of reducing greenhouse gases would obviously be money spent anyway; and it could be spent in ways that reduced the warming.

Only any incremental costs of building new cars or buildings so they are energy-efficient or installation of alternative energy systems should be chalked up against the costs of saving the environment; and dollar savings, such as lower fuel costs and those from less wear and tear on roads from lighter vehicles, should be included on the benefit side. Perhaps the final irony is Nordhaus's estimate that an efficiently phased-in conversion over several decades would cost the world over $300 billion— about equal to one year's United States military budget. What is more directly relevant, assuming three decades for the conversion, the annual cost is the same as that mentioned above: $10 billion, or one-fifth of the United States oil-import bill.

Curiously, while Nordhaus is nervous about shifting from fossil fuels to other energy sources to deal with the potential greenhouse threat, he views with equanimity the option of carrying out vast geochemical experiments to counter it: "Possibilities include shooting particulate matter into the stratosphere to cool the earth, altering land use patterns to change the globe's reflectivity, and cultivating carbon-eating organisms in the oceans." The costs of such programs (ignored by Nordhaus) could, of course, be much greater than the costs of slowing the greenhouse buildup—without including the potential costs of trying to cope with the results of blindly tampering with the little-understood systems upon which civilization is totally dependent.

One proposal, known as the "Geritol solution," is to spread an iron solution in the seas to encourage the growth of marine algae now limited by a lack of that nutrient.[120] Photosynthesis by the algae would soak up CO_2, and when the algae died, some would drift downward in the oceans and sequester in deep-water carbonate deposits some of the carbon extracted from the atmosphere.

It sounds good if you say it fast, but a closer look reveals high uncertainties and high costs. It is not clear what the next limiting nutrient for the algae would be; adding iron might not create algal blooms as large as expected. Algae that live in the iron-poor open ocean are adapted to the lack of iron; iron fertilization would doubtless cause them to be replaced by algae that could outcompete them if iron were present in abundance, with unknown consequences for oceanic food webs—consequences that could seriously damage fisheries. Experimentally fertilizing 600 square miles of ocean (a tiny portion of the area needed to make a significant dent in the problem if the system worked) would require $2.5 billion worth of special barges to haul the iron solution. A flotilla of hundreds of ships and barges operating 365 days a year would be required to spread the fertilizer, and when the costs of labor, maintenance, fuel, and producing the iron solution are considered, the benefits of the kinds of energy-conservation programs Nordhaus considers unwise look very attractive indeed.[121]

The Bush administration, influenced by dollar-dominant (and natural systems-ignorant) analyses of the sort done by Nordhaus, has remained recalcitrant on the global warming issue, just as it has refused to face the critical population component of environmental problems. At the July 1990 economic summit, European nations put pressure on the administration to join the effort to limit greenhouse-gas emissions, to no avail. John Sununu, chief-of-staff to the "environmental President" and architect of many of the administration's environmentally destructive positions,[122] then responded: "The day America puts caps on greenhouse gas emissions is the day America stops growing."[123] This is a good example of a "spherically senseless" statement—one that is stupid when viewed from any angle.

But perhaps the most egregious mistake the Bush administration made through 1990 was its utter failure to take advantage of the Persian Gulf crisis to declare a "war for energy efficiency." It was a golden opportunity to start weaning Americans from their big cars and other energy-inefficient devices, but the "environmental President" not only let it pass, but also publicly bragged that he would keep right on driving his gas-guzzling speedboat and urged other Americans to do the same. As late as January 1991, Bush, Sununu, Richard Darman (head of the Budget Office), and Michael Boskin (chief of the Council of Economic Advisors) gutted a proposal by members of Bush's cabinet to institute energy conservation in the administration's energy policy bill.[124]

Clearly, despite Sununu's silly statement, it would be possible to keep the United States economy growing (should that prove desirable) *with* caps on greenhouse-gas emissions. Finding and deploying alternative energy sources alone would generate substantial economic activity. Conserving energy after the energy "crises" of the 1970s was hardly a drag on the American economy—increased efficiency helped to cut economic losses caused by higher oil prices. Further progress is likely to bring more profits, not fewer, as well as improve America's competitive position in world markets. Similarly, stopping forest destruction (largely a loss to the economy and a cost to taxpayers today) and accelerating reforestation would benefit the economy.

Indeed, virtually all the steps required to reduce greenhouse emissions would have positive economic effects, even in the medium term, if the market distortions created by inefficient government regulation and, especially, the failure to internalize externalities were corrected. The conservative course would be to get on with ameliorative measures immediately in our domestic economy and to begin working with other nations toward equitable solutions.

Global warming is probably the greatest single environmental threat to the security of the United States and all other nations. Because it arises from a variety of sources and is caused by a multiplicity of culprits, global warming also will be the most difficult problem to combat effectively. Greenhouse-gas emissions, moreover, are intricately intertwined with other major environmental and economic dilemmas.

Nevertheless, the technologies needed to begin reducing greenhouse-gas emissions exist and are largely known. If the world's nations agreed to make the effort, global CO_2 emissions could be reduced by a quarter within fifteen years, according to a report by an international team of experts.[125] They advocate setting targets of holding temperature rises below 0.2° per decade (a rate that would probably allow adaptation by farmers and ecosystems without undue disruption) and sea-level rises below one to two inches per decade. They also propose setting a limit on the total CO_2 buildup of 400 to 560 ppm (the 1991 concentration is about 355 ppm), corresponding to a total postindustrial average temperature rise of 2° to 3.8° F, and a total sea-level rise of no more than 20 inches. Meeting such targets no doubt is possible, but would require a level of global cooperation and dedication far beyond any historical precedent.

While technological changes might help a great deal in making initial reductions in greenhouse emissions, here again we see the underlying problem of the excess scale of the still-expanding human enterprise: too many people demanding too much from the planet. Humanity is unlikely to avoid a greenhouse catastrophe unless it confronts and comes to grips with its fundamental cause.

Ozone: A Cautionary Tale

CHAPTER Four

Humanity has had the hubris to accept Linnaeus's name for itself, *Homo sapiens*. But as the global warming debate has shown, it's not clear that we deserve the formal label of "wise man." Indeed, pure luck may have saved civilization from a catastrophic threat closely related to global warming— ozone depletion. Luck appeared in the form of the insight and determination of just two scientists: Sherwood Rowland and Mario Molina. Perhaps "lucky man" would be a more appropriate label for our species—if still a bit sexist. How we got lucky in the case of ozone is an instructive tale.[1]

In the early 1970s, Sherwood Rowland, a distinguished chemist, had a list of research projects that he hoped to tackle. When Mario Molina joined his laboratory as a postdoctoral researcher, Rowland was pleased with his choice from the list— to figure out the fate of a group of long-lived synthetic molecules. The molecules belonged to a group of highly stable chemical compounds called chlorofluorocarbons (CFCs), constructed of atoms of chlorine and fluorine (and sometimes hydrogen) hooked onto a backbone of carbon.

The CFCs were best known to the public under the Du Pont trademark name "Freons," and were famous as a safe, nonflammable refrigerant. CFCs were invented by a Du Pont chemist in 1928, and it was considered a great triumph of chemistry when Freon replaced the sulfur dioxide or ammonia then used as the working fluid in refrigerators. CFCs were later used in automobile air conditioners and eventually turned out to be ideal, nontoxic propellants in aerosol cans.[2] Their insulating properties also proved them to be fine blowing agents for foam plastics used in packaging, building insulation, and styrofoam plastic cups. Later they also found an important role as cleaning agents for electronic circuitry.

Late in 1973, Molina went to work, and just before Christmas his calculations indicated a startling result. CFCs are almost inert in the troposphere (the mixed portion of the atmosphere below an altitude of 50,000 feet or so) and so would slowly drift upward until they reached the mid-stratosphere (about 100,000 feet). There they would be broken down by short-wavelength ultraviolet radiation of the sun—radiation that doesn't reach the lower atmosphere in large doses because it is screened out by the ozone layer. When the CFCs broke down, they would release atomic chlorine;[3] chlorine then would react with ozone (O_3), converting it back to the common form of oxygen (O_2). Unfortunately, the chlorine itself would not be incorporated into inactive chlorine compounds in the chemical reactions, but would appear again as a reaction product and be available to react with other ozone molecules. Thus it would function as a *catalyst*, a single chlorine atom participating in the chain-reaction destruction of numerous ozone molecules.

Ironically, if Rowland and Molina were correct, once the CFCs rose above the ozone shield, they could start chemical reactions that would thin the shield beneath which they had migrated upward. This thinning of the ozone layer could precipitate a vast tragedy for Earth's living beings, including *Homo sapiens*, by allowing increased amounts of deadly ultraviolet-B (UV-B) radiation to reach the Earth's surface.

Rowland and Molina were extremely careful in checking and rechecking their calculations, since the conclusion they led to seemed intuitively preposterous. A major use of CFCs in the early 1970s was in the booming aerosol-propellant industry. Could people aiming spray cans of deodorant at their armpits actually be destroying a feature of the atmosphere that had been created over many hundreds of millions of years? Could life on land truly be threatened by chemicals routinely discarded when automobile air conditioners were repaired? It was difficult to credit, but that was what the numbers said. So Rowland and Molina, after having their calculations checked by colleagues, "went public" in 1974 with a paper in the prestigious scientific journal *Nature*.[4] Soon thereafter, Molina and Rowland drew the only sensible conclusion from their work and recommended that the use of CFCs be banned. To them the costs of closing down an $8 billion industry in the United States were trivial compared with the risk to all life represented by the possibility of significant ozone depletion.

It is now clear that the Molina-Rowland paper was one of the great scientific contributions of all time in terms of its importance for the future of civilization. The two scientists' calculations and recommendation were not, as a naive observer might expect, greeted with a combination of gratitude and prompt action. Instead, there ensued more than a decade of scientific debate, political dispute, and, on the part of the chemical industry, foot-dragging and outright stonewalling. The Du Pont corporation, a major CFC producer, claimed that the Molina-Rowland calculations were speculative and controls were unwarranted—although Du Pont spokesmen said the company would cease production of CFCs if they proved to threaten the ozone shield.[5]

Most of the history of that period need not concern us. As one might expect, the manufacturers of CFCs persistently attacked the Molina-Rowland work. But the two scientists did outstanding jobs of both defending their results and explaining the significance of the calculations to other scientists, government bodies, and the press.[6] Rowland, as befitted the senior member of the team, was especially on the firing line, and he acquitted himself very well. A small group of scientists, a handful of politicians, and several environmental organizations rallied to Molina's and Rowland's support, however. So did the American public. While scientists and politicians haggled and industry scoffed at the CFC-ozone connection, consumers learned about the threat from aerosol sprays and stopped using them. In the first half of 1975, shipments of aerosol cans were off some 25 percent.

The Tide Turns

Most scientists were aware of the potential for disaster in the depletion of the ozone layer by the mid-1970s. Alarm about the ozone shield had first been raised in the early part of the decade when it was calculated that nitrogen oxides injected into the stratosphere by a large fleet of high-flying supersonic airliners (then proposed to be developed in part with government funds) could lead to a substantial ozone thinning. The scientific community was thus sensitized to the vital importance of the shield, and the stage seemed set for action when, in September 1976, a committee of the United States National Academy of Sciences confirmed the Molina-Rowland calculations. Shortly thereafter, the Environmental Protection Agency and the Food and Drug Administration proposed to phase out the use of CFCs as aerosol propellants, and CFC use in aerosols was banned in the United States in 1978.

But the story was far from over then. Scientific confusion and the arrival of the antienvironmental Reagan administration made it extremely difficult to progress toward the near-total ban on CFCs for which Rowland, Molina, and other prudent scientists were pressing. The National Academy of Sciences published several more reports with differing, but all significant, estimates of the amount of ozone depletion to be expected. A critical chemical compound predicted by the Molina-Rowland calculations to be produced as part of the reactions started by CFCs reaching the stratosphere was actually found there in the late 1970s. Nonetheless, the fight to close down the CFC industry remained stalled. Between 1981 and 1983, the EPA rolled over and played dead under the disastrous stewardship of Anne Burford. She may have gained herself a small niche in the history of unintentional humor by writing later, "Remember a few years back when the big news was fluorocarbons that supposedly threatened the ozone layer?"[7]

Then in 1985 came stunning news. That year Joe Farman of the British Antarctic Survey announced the presence of an ozone "hole" over Halley Bay in Antarctica. During the Antarctic spring, the survey's ozone-measuring instrument showed that there was a decline of more than 50 percent in the amount of ozone overhead. Instruments on NASA's *Nimbus-7* satellite quickly showed that depletion of the layer was widespread above the southernmost continent. Satellites had failed to detect it previously, in part because they generated so much information that scientists were far behind in analyzing it, and in part because the computers were programmed to discard as an error any data showing so much depletion; such an extreme drop was deemed impossible. Worse, it appeared the spring ozone declines had been occurring since the mid-1970s. The Antarctic ozone hole amounted to a "smoking gun" that almost everyone could perceive, and things got moving again.

The scientific community was shocked and frightened, and an expedition was quickly organized to go to Antarctica in August 1986 to see if the cause of the hole could be determined. The National Ozone Expedition, sponsored by the United States National Science Foundation, was led by a brilliant young atmospheric chemist, Susan Solomon from NOAA (National Oceanic and Atmospheric Administration). The hurriedly assembled expedition gathered critical data supporting the notion that chlorine from CFCs was playing an important role in creating the ozone hole. In 1987, the Antarctic Airborne Ozone Expedition was mounted to check the Solomon group's findings. NASA pilots flew very hazardous missions in a Lockheed ER-2 aircraft (a version of the U-2 spy plane) and in a DC-8 in order to gather the needed data. At the end of the expedition, there was little doubt that chlorine, carried aloft by CFCs and spread throughout the stratosphere, was destroying the ozone.[8]

The hole over Antarctica was so large and deep because of the presence in winter of polar stratospheric clouds, made of ice crystals, that can incorporate important compounds when they form and permit reactions to occur on their surfaces that will not occur in the atmosphere itself.[9] The clouds remove nitrogen oxides, which tend to stop the chain reaction, and they expedite production of chlorine molecules from chlorine compounds. Sunlight, returning as winter ends, converts the molecules into atomic chlorine, which destroys ozone wholesale in the absence of nitrogen oxides.

The consensus of the scientific community was by then, and is today, that Molina and Rowland clearly had been right all along. Indeed, it now appears that some 10,000 to 100,000 ozone molecules are destroyed for each chlorine atom that is released into the stratosphere by gradually ascending CFC molecules.[10] Depletion can also be caused by bromine, a close chemical relative of chlorine, which can reach the stratosphere carried by halons—stable chemicals used in fire extinguishers, but manufactured in only one-thirtieth the quantity of CFCs. Rowland, just after he and Molina had double-checked their calculations, had told his wife, "The work is going well, but it looks like the end of the world."[11] But for his efforts, it might have been.

━━ The Costs of Ozone Depletion

UV-B radiation has many adverse effects on people and other organisms, as well as on the natural and agricultural ecosystems that support them. The best-known and most frequently discussed impact of increased UV-B flux at the surface is its direct threat to human health. The importance of ultraviolet radiation, especially UV-B, in causing skin cancers has been well demonstrated experimentally; and depletion of the ozone shield would indeed cause a substantial increase. A rough estimate is that every 1-percent decrease in the ozone layer would lead to a 3-percent increase in nonmelanoma skin cancers.[12] These cancers are extremely common, but rarely fatal (mortality rate slightly below 1 percent). In the United States, a 10-percent reduction in the ozone layer would result in an estimated 160,000 additional cases of nonmelanoma skin cancers annually.[13] Nonmelanoma cancers alone resulting from a 10-percent decline by 2050 could kill over 10,000 Americans now alive (assuming no change in American habits of sun exposure).

Melanomas are rarer, but that form of skin cancer is frequently fatal. The cause of melanomas is less well understood than that of the other skin cancers, but their incidence appears to be strongly associated with patterns of exposure to ultraviolet (getting heavily tanned or burned during infrequent vacations appears to be the most dangerous pattern).[14] If ozone were depleted only 10 percent by 2050, there would be something on the order of 100,000 additional deaths among Americans born around the middle of the twenty-first century.[15]

Cataracts, clouding of the lens of the eye, can also be induced by UV-B exposure. The Environmental Protection Agency has estimated that a 10-percent ozone depletion by 2050 would result in some 4 million cataracts among Americans born between 1986 and 2029. Another worrisome health threat is presented by UV-B's general suppression of the immune response, which can affect human beings through exposure of the skin and peripheral blood.[16] The result may be increased susceptibility to serious diseases such as leishmaniasis, measles, herpes, tuberculosis, and leprosy, and to substances that cause allergic reactions. As two internists put it: "Immune suppression by ultraviolet light could predispose vulnerable populations to infection illness, particularly if compounded by poor sanitation, crowding, and malnutrition. In the presence of immune suppression, even mild illness could become serious, and serious illness fatal, as demonstrated by the lethality of opportunistic infections in patients with . . . AIDS."[17]

The direct health threats to people from ozone depletion may turn out to be trivial, though, compared to indirect threats of other kinds. About two thirds of the over 200 kinds of plants (mostly crops) that have been tested are sensitive to UV-B. Effects in sensitive plants included reduced stem and leaf growth, lower dry weight, reduced photosynthetic activity, and reduced quality (lowered protein and oil content).[18] Some of the most sensitive plants were legumes and cabbages. One study indicated that an increase of 16 percent in UV-B flux on average would substantially reduce soybean yields (by an average of 10 percent), but different strains of bean plants reacted very differently, and UV-B actually seemed to benefit some, suggesting the possibility of developing new UV-B–resistant strains.[19] Grains fortunately proved to be less sensitive, but rice production could suffer because of nitrogen deficiencies, since the microorganisms (cyanobacteria) that fix that nutrient in rice paddies are very sensitive to UV-B.[20]

There is also evidence that some tree species are more sensitive to UV-B than others and reason to believe that some pollinating insects that orient by ultraviolet light might be negatively affected. Essentially nothing is known about how an increased UV-B flux might alter relationships between plants and plant-eating animals—or indeed what effects it would have on natural terrestrial and shallow-water ecosystems in general.

With so little known, it is difficult to make predictions, although what is known doesn't inspire complacency. Much, of course, will depend on the amount of ozone loss; much on the still poorly understood responses of plant populations. While it is conceivable that strains of crops can be bred with reduced sensitivity to UV-B, the speed at which this can be done and the possible prices to be paid due to reduced pest resistance and lower yields can only be guessed at present. It seems clear that, especially in the short term, substantial depletion of the ozone layer could lead to significant declines in agricultural productivity.[21] And even small declines, say a few percent below what otherwise would be the production level, could have very serious consequences in a world that seems fated to have increasing difficulty in feeding the entire human population for the indefinite future.[22]

Even more ominous, no one can predict how enhanced levels of UV-B will interact with rapidly changing climates, acid precipitation, surface air pollution, and other insults to which agricultural ecosystems are certain to be subjected over the next half-century. In addition, natural ecosystems are likely to be similarly disrupted as these assaults change relationships among species. One can guess that rapid environmental changes, and possibly increased UV-B itself, will favor "weedy" species over those more valuable to humanity; weedy species are specialists at living in disturbed environments. Such disruptions also are likely to impair the ecosystem services that support agriculture.

An increased UV-B flux would have deleterious effects on life in the already badly stressed oceans as well. The effects of UV-B have been detected as deep as 60 feet in clear water and 15 feet in murky water, and that radiation has been shown to damage larval fishes, crabs, shrimps, other small animals, and, most important, phytoplankton (tiny aquatic plants).[23] Increased UV-B could cause fundamental shifts among species in marine phytoplankton communities, which could have serious impacts on organisms such as fishes that live at higher levels on marine food chains—and that human beings harvest.[24] There could also be a general decline of oceanic productivity.[25] In that case, reduced phytoplankton populations might lower the ocean's ability to absorb CO_2, thereby exacerbating global warming.[26] In short, humanity has been taking a gigantic gamble with the ozone layer by continuing to manufacture and release CFCs; if we're very unlucky, the consequences could be catastrophic.

Saving the Ozone Shield

Sadly, though, the scientific consensus on the dangers of ozone depletion still has not been converted into the immediate international ban on the manufacture of CFCs that is needed. But some encouraging progress has been made on the international front; and how that progress was achieved is worth considering, since it may serve as a model for international negotiations on other global environmental issues.[27] Genuine negotiations started just as the ozone hole was breaking into the news in 1985, with the Vienna Convention for the protection of the ozone layer. The convention, sponsored by the United Nations Environment Programme (UNEP) set a precedent by establishing an international framework for dealing with a *future* environmental problem. It committed its signatories to take steps to prevent damage to human and environmental health from the effects of thinning of the ozone shield, encouraged research into atmospheric processes, and urged scientific, technical, and legal cooperation, along with a free flow of pertinent information. The need to adopt future protocols actually to protect the ozone shield was recognized, but it was not possible to get agreement on them at Vienna.[28] Indeed, many nations delayed ratification, waiting to see what control measures would eventually be adopted.

The appearance of the ozone hole changed the situation dramatically. Increasingly, world leaders began to be convinced of the seriousness of the threat, and a series of workshops and meetings was held to consider various schemes for controlling CFCs, looking toward a major conference of plenipotentiaries (delegates legally authorized to act on behalf of their governments) to be hosted by the Canadians in Montreal in September 1987.[29] Although they benefited from the skilled leadership of, among others, Ambassador Richard Benedick of the United States, delegates at these preliminary sessions had difficulty reaching agreement as nations jockeyed for position. Debate centered on such questions as whether controls should be instituted on production capacity or on production itself, and whether CFC production should be eliminated, reduced, or merely frozen at current levels. A further problem was presented by the huge disparity between levels of use in rich and poor nations. Per-capita usage of CFCs in developing countries is one-tenth or less than in rich nations. Governments in the poor nations opposed controls that might limit their aspirations to provide more refrigeration facilities for food (thereby reducing the wastage that now contributes to food shortages).

The argument was made that freezing production of CFCs at 1986 levels and waiting for more scientific certainty before deciding whether to go further was sufficient action to protect the ozone layer. That argument was rejected for two reasons. First, when given the same input data on projected emission rates, the various models gave the same basic answer: unless both CFCs and halons were regulated, ozone depletion would continue. Furthermore, a mere freeze would be insufficient unless it were global—which meant that either poor nations would have to forgo increased use, or rich ones would have to reduce their use significantly in order to let the poor have more refrigerators.

The second and more important reason for rejecting a "freeze-and-wait-and-see" approach was more basic. Even if the computer models were wrong, the risk could not be taken. Because of the long residence times in the atmosphere of CFCs and halons, it might well be too late to avoid a ravaged ozone shield by the time a high degree of certainty was achieved. From the perspective of elementary risk analysis (see Chapter Eight), CFC and halon production should have been sharply cut back much sooner.

With the Antarctic ozone hole forming a grim backdrop to the proceedings, the working groups agreed to move toward a formula for reducing emissions of CFCs and halons. In September 1987, 23 nations signed the Montreal Protocol, which mandated a gradual reduction of CFC and halon emissions worldwide to 50 percent of the then-current levels. The document was complicated, since the reduction was to occur in phases, starting with a freeze, moving in 1994 to production and consumption at 80 percent of 1986 levels, and then to 50 percent of those levels in 1999. The protocol controlled the import of CFCs from nonsignatory countries, but did not prohibit the import of devices containing CFCs, although some restrictions would come into effect later.

The situation of developing nations was recognized by a provision allowing a grace period of ten years in compliance to "meet basic domestic needs" as long as the nations consumed annually less than two-thirds of a pound (0.3 kg) per person. Average global use in 1985 was about 0.2 kg per capita; the United States and other rich nations used 1 to 2 kg per capita.

One of the most important provisions in the Montreal Protocol is for a review every four years of whether the controls are doing the job of protecting the ozone layer. Each review is to be based on expert scientific, environmental, economic, and technical evaluations.

That the review provision was badly needed is clear, as there is now a scientific consensus that the proposed 50-percent reduction (which for technical reasons would reduce annual emissions of CFCs from about 1.32 million tons to 0.78 million tons) will not be sufficient to stabilize the ozone layer. Indeed, it appears that at least an 85 percent reduction in ozone-depleting chemicals will be required to keep the ozone shield at its current thickness.[30] There is evidence, especially in the behavior of the Antarctic ozone hole and the observed thinning over the Northern Hemisphere, that models may be underestimating the rate of ozone destruction.[31] Furthermore, two important chemicals that attack the ozone layer are not covered in the protocol. One of these, methyl chloroform, is already manufactured in larger quantities than any CFC. Although its potential for destroying ozone, molecule for molecule, is only 15 to 25 percent of that of CFCs, if its production continues to grow unregulated, it could account for about a third of the ozone-destroying chlorine in the stratosphere by 2075.[32]

Industry continued to stall even after the Montreal Protocol was signed. In February 1988, three United States senators asked Du Pont to live up to its pledge to discontinue use of CFCs if they were shown to be a threat to the ozone shield. On March 4, the chairman of Du Pont refused, claiming that "scientific evidence does not point to the need for dramatic CFC emission reductions." But later that month, evidence of ozone thinning over the Northern Hemisphere forced Du Pont to announce belatedly that it would cease production of CFCs as soon as substitutes became available. The decision ended more than a decade of disgraceful stalling—made all the more disgraceful for Du Pont since CFCs accounted for less than 2 percent of that giant company's earnings, and it had long been in a superb position to spearhead a phase-out and develop alternatives.[33]

Further evidence that global ozone depletion was more severe than had previously been thought began to appear in 1990.[34] A Norwegian geophysicist, Ivar Isaksen, reported that balloon observations made in the lower stratosphere showed a 10-percent ozone reduction over the middle latitudes of Europe and North America.[35] Measurements made at 12,000 feet in the Alps indicated about a 1 percent increase per year in the UV-B flux since 1981.[36]

In early 1990, there was still some disagreement on the urgency for phasing out use of CFCs. The European Community wanted a 50 percent cut in production between 1991 and 1992, an 85-percent cut in 1995-1996, and an end to production between 1997 and 2000. The United States, Soviet Union, Japan, and several other countries wanted a somewhat slower reduction, but one that still ended production by 2000.

Evidence that the seriousness of the problem was finally recognized almost universally came when the Bush administration, which has been recalcitrant on most critical environmental issues from population control to global warming, reversed its position and decided to contribute to a new international fund to help poor nations deal with the technical and financial problems of substituting for ozone-destroying chemicals. The White House chief of staff, John Sununu, had feared that "a new fund would set a precedent for expensive new foreign aid programs on the environment."[37] Pressure from other nations, Congress, environmental groups, and even CFC manufacturers persuaded the President to change his mind. The irony, of course, is that some form of "expensive new foreign aid programs on the environment" almost certainly will be required of all rich nations if they are to remain rich, or even nations.

■■■ CFC Phase-out

The Bush administration's decision cleared the way for 93 nations to agree in London in June 1990 to terminate production of CFCs, halons, and carbon tetrachloride (another ozone-threatening chemical) by the end of the century. The participants also agreed to cut production of methyl chloroform by 70 percent by 2000 and cease production by 2005. An exception was made for developing nations, which will not have to stop CFC production until 2010. Best of all, both China and India are expected to ratify the agreement, even though India was bitterly critical of the United States for its "economic imperialism" in pushing poor nations to forgo CFCs without guaranteeing adequate assistance in switching to substitutes. A fund was established to provide such assistance, and the United States agreed to contribute $40-60 million over the first three years, during which the fund will disburse $240 million.

Meanwhile, the search for and evaluation of substitutes for CFCs has been proceeding steadily. Replacing them will be costly, and in some cases it appears that the substitutes will be less satisfactory in immediate applications. For example, the principal substitutes for CFCs as aerosol propellants will be hydrocarbons such as butane/propane mixtures, which have the disadvantage of being flammable (but the active ingredients in most CFC-propelled aerosols were dissolved in alcohol and thus also moderately flammable). Many aerosol uses, however, are essentially frivolous and give consumers less active product for their money; so society would benefit by the disappearance of aerosols in those uses. A classic example is in deodorants, where stick and roll-on varieties appear to be superior in every dimension. In other applications, pump dispensers substitute quite adequately.

For blowing polyurethane foam (used in insulation), a hydrocarbon can be used, or one of the new hydrochlorofluorocarbons (HCFCs),[38] or a new carbon dioxide technology that makes a satisfactory foam for some uses, such as foam furniture cushions, in which insulating properties are not important. For refrigeration, hydrofluorocarbons (HFCs) show great promise.[39] HFC-134a, especially when blended with a small amount of HCFC, comes close to being a "drop-in" substitute for CFC-12 (the most widely used CFC) as is likely to be found. It is not, however, compatible with the lubricant used with CFC-12 in refrigeration systems and will have a different effect on other system components. Not only will a new lubricant have to be developed, but systems using one refrigerant cannot be topped off with the other. Furthermore, it is estimated that in 1994 a pound of HFC-134a will cost about $3, somewhat more than the cost of the CFC it would replace,[40] although the gap is being closed by a substantial tax on CFCs designed to discourage their use, and unit costs can be expected to drop as production increases.

On the other hand, corporations at the forefront of chemical technology stand to make substantial profits from producing substitutes. In 1990 Du Pont, which for so long refused to recognize the threat posed by CFCs and was so convinced that Molina and Rowland were wrong that it gave up the search for substitutes around 1980, announced that it would build four plants to produce HFCs.[41] Imperial Chemical Industries in Britain won't go broke either, but will also reap the rewards of its investments in research and development.

Needless to say, the possible interactions of CFC substitutes with the ozone layer and their greenhouse characteristics are both of great concern. The HFCs have no ozone depletion potential (ODP), since they lack chlorine. The HCFCs have been calculated generally to have only a tenth or less of the effect on ozone as the worst CFCs.[42] Similarly, most of the replacement compounds would contribute only about a tenth as much to global warming, molecule for molecule, as the CFCs. Reassuring as those findings are, if the scale of use of the new compounds were permitted to grow too large, they could still cause great trouble.[43] It is clear that all CFC substitutes should be closely monitored both for their atmospheric effects and for the quantities produced. Finally, despite the success of the final agreement, the concentration of CFCs in the atmosphere will continue to rise as long as they continue to be produced, and their long residence times guarantee that they will continue destroying ozone and contributing to global warming for decades, perhaps centuries to come.

▆▆▆ Lessons

Several things seem obvious from the history of the affair begun by Molina and Rowland. First, as things now stand, with the scale of human activities so large, serendipity will play an increasingly important role if civilization is to survive. If Sherry Rowland hadn't become curious, the background might not have existed to interpret Joe Farman's observations quickly. Another decade of delay in controlling ozone-destroying chemicals could have been lethal. Much more effort put into monitoring the state of our life-support systems could reduce the bad luck factor. It also would be relatively inexpensive (compared to fixing well-advanced global problems), especially if action were taken promptly on the basis of incoming data.

But getting fast international action, as we have seen, is exceedingly difficult, even when scientists understand quite well what is happening. Pathetically little has been done, for example, to arrest the global hemorrhage of biodiversity. No substantial steps have yet been taken to slow the flow of greenhouse gases into the atmosphere, except for the Montreal Protocol and the subsequent move virtually to eliminate production of ozone-depleting chemicals by 2000, steps that were not taken in response to the potential of those chemicals to enhance global warming. Population control is not even a real political issue in nations like the United States, which make the biggest contributions to the most serious problems. And, of course, the subject of restricting the scale of human activities is unimagined in political circles, even though the need to do so is manifest.

All these problems are gigantic, pervasive, and complex compared to that of ozone depletion. Their causes are much more diffuse and often poorly understood, and instituting the necessary cures will gore infinitely more oxen. In short, the long, grim struggle of Rowland, Molina, and many others to solve the ozone dilemma (if it is truly solved) was a comparative cinch. A British scientist, R. Russell Jones, summed up the ozone situation very well: "For my part, I have no doubt that the issue of stratospheric ozone represents the litmus test of man's ability to prevent the ultimate degradation of Planet Earth. If this relatively simple problem cannot be solved, how can mankind hope to survive the vastly more complex problems that will arise in centuries to come?"[44] How, indeed!

Yet, more than fifteen years after action should have been taken, it is just beginning to appear that the ozone problem may be solved[45]—providing, of course, that compliance with the new protocols is effective *and that atmospheric chemistry does not harbor any more dangerous surprises.* Humanity is not necessarily out of the woods yet. The latest calculations are that, under the revised Montreal Protocol, the stratospheric concentration of chlorine will peak early in the twenty-first century at almost six times the pre-CFC level, and will not drop back to the level at which the ozone hole was generated until around 2050. Joe Farman put it very well: "We are condemned to 60 years of the unknown."[46] The discovery in 1991 that ozone loss over the United States, Europe, and much of Asia had been proceeding twice as rapidly as expected was ominous news.

Humanity must live with those six decades of uncertainty because a relatively small and concentrated industry and complacent governments (in the absence of a well-informed, active electorate) managed to prevent necessary steps from being taken for more than a decade.[47] Still, there is no question that the Montreal Protocol and its follow-up are reasons for some optimism. Scientists and governments, encouraged by informed public opinion, eventually put together a precedent-shattering agreement that was sensitive to special conditions, attempted to distribute burdens equitably, and established an ongoing process for monitoring and negotiation.[48] Nonetheless, the ozone "solution" cannot form a complete model for the future. The day must soon come when potential environmental threats can no longer be presumed innocent until "proven" guilty.

Science, unfortunately, never "proves" anything (although it can *disprove*). All it can do in evaluating most environmental threats is to give if-then statements accompanied by reliability estimates. They are of the following form: "If X tons of CFCs are released over the next fifty years, there appears to be about a 75-percent chance that the ozone layer will be thinned by an average of 25 percent." Then society presumably can decide, when fully informed of the best estimates of the impact (and costs of avoidance) of such a thinning, what steps to take.

But as we have seen with the relatively simple ozone problem, this system works slowly and extremely imperfectly at best. Scientists often disagree strongly with one another. It takes a great deal of time for society to inform itself even partially, and those likely to suffer economically will resist any policy change. Politicians are first uninformed and then they vacillate, torn by conflicting interests. Meanwhile, the environmental damage accumulates.

The solution to our intertwined environmental problems, if one can be attained, clearly must take another form. *Prevention* is the only safe way to deal with global environmental problems. That entails, as we will continue to repeat, a reduction in the scale of human activities. It also entails making human activities, regardless of their scale, as environmentally benign as possible. Often that will be easier than reducing human numbers (lowering P in the $I = PAT$ equation) or curbing consumption (lowering A). After all, everyone agrees that it is desirable to reduce the impacts of technologies on the environment (to lower T). And part of that reduction must consist of avoiding the deployment of technologies, such as CFCs, that raise T.

To do so, ways must be found to evaluate beforehand and regulate all large-scale technological activities that are undertaken—like the introduction of certain substitutes for CFCs. Such evaluations will not avoid all unpleasant surprises, but it should reduce their frequency and help ameliorate their impacts. As the human population will inevitably continue to grow in the short term, that sort of regulation will be increasingly important, even though it will impose costs on society. We cannot escape the iron grip of $I = PAT$.

Pollution: Dead Trees and Poisoned Water

When people think about environmental problems, they usually think first of air and water pollution, and of course these are serious problems. No two things are more critical to human life than air and water, and many of the substances civilization adds to air and water are directly toxic to human beings as well as other living things. The amounts of pollutants released into the environment and their effects have increased almost explosively in the second half of the twentieth century. Between 1950 and 1985, the annual production of synthetic organic chemicals in the United States increased nine-fold, from 24 billion to 225 billion pounds.[1] Tens of thousands of these compounds, from food additives and medicines to plastics and pesticides, have been incorporated into the economy and released into the environment. Of about 70,000 chemical compounds used commercially today, roughly half are considered potentially or actually hazardous to health by the United States Environmental Protection Agency.

We will not, however, dwell here on the cancer, heart and lung diseases, and other ailments caused by these toxic substances. Nor do we dwell on the gross threats to health from water-borne disease—an especially serious problem for Third World children, millions of whom have no access to clean drinking water. The direct human toll from pollution can be measured in millions of early deaths and tens of billions of days of illness annually. No one knows the exact burden of death and illness directly attributable to pollution, but most people are aware that it is substantial. Our focus will be on the less familiar, but ultimately even more dangerous indirect threats that pollution represents: its impact on ecosystems, humanity's life-support apparatus.

■■■ Air Pollution

Air pollution was the first environmental problem to get broad coverage in the American media, when the increasingly smoggy atmosphere over Los Angeles became the butt of comedians' jokes in the late 1940s. The direct health effects of air pollution still capture most of the public's attention. In 1990 it was announced that more than half of all Americans live in cities where ozone levels exceed permissible standards and a third are frequently subjected to dangerous levels of carbon monoxide.[2] In response to the need to conserve energy in the 1970s, offices and homes were designed to be airtight, thereby unexpectedly creating a new source of air pollution, as toxic gases and radioactivity built up indoors. Indeed, for millions of Americans, these may be more important threats to health than the more visible pollutants belched from smokestacks and tailpipes.[3] Because of many uncertainties,[4] it is difficult to calculate how many people die in the United States each year primarily because of air pollution, but plausible estimates range up to 70,000.[5] Millions more undoubtedly suffer illness and lose work time.

Equally serious problems exist in many developing countries, where essentially no emission controls are put on industry or vehicles. Some cities in developing nations have worse smog than Los Angeles has.[6] In rural areas, indoor air pollution from inefficient cooking stoves in poorly ventilated huts places a huge load of illness on populations.[7]

Living in the polluted air that often envelops Los Angeles and most of the planet's other large cities undoubtedly is not conducive to robust health or long life. And the chronic respiratory problems that people—women and children especially—are subjected to in poor regions surely does not increase their productivity or ability to improve their lives. Even so, if the direct health effects were the only consequences of air pollution, humanity could carry on for a long time, a little sicker and a little shorter-lived because of it, but in no great strife.

The substances that humanity injects into the atmosphere pose more serious threats to the future of civilization through their effects on natural and agricultural ecosystems. We have already seen the potential for disaster inherent in the atmospheric pollutants that enhance the greenhouse effect and destroy the ozone layer. Air pollution also directly damages agriculture and natural ecosystems, primarily by injuring and killing plants.[8]

━━━ Acid Deposition

Perhaps the most serious effect on ecosystems of air pollution is acidification, mainly through acid rain. The connection between air pollution and acid rain was first made in 1852;[9] in 1911 inhibition of the nitrogren cycle in soil by acid around the city of Leeds was reported.[10] In 1959 a Norwegian scientist made the connection between losses of fishes from freshwater lakes and acid deposition, and in 1968 Svante Odén first pointed out the potential widespread damage to ecosystems from the movement of acidifying air pollutants across national boundaries. The problem attracted public notice in the United States in the early 1970s, when ecologist Gene Likens and his colleagues pointed at fossil-fuel combustion as a cause of acid rain in eastern North America.[11]

Acid deposition traces to elevated levels of oxides of sulfur and nitrogen in the atmosphere as a result of releases from power plants, automobile exhausts, and the burning of forests, savannas, and grasslands.[12] These are transformed into sulfuric and nitric acids in the atmosphere, which acidify rain or fog, or form in material deposited dry on leaves or other surfaces.

Acidification has diverse impacts on ecosystems. Most organisms are closely adapted to their environments, including their acidity (or "pH"), and are sensitive to changes. In soil, drastic alteration of the pH causes marked changes in the composition of the soil flora and fauna. Heavy metals and aluminum are mobilized in acidic soils and can reach toxic levels. Aluminum in the soil may be taken up by plant roots, or it can be washed into streams and lakes, where it injures aquatic organisms.

Many lakes in northeastern North America, Britain, and

Scandinavia are already heavily acidified and have lost much of their former animal life, including valuable fish species such as trout. As a stopgap measure, putting lime in lakes to reduce their acidity has been attempted in Norway and Sweden (where more than half the lakes have become fishless) and in Britain and North America. The treatment has proven very unsatisfactory for various reasons. One is that outflow streams are neutralized by lime applications, but inflow streams are not. Inflow streams often are where the fish breed—and breeding failure is the main cause of the loss of fish populations. Liming also has negative effects on many other elements of the ecosystems of lakes and adjacent areas.[13]

Acidification can severely depress populations of amphibians; like fishes, they have difficulty reproducing in acidified lakes and streams. Indeed, acidification may be one cause of the observed decline of amphibian populations around the world.[14] But the pervasive impacts of acidification in ecosystems recently have been underlined by its implication in causing declines of bird populations.[15]

As might be expected, the first effects appeared to have been mainly on reductions in birds' food supplies, especially birds whose sustenance comes from aquatic systems. Declines in populations of loons that tried to raise their young on fish-poor acidified lakes in North America have been reported, and acid precipitation is suspected of contributing to the decline of the American black duck as well. In Britain, surveys have shown breeding dippers to be absent from acidified sections of streams, sections that also contain few of the larval insects upon which dippers (which forage underwater in fast-flowing streams) feed.

Direct acid damage to breeding bird populations may now have been detected in northern European forests. In the Netherlands, resident titmice, nuthatches, and great spotted woodpeckers all showed eggshell thinning in the 1980s.[16] Acid precipitation seems to have an effect on birds reminiscent of that seen earlier from overuse of chlorinated hydrocarbon pesticides such as DDT. The pesticide toxification and more recent acidification of the birds' habitat both cause birds to lay thin-shelled eggs that soon are crushed by the incubating parents.

In the late 1970s and early 1980s, forests in Europe started showing signs of a new kind of damage so serious that in Germany it was christened *Waldsterben*—forest death.

Forests throughout central Europe have now shown symptoms—
a yellowing and dropping of needles or leaves, often followed
by death of the trees—and in many areas they are well advanced. The damage to trees follows a long period of slow soil
acidification, which directly injures tree roots and may reduce
the availability of essential nutrients such as magnesium.[17] Direct injury to leaves by airborne acids may also contribute to
the syndrome.[18] To complicate the diagnosis, the problem apparently was exacerbated in the early 1980s by a drought in
Europe. Some recovery was noted when the drought ended in
1985, but forest biologists predict that the next drought will
cause even more severe damage.[19]

Signs of forest deterioration have also been detected in
the forests of the eastern United States.[20] The deterioration has
been associated with air pollution and acid deposition because
the degree of damage at different altitudes correlates with the
concentrations of air pollutants at those altitudes.[21]

The cause of the damage to forests remains controversial;
besides air pollution and acid precipitation, blame has been put
on climate changes, as well as attacks of insects, plant diseases,
or other predators or parasites, and other sources of natural
stress. Various tests to determine the reasons for forest damage
have not proven definitive; often the apparent causes differ from
one place to another, despite the superficial similarity of symptoms and the widespread nature of the problem.[22]

The most likely answer is that a complex set of interacting stresses, varying in relative importance or effect from
place to place, is responsible for the syndrome of forest decline.[23] Natural conditions or events (such as a drought or a
pest outbreak) may play predisposing or triggering roles in producing symptoms. But many or most of the stresses contributing
to forest decline probably arise in one way or another from human
activities.

The vulnerability of an ecosystem to acidification depends heavily upon the capacity of its soil to neutralize acids.
Areas with highly alkaline soils are much less susceptible than
those with neutral or acidic soils. The most serious problems
appear where high emissions of sulfur dioxide and nitrogen oxide
coincide with susceptible soils. Such vulnerable areas today
include eastern North America and central and southern Europe.
Sensitive areas that are likely to see great increases in emissions

include southern China (northern China, where most of Chinese industry is now located, is less prone to damage because of its alkaline soils), equatorial Africa (especially Nigeria), southeastern Brazil, northern Venezuela, and southwestern India.[24] Decision makers would be well advised to avoid locating future industrial facilities or fossil-fueled power plants in those areas or upwind of them.

The deadly effects of acidification on sensitive organisms often become apparent only after long periods of acid deposition, as the buffering capacity of soils and watersheds is used up. As a result, the number of lakes throughout the world that are sensitive to acid rain but that have not yet become acidic far exceeds the number that are now acidified (another example of the time-lags that often obscure the potential seriousness of environmental problems). Many of these sensitive, but as yet undamaged lakes are found in regions where acid rain occurs. Thus the dead or dying lakes in eastern North America may only be the tip of the iceberg. Clearly it is critical to determine whether, and on what time scale, these sensitive lakes will become acidic.

Two University of California ecologists, John Harte and his colleague James Kirchner, have tackled that very problem. They developed a mathematical model, based on properties of the soil in watersheds above lakes and records of the pH of local precipitation, that could be used to predict the rate at which lakes will become more acid in the future. The model has been validated by comparison with historic data from lakes in which increasing acid levels had been monitored, and now has been applied to a dozen sensitive lakes in Colorado, the Adirondacks, and New England. The results suggest that the iceberg is, indeed, a big one. Many lakes that are not yet acidic are likely to become so in coming decades if current levels of acid rain continue.[25] Ecosystems in trouble from acid deposition may seem healthy for years, then suddenly begin to collapse. By then it may be too late to reverse the insidiously accumulating effects of acidification.

Even though the precise effects of acid deposition on forests remain uncertain, one regionally dispersed air pollutant, ozone, has been unequivocally shown to damage forests and crops.[26] Vegetation is also sensitive to sulfur dioxide, the widespread air pollutant released in coal burning, smelting of ores, and refining of petroleum (and also largely responsible for acid rain).[27] In some areas close to industrial stacks, sulfur dioxide has killed all the vegetation. Various other components of air pollution are also known to harm plants. In situations where one dominant pollutant is clearly responsible for observed damage (and laboratory tests can confirm the effects), there is little controversy.

But sorting out the precise effects of acid precipitation is so difficult because forests, other natural vegetation, and crops in many places are subject to a variety of airborne chemical stresses. There doubtless are synergisms (interactions in which the combined effects of two or more elements are greater than the sum of their effects taken separately; that is, the pollutants in essence act to worsen each other's effects) among the direct impacts on flora and fauna of the airborne pollutants. These synergistic effects in turn almost certainly synergize with other stresses such as climate change and increased UV-B flux. And plants weakened by some or all of these damaging impacts are less resistant to "normal" natural stresses such as plant diseases, insect attacks, and adverse weather.

The consequences of such combined assaults have been most dramatically displayed in formerly communist central and eastern Europe, where pollution controls have been virtually nonexistent.[28] In Czechoslovakia and the former East Germany, almost 500 square miles of forest have disappeared or been reduced to stands of dead and dying trees, and lesser damage is visible throughout central and eastern Europe. The "best" situations are in Hungary, where only 25 percent of the forests are damaged, and Bulgaria, where the damage level is approaching 50 percent.[29] Where forests are most damaged, disruption of the hydrologic cycle has been severe, resulting in widespread flooding and erosion in the spring and water shortages in summer.

Yet, although eastern Europe's problems resulted from decades of gross environmental irresponsibility, comparable damage has been seen in parts of western Europe, despite its better record of environmental management. Indeed, the ability of the pollutants to cross borders with impunity has resulted in a great deal of finger-pointing and blame between European nations as to the sources of trouble.[30] Similar tensions also damaged relations between the United States and Canada during the Reagan years, when the United States administration stoutly refused to admit that acid rain could be a problem.[31] Better relations since George Bush took office are largely due to his receptivity to addressing the acid-rain issue.

Acid precipitation is more than an ecological disaster; it carries huge economic consequences as well. The direct economic losses resulting just from forest damage in all of Europe were conservatively estimated by the International Institute for Applied Systems Analysis (IIASA). By IIASA's reckoning, acid rain will cost the continent some 118 million cubic meters of wood, worth about $30 billion in 1990, per year for the next century if emissions of sulfur and nitrogen compounds are not sharply reduced.[32] That cost amounts to two and a half times as much as European governments have so far agreed to spend annually to abate air pollution.[33] As fossil-fuel use continues to rise and nations continue to drag their heels on pollution abatement, the destruction of Europe's forest ecosystems may be a sad harbinger of what awaits most of the globe as the human enterprise keeps expanding.

The appalling destructiveness of acid precipitation and the insidious way its effects develop are worrisome enough. Even more so are possible interactions between acid rain and global warming, an area of trouble that is only beginning to be understood. Uncertainties abound, but preliminary assessments give scant room for complacency; the effects of each can worsen those of the other, although in a few cases, they may have mitigating effects.[34] Acidification disrupts ecosystem processes and causes losses of species, thereby making the systems more vulnerable to climate changes (such as Europe's 1980s drought), which in turn can intensify the effects of acidification. Deposits of nitrogen compounds in air pollution may also lead to accelerated releases of nitrous oxides from soils while reducing soils' ability to absorb methane, thus increasing the buildup of both potent greenhouse gases.[35]

Some good news is that nitrogen deposits might slightly "fertilize" waters over the inner continental shelf and increase the uptake of carbon dioxide by phytoplankton. Sulfurous particles emitted by plankton or present in air pollution may "seed" clouds that help shade and cool the planet.[36] Much more needs to be learned about these interactions; but on the whole (despite any possible counteracting effect on global warming) they add urgency to civilization's need to make substantial cuts in emissions of the acidifying pollutants.

As with other widespread problems, the fundamental long-term solution to problems of air pollution is a reduction in the scale of human activities. Cutting per-capita emissions in half while the population doubles leaves no improvement. People, even subsistence farmers, inevitably affect the atmosphere by adding heat, combustion products, dust, gases, and other substances to it. Industrial societies add a daunting variety of chemical compounds. Even seemingly minuscule sources, such as solvents evaporating from drying paints and aerosol propellants from underarm deodorants, have proven to be significant air pollutants when used by tens of millions of people in an urban area.[37]

There is no question that substantial gains can be made in abating air pollution, and those gains will be crucial in the face of projected growth in both population and per-capita economic activity worldwide.[38] Serious efforts must be made to shift society away from reliance on the most damaging technologies, as well as to reduce the wasteful components of per-capita consumption and hasten the reversal of population growth. Air pollution, even more than many other environmental problems, is the result of energy use, especially combustion of fossil fuels. Without direct efforts to curb the volume of pollutants released per unit of energy mobilized, it is doubtful that even achieving the per-capita energy-use levels of the Holdren "optimistic" scenario would suffice to prevent a continued undermining of Earth's life-support capacity.

In the United States, a long-overdue new amendment to the Clean Air Act was passed in 1990, which will considerably strengthen the nation's ability to reduce air pollution. For the first time, the act specifically addressed the pollutants responsible for stratospheric ozone depletion (CFCs) and acid deposition (sulfur dioxide and oxides of nitrogen).[39] Production of CFCs in the United States must be ended by 2000, and other ozone-depleting chemicals are also on phase-out schedules. Sulfur dioxide emissions from the nation's most polluting utilities must be reduced by half by 1995 and total emissions from all power plants halved by 2000, after which date a cap on total emissions will apply. Oxides of nitrogen emissions from stationary sources and vehicles are to be reduced by half or more by 2000. While these reductions will not be sufficient to reverse acidification processes in sensitive areas, they will significantly slow them.[40] Further reductions later may be achievable through improved technology and a shift away from dependence on fossil fuels, especially coal.

The 1990 Clean Air Amendments for the first time introduced at the national level a tradable emissions-permit system to abate air pollution. The advantage of tradable permits is that pollution from fixed sources such as power plants and factories can be abated without unduly penalizing sources that lack or cannot afford newer control technology. In order to maintain a desired standard of air quality, permits are issued by the government or a regulatory agency allowing each firm a share of permissible release based on its production. Those firms that can maintain or increase production of goods while reducing emissions below their allotments can sell their unused permits to firms that cannot.

A tradable permit system has been used successfully to control air pollution in California for several years. The system has benefits similar to those resulting from a pollution charge levied on specific rates or quantities of pollution, such as the proposed carbon tax (Chapter Three). The total costs of achieving a given reduction in overall pollution are minimized, and incentives are provided for industries to develop and install more effective pollution-control technologies.[41] Meanwhile, as the United States begins implementing tradable permits, pollution charges—a tax on specific quantities or rates of pollution, such as a carbon tax—are gaining popularity in Europe.

The Clean Air Act also tightened up controls on automobile emissions, but not as much or as quickly as environmentalists wished. They were disappointed that another bill to mandate substantially increased fuel efficiency in cars failed to pass, nor has any legislation that emphasizes alternate means of transportation even been on the agenda. Yet, although the 1990 Clean Air Amendments were passed a decade late and were somewhat weaker than they should have been, the United States is still generally ahead of other industrial nations in regulating air pollution.[42]

A slow step in the right direction is better than none, but tinkering with technologies within a framework of an inherently unworkable overall consumptive pattern—that is, the

American automobile-centered urban transport system—can take us only so far.[43] It's another classic case of suboptimization—doing in the best possible way something that shouldn't be done in the first place.

▬▬ Water Pollution

There are two major environmental aspects to human use of fresh water. The first is the classical area of "water pollution"—contamination by human beings of both fresh and salt waters and the direct consequences of that contamination to human beings. The second is that water use and pollution have led to widespread degradation of ecosystems and ecosystem services. As in the case of air pollution, effects on ecosystems are likely to be the most serious and long-lasting, although public attention is, quite understandably, more focused on human health effects. Misuse of water resources has been responsible for much ecosystem disruption; we discuss that topic more fully in the next chapter, and here we will mention some problems associated with pollution.

That the billions of dollars spent in the last two decades in the United States to clean up surface drinking-water sources—rivers and lakes—have failed to eliminate pollution is obvious just from sampling headlines around the nation: concerns about contaminated fish in the Great Lakes and the Everglades, toxics in New Jersey, dioxin downstream from paper mills, and so forth.[44]

An even greater health concern today is pollution of groundwater, for two reasons. First, one out of every two Americans depends on well water for drinking; and second, once contaminants reach groundwater, they are extremely hard, if not impossible, to remove. But a variety of highly toxic substances have been finding their way into our wells, from leaking landfills and dumps, from mines and farms, and even as a result of acid rain.[45]

Such problems are also widespread elsewhere in the developed world, especially Western Europe.[46] In the Netherlands, a major source of trouble is the manure produced by intensive livestock production, which causes numerous problems. Ammonia is a major air pollutant and contributes to acid rain; toxic levels of nitrate and phosphate leach into groundwater and cause eutrophication (overfertilization leading to oxygen depletion) of surface waters; and heavy metals from feed additives are accumulating in the soil, poisoning soil flora and fauna, and reducing crop productivity. As in other realms, the appropriate answer to pollution of groundwater is prevention. Where the escape of pollutants has already occurred, cleanups are essential, and the quicker the better.

Water pollution's effects on ecosystems, however, are mainly through surface water systems. The Great Lakes region is a case in point for North America. Long a dumping ground for industry and mining, still the drainage for much of the Midwest's intensive, chemical-based agriculture, and, curiously, a continent-wide sink for many airborne toxic substances, including PCBs and pesticides, the Great Lakes system is still endangered.[47] The effects on fishes and wildlife suggest that the region's toxic brew of pollutants may pose a threat to the human population that uses the water and eats the region's fish, game, livestock, and dairy products. Despite substantial efforts to clean up the region's lakes and rivers—the Cuyahoga no longer bursts into flame, and Lake Erie has been pronounced alive—the outlook is for continued deterioration unless a massive renewed commitment is made jointly by both the United States and Canada to reverse the trend. Fortunately, such a reversal is both technically and economically feasible; here the problem is that the consequences to both the region's ecosystem and human health are sufficiently subtle that action may be delayed while the problems are studied—and unnecessarily perpetuated.[48]

■■■■ Pollution of Land: The Waste Crisis

Even land is becoming polluted, as the standard place for disposal of urban and industrial wastes. Society digs a hole, stuffs in the garbage, and forgets it—until years later a Love Canal seeps out as a grim reminder. Perhaps the ultimate in this kind of land abuse is the hundreds of millions of gallons of radioactive and toxic wastes buried in pits and leaking metal tanks by the Department of Energy (and its predecessor agencies) at the sites where nuclear weapons have been made for over forty years.[49] The costs of cleaning up all the sites have been variously estimated at upwards of $150 billion.

No American cities have waste-disposal problems quite so horrendous, but they are serious enough. The nation is dotted with tens of thousands of "Superfund" sites—abandoned dumps eligible for enforced cleanups under the national Superfund law set up for the purpose. The majority of these sites have barely been identified, let alone properly evaluated or cleaned up. Most contain a witches' brew of toxic and hazardous substances. Leaking dumps not only are a threat to people using the land later, but chemicals and heavy metals such as lead, mercury, cadmium, and arsenic may leak downward to aquifers, poisoning drinking water from wells, or find their way into surface water supplies.[50]

Perhaps the biggest problem in waste disposal is finding space for it. Conflicts have taken place over waste-disposal sites between cities and even between nations; about 10 percent of the European Community's roughly 25 million tons of waste per year are passed from one country to another for disposal.[51] Developed countries produce such excessive amounts of garbage and toxic wastes that they have taken to paying poor nations (whose disposal problems are comparatively small; they cannot afford to waste materials) to accept them. Unfortunately, the people in less developed nations, desperate for foreign exchange, often are unaware of the hazards in the cargo they have agreed to take.[52]

The waste crisis in the United States is only a symptom of our basic environmental problems, but it is an instructive symptom. First, it shows $I = PAT$ in action. From 1960 to 1986, the American population grew by 34 percent, but the amount of solid waste generated skyrocketed by 80 percent.[53] This means that the average person threw away 34 percent more trash ($1.34 \times 1.34 = 1.80$) in 1986 than in 1960—that is, half the increase in trash (I) was caused by population growth and half can be ascribed to $A \times T$—increased affluence or changes in "technology" (more wrappings, more unrecyclable plastics, and so on).[54]

Recycling has been pushed by the environmental movement for two decades, but only in the late 1980s did it become a reality for many American cities, under the impetus of over-stuffed landfills. Even so, only about 11 percent of American trash was recycled by the end of 1989.[55] Some of the "solutions" to this dilemma are nothing of the sort. Biodegradable plastics will help in the long term to reduce the volume of materials in landfills by perhaps 10 percent if widely used. We do not know, however, what the effects will be of their dusty residue, but that is almost a trivial point. If by technological tricks and regulatory efforts the volume of garbage per person could be reduced by half, that would simply double the lifetimes of landfills and still leave us with a crisis early in the next century. And, to the degree that the population continues to grow, reductions in per-capita waste disposal will fail to reduce the total waste flow.

It is crystal clear that dramatic reductions in the *production* of materials that will be transformed into wastes must be made soon, along with brand-new ways of handling the ones we can't avoid producing. Exactly how these changes are to be brought about cannot be specified here, but the following are the sorts of things that will be required.[56] Prices of materials used in packaging should be raised, perhaps by direct taxes (or perhaps even by natural resource depletion quotas)[57] that greatly increase the costs of raw materials used in packaging, until every producer's goal is to minimize the packaging that accompanies any item. The cost of cutting trees to produce pulp for paper should be raised to the point where paper recycling becomes profitable, or taxes should be used to subsidize recycled paper, or both. New technologies should be encouraged to replace the horrendous amount of paper that now goes into newsprint or is used in offices. Paper now makes up over 50 percent by volume of the material in landfills and almost half by weight. Of the paper in municipal solid waste, packaging and newsprint each account for about a fifth.[58]

Some changes, of course, would be viewed as inconveniences, but not being served on throw-away plates in fast-food restaurants would provide more jobs for busboys and dishwashers; one can quickly get used to carrying string or canvas bags to supermarkets; and electronic newspapers could be immensely more convenient than wrestling with giant piles of paper. And the escalation in garbage-collection bills might even slow a little.

Ocean Pollution

Finally, we should say a few words about what is happening to the greatest of water resources: the oceans and the valuable ecosystems they contain. An enormous amount of attention was focused on pollution of oceans by the great Alaskan oil spill of 1989, one of three large spills that year.[59] The Alaskan spill was the largest in history up until then (11 million gallons, coating some 4,000 miles of shoreline),[60] and the most controversial and costly to clean up. Unfortunately, the legalities of the situation led to suppression of some scientific findings and hindered the usual collaboration and exchange of information between scientists who were determining the oil's effects on the rich flora and fauna of Prince William Sound. Legal considerations also caused studies to be focused on short-term rather than long-term consequences.[61]

There is no question, though, that hundreds of thousands of marine birds were killed in the first few months, as well as thousands of sea otters and fishes.[62] Large but uncountable numbers of seals, porpoises, and whales are assumed to have been injured or killed. Among the animals exposed to toxic chemicals were the human cleanup workers.[63] Concern was high about the impact on Alaska's valuable fisheries, especially salmon,[64] but the impact will not be clear until 1991, when salmon hatched in 1989 first return to the streams where they will breed. In the longer term, breeding failures are likely in bald eagles and other birds that may be affected indirectly through buildups of toxins in the food chain.

The extent of damage from oil spills depends on many factors including: the kind and amount of oil spilled, the ecosystem into which it is dumped, the climate, and wind and current conditions.[65] But in sensitive and highly productive ecosystems, such as exist in Prince William Sound and in many coastal areas where spills are most likely to happen, damage can be devastating in the short term and may well have lasting effects. One well-studied oil spill of 2 million gallons off the coast of Panama in 1986 was found to have killed reef corals and intertidal flora and fauna over an extensive area.[66]

If ever there were a situation where a smidgen of prevention is worth a ton of cure, it is with oil spills. The Alaska spill may at last force the industry to use double-hulled tankers (the $2 billion that Exxon had to pay to clean up Prince William Sound inadequately would have bought a lot of tanker hulls!). Such preventive measures are available for offshore drilling facilities as well, not all of them requiring costly new technologies. One analyst has pointed out that much greater safety can be achieved by reducing the human error factor through better organization and planning at a far lower cost than new high-tech fixes—a lesson that undoubtedly could apply in many environmental arenas.[67]

While the Alaskan spill was spectacular, smaller spills occur all the time—some 10,000 in the year following the day the *Exxon Valdez* hit the reef (at least three of which were in the million-gallon range).[68] Not all of these spills are into the oceans; probably more birds have been killed in oil-collection ponds associated with oil facilities than are killed in spectacular ocean spills. The steady rain of small spills, along with a constant low-level flow into the seas from tanker operations that do not get recorded as "spills" and oil flowing from natural submarine seeps, exert continuous toxic pressure on both oceanic and coastal ecosystems.

The oceans, of course, are constantly being assaulted along their shorelines by a flow of much more than petroleum: sewage, industrial wastes, and agricultural runoff, including insecticides and herbicides, from the land. Beaches are commonly closed for health reasons in industrialized nations, and worldwide they are repositories for tons of plastic debris and gobs of oil. In 1988, beaches in the eastern United States were closed repeatedly because of contamination from medical wastes.[69] Poisonous algal blooms and eutrophication of bays and estuaries have led to large-scale fish kills; pollution has forced closing of shellfish beds in California, Washington state, the Gulf coast, and Chesapeake Bay. San Francisco Bay and other bays on the West Coast are contaminated with mercury, zinc, and other heavy metals, along with all the usual industrial effluents, sewage, and farm chemicals.[70]

Even on the remote Pribilof Islands of the Bering Sea, the beaches are littered with bits of plastic fishing net, crab trap floats, detergent bottles, and the like.[71] When crab traps lose their floats, they become crab killers, sitting on the bottom and continually rebaiting themselves with new victims, whose decaying bodies attract more crabs. The rich fauna of the waters around the Pribilofs are being wasted in various ways, including indiscriminate hunting for sport. One of the world's last great fish stocks, a pollack fishery (in the words of a local ecologist) "is being strip-mined."[72]

Some fishing operations now use drift nets up to 30 miles long. Not only do these enormous nets, pioneered by Japanese and Taiwanese fishers, essentially vacuum the sea of fishes; they also trap and kill thousands of seabirds and marine mammals. Even worse, the nets are made of nylon, so when lost or discarded they are virtually immortal, sweeping through the seas and killing fishes and other animals in perpetuity. The impacts on fishery stocks after a few years of drift-netting can be devastating. Drift nets have been banned in the South Pacific, and Mediterranean nations are trying to ban them from their sea.[73]

No one is sure just how much oceanic pollution is affecting oceanic fish production, although deformed fishes are commonly caught in highly polluted areas.[74] Since most of the productivity of the seas is concentrated near the shores, the impact on coastal fisheries is surely substantial. China has suffered about a one-third decline of fisheries off her coasts because of enormous discharges of domestic and industrial sewage into coastal waters.[75] Fish kills and mysterious outbreaks killing marine mammals have been recorded everywhere from Japan's Inland Sea to Brazil, the North Sea, and Europe's shores.[76]

The Mediterranean Sea illustrates the problems in concentrated form: severe overfishing and pollution by effluents from cities, resorts, factories, refineries, and farms have brought the sea to the brink of crisis. Efforts to control pollution have been complicated by the difficulty of coordinating twenty nations whose relations are not always cordial and which vary greatly both in causing the problems and in their ability to finance cleanups.[77] Nonetheless, progress has been made, and a compact called the Nicosia Charter, an agreement to achieve "an environment . . . compatible with sustainable development" by 2025, was signed by eighteen nations in May 1990.

In view of the multiple assaults on oceans, it seems unlikely that fisheries production can be increased much further. As the human population grows and pressure on fish stocks keeps rising, it is more likely that production will shrink.[78] Furthermore, no one is certain just how much abuse the oceans can take until they go into a downward spiral and "die"—losing much of their biodiversity. Experts in marine ecology generally believe that the oceans are tough and resilient but that once they are pushed too far, they could enter a degraded state from which it might take millions of years to recover. As pioneering restoration ecologist John Cairns put it: "It appears highly probable that the vast oceanic ecosystems are quite fragile . . . and are protected primarily by their vastness and the resultant dilution of all potentially deleterious materials. Should an entire ocean be damaged, the time required for recovery staggers the imagination."[79]

Addressing Pollution

The secret to abating pollution of air, water, and land, of course, is not to emphasize cleaning it up but to prevent its emission in the first place. It is always much more difficult (and often impossible) to remove pollutants from the environment than it is to keep them from being released. Two basic tasks are involved in preventing releases: the first is to design and develop technologies that either do not produce the pollutants at all or that will prevent their release; the second is to provide incentives to develop and deploy those technologies. The first task is largely a technical one, the second a political and economic problem. Society's responsibility is to see that existing cost-effective pollution-preventing technologies are universally used, to encourage the improvement of those technologies and the development of new ones, and, wherever possible, to work for changes that reduce the overall need for those technologies.

Technologies are already in hand to remove many pollutants from the effluents of smokestacks, exhausts of automobiles, and waste-water discharge pipes. Scrubbers that remove sulfur from power plant stacks, catalytic converters that remove carbon monoxide and oxides of nitrogen from automobile exhausts, and various kinds of sewage and waste-water treatment plants are examples. The 1990 Clean Air Act and most of the earlier environmental legislation of the United States leans heavily on such remedies.

The superior approach of preventing the production of troublesome pollutants in the first place is too seldom taken, largely because incentives have been lacking to do so.

Economically speaking, the pollutants are externalities—the cost of cleaning them up has fallen on society, not the polluter. Laws like the Clean Air and Clean Water acts have been partly successful in "internalizing" cleanup costs, but have yet to motivate much effort toward preventing the creation of pollutants—what ecologist Eugene Odum calls "input management."[80] A truly beneficial case of input management was the removal of lead from gasoline. That particular improvement was made for a less compelling reason—the lead clogged up catalytic converters—but subsequent studies have shown lead to be an even more serious threat to the health and mental development of children than was realized at the time.[81]

In some situations, a simple prescription for internalizing pollution control might greatly alleviate a problem with relatively little enforcement effort. An example would be to require any individual, firm, or government unit that extracted fresh water from and released waste water into a river or stream to place the waste-water outlet *upstream* of its freshwater intake. Variances could be applied for where either ecological or economic factors made enforcing the regulation unnecessary or too costly.

One mechanism suggested for limiting the profligate use of resources by the United States could, if instituted, greatly reduce air and water pollution. That is the placing of "depletion quotas" on several hundred natural resources—especially non-renewable ones. This proposal, made by economist Herman Daly, would place strict limits on the total amount of each resource that could be extracted or imported by the United States annually.[82] Shares of the right to deplete within the quota of a given resource for a given period would be sold at auction by the government and would then be freely tradable. As the United States economy expanded, the prices of the resources would rise.

Depletion quotas (and higher resource prices) could have many beneficial effects, including encouragement of resource conservation through recycling and the placing of a premium on the durability of manufactured articles. But in the context of this discussion, their most beneficial effect would be pollution abatement. If raw materials were scarce, then those that could be extracted from smoke by scrubbers in stacks, retrieved by waste-water treatment, or otherwise retained would increase in value and thus encourage pollution controls.

The problems with such a scheme, however, could be serious. They would include those of bureaucrats setting the quotas, possibly with the sort of arbitrariness that helped cripple communist economies. Another would be the likelihood of black markets flourishing, especially if other nations did not institute similar quotas. Most economists would prefer to attack effluent problems directly, with taxes and subsidies, which would probably create less of a regulatory mess and could be tied more directly to output. But even a debate over the pros and cons of depletion quotas could have a beneficial educational effect for society. The depletion quota approach, unlike trying to control outputs or cleaning up the environment after the outputs are out, emphasizes both the problems of scale and the advantages of limiting physical inputs to the economic system.

Under the relentless pressure of human expansion, some of our most essential resources—air, land, and water—are being subjected to unprecedented deterioration. While that abuse endangers human health directly, it also threatens that other precious resource, biodiversity. The toxification of air and water is also intimately connected to other abuses of land and freshwater resources that are undermining Earth's ability to support humanity, and which now merit our attention.

Use and Abuse of Land and Water

Land and fresh water are resources in increasingly short supply on an overpopulated, finite planet. Both are being subjected to rising pressure for diverse uses from producing food to supporting burgeoning urban populations and industry and even disposal of wastes. In the process, these resources—supposedly renewable—are often degraded by mismanagement or by pollution. The amount of undisturbed land available to maintain natural ecosystems and the vital services they provide keeps on shrinking as the human population grows. Freshwater resources also are being diverted away from those ecosystems or poisoned by contaminants generated by humanity, often as a result of land-use practices.

■ Land Resources under Pressure

The scale of the human enterprise is now so vast that nearly all of Earth's desirable land has already been put to use. Humanity grows crops on about 11 percent of Earth's land surface (excluding Antarctica and the Greenland icecap), and uses nearly 25 percent as permanent pasture.[1] Perhaps 2 percent more is occupied by urban development: homes, offices, industry, roads, airports, and the like. About 30 to 40 percent of the land is covered by forest and woodland, and most of that is used by people, often heavily.

So people are now intensively using at least 40 percent of Earth's land surface, and less intensively another 25 to 30 percent. Most of the one-third of Earth's land that people haven't seriously exploited is inhospitable desert, tundra, high mountains, or regions of permafrost—areas of little or no biological productivity. A small fraction of the unused land is inaccessible forests both in the humid tropics and the far north—and how long they will remain relatively untouched is open to question.

The global level of land use not unexpectedly correlates with the direct human consumption, indirect diversion into alternate systems, or conversion to systems of lower productivity of some 40 percent of the world's net primary productivity (NPP) on land.[2] The degree of land takeover helps to explain why competition for suitable land to meet various needs is becoming intense in many regions of the world. In particular, there is a growing shortage of high-quality agricultural land, and much of what exists is being rapidly degraded.

▬ Agricultural Land

Agricultural land, among the most essential of natural resources, and in some ways one of the most undervalued, is under increasing pressure, as the century nears an end, to produce ever more food for ever more human beings. At the same time, demand for land for other uses such as urban development, roads, and waste disposal is also on the rise. Yet less and less suitable new land remains available to open for farming in most regions of the world. The few exceptions include Latin America, where skewed land-ownership patterns cause relatively inefficient land usage; the United States, where abundant food production has allowed some inefficiency in land-use patterns; and some parts of Africa, where diseases carried by tsetse flies have prevented people from establishing settlements and farms.

Much of the existing farmland, moreover, is subject to a variety of damaging assaults that undermine its capacity to continue producing food. Croplands are exposed to air and water pollutants of various sorts, many of which are known to injure plants. Widespread acid deposition damages crops in many areas; increased UV-B radiation from depletion of the stratosphere's ozone shield may take its toll as well. The process of irrigation (discussed below) often damages land.

Cultivation itself, especially the machine-and chemical-intensive farming generally practiced today, can lead to a slow deterioration of land and loss of fertility, mainly by accelerating the erosion of soil by wind and water. Nutrients are lost in the eroding soil, in runoff waters, and in water leached through the soil profile. Moreover, the loss of organic matter by microbial activity may not be offset by inputs of organic matter to the soil, since conventional agriculture allows for little or no incorporation of plant material into the soil. Recently introduced no-till practices could go a long way toward preventing erosion and runoff as well as toward maintaining soil organic matter (but at the cost of increased herbicide use).

Soil erosion is a worldwide scandal; an estimated 24 billion tons of soil are washed into the oceans each year from the world's croplands and grazing lands.[3] Declining productivity due to soil depletion has been noted in many regions, including the American Midwest.[4] One-third or more of the topsoil in the Corn Belt has been lost in the 150 years since it was first plowed.[5]

Seeking foreign exchange to balance import expenditures (largely for imported oil) in the 1970s and early 1980s, the United States government encouraged farmers to maximize their production, to the neglect of older, soil-protecting policies dating from the 1930s. The inevitable result was an acceleration of soil erosion and noticeable declines in crop yields in some areas. For every inch of topsoil lost, wheat and corn yields fall by roughly 6 percent (everything else being equal). In 1977, about six tons of soil were lost for every ton of grain harvested by American farmers.[6]

Increased fertilizer applications can mask depletion of soil for a while, but sooner or later, the losses show up as falling yields. When the land is so depleted that farming has become unprofitable, restoration of the soil becomes a very slow and difficult process if it is possible at all; an inch of topsoil requires perhaps 500 years to form, and at least six inches are needed for crop production.

Fortunately, recent United States farm policies that included soil-conservation incentives have helped to stem the horrendous losses and put American agriculture on a somewhat more sustainable basis. The problem is far from fully remedied, though; the United States is still losing more than 2 billion tons of soil a year.

Every continent had suffered significant land deterioration by the late 1970s, including severe degradation on areas ranging from 6 to 17 percent of the agricultural land, according to a study by agronomist Harold Dregne.[7] Since 1981, the world's total cropland base has been shrinking as land has been taken out of production in many regions, including the United States, the Soviet Union, China, India, and sub-Saharan Africa. Some of the abandoned farmland was of marginal quality and should never have been plowed, and some has been taken over for other purposes. But tens of millions of acres of cropland have been abandoned after deteriorating to the point that farmers could no longer make a living from it. A million acres per year were abandoned in the Soviet Union alone from 1977 to 1987. By 1988, worldwide acreage planted in grain had fallen by some 7 percent from the 1981 maximum.[8]

The loss of productive land around the world is clearly a threat to the increased food production that will be needed to support the human population over the next several decades, as it is swelled by billions more people. Most land deterioration is preventable, with care and appropriate management. But economic pressures often override conservation concerns, especially in poor nations where populations are growing rapidly and farmers' concerns focus on the current crop, not harvests ten or twenty years ahead. Once again, capital is being consumed, undermining the income-producing resource.

▬ Desertification

Human-caused land deterioration in arid and semiarid lands is known as desertification. A great deal has been written about desertification, especially in Africa and Asia, and there is some confusion about which forms of land degradation should be included in the term.[9] Nevertheless, desertification processes are under way on every continent.[10] The United Nations in 1984 estimated that more than one-third of Earth's ice-free land in regions supporting more than a billion people is vulnerable to desertification, and that three-fourths of that land has already suffered at least a moderate degree of degradation.[11] Around the world each year, an estimated 50 million acres of land (an area as large as Spain) reach the point of being unprofitable for farming or grazing, and nearly a third of that land has been irreversibly turned to desert. While others dispute these ominous figures on the extent of desertification, no one contends that the problem is not real.[12] Nor is there any doubt that people are impoverished by desertification of the land they depend on for their livelihood.

The basic causes of desertification are overuse and abuse of relatively fragile soils and ecosystems: overgrazing, overcultivation, poorly managed irrigation, and deforestation. Consequences include accelerated soil erosion, loss of vegetative cover and a resultant change in local climate, increased floods and droughts, and reduced crop or livestock productivity. A frequent side effect in arid developing regions is a severe and growing shortage of fuelwood (about which more below).

At least 40 percent of Asia's land is at high risk, much of it in the environmentally battered Soviet Union. An even higher fraction applies to Africa. Not all the damage is recent; the Mediterranean basin and much of the Middle East, India, and China have been subject to desertification processes since antiquity.

Desertification, however, is far from restricted to places like the Sahel, central Asia, and northeastern Brazil. Much of the semiarid western United States was converted from rich grassland to desert by severe overgrazing, primarily by cattle, in the last half of the nineteenth century.[13] Serious overgrazing persists in much of the semiarid West, although the range has partially recovered from the past extreme abuses after a century of reduced pressure. The benefits accruing to the nation from livestock operations in the region are negligible, and the costs are huge (if largely unnoticed). The main beneficiaries are some 13,000 politically powerful cattlemen in the western United States. Each state has the same number of senators, and western senators have been responsive to the cattle industry's powerful lobby, the National Cattleman's Association. Those western senators do very well trading votes with eastern senators on an issue that means little to representatives of urbanized areas.

As a result, not only has much of the West been desertified by overgrazing, but the overgrazing is concentrated on federal lands (owned by all of us). The grazing fees on public land are about one-fifth of those that would be charged for grazing cattle on private lands, and little effort is made to oversee the grazing by bureaucrats in the Bureau of Land Management (BLM, sometimes called the Bureau of Livestock and Mining) and the Forest Service.[14] The bureaucrats have long supported the Forest Service's "land of many abuses" policies and cooperated with the livestock industry. Like most Americans, they probably would not even recognize that the range is overgrazed because it has been overgrazed for so long.[15]

To limit the destruction, new "Change on the Range" grazing guidelines for public lands were recently issued, which may bring some improvement if they can be enforced. One Forest Service district ranger in Idaho, Donald Oman, was courageous enough to try; he was subjected to repeated public death threats from local ranchers, who were unaccustomed to interference with their grazing practices.[16]

The benefits the nation derives from cattle operations in these areas are scant, to say the least. Less than 3 percent of the beef marketed in the United States comes from public lands in the arid West. Many of the operations are marginal at best and would not survive without the subsidized access to public land. American demand for beef, which has been dropping for more than a decade anyway (largely because of health concerns), could easily be supplied by states east of the Rockies and the few relatively water-rich areas of the West, without degrading fragile lands.[17]

The costs to the nation of desertification are very high, but generally unrecognized. Removal of vegetation in arid regions promotes soil erosion when the scarce rains do come; instead of soaking in and recharging aquifers, the water simply runs off in gullies and eventually ends up in the sea. Cattle damage the critical riparian (stream-side) areas by eating vegetation, trampling streambanks, and removing shade. Fishers and hunters have less quarry because of despoiled streams and competition by cattle with native deer and antelope for food and water, while hikers and campers find their outdoor recreation areas befouled by cow pies. Even towns seeking pure, dependable water supplies must compete with the cattle.

The situation in Arizona is so bad that it has been suggested that its license plates' motto should read "The Cow Pie

State." In the Sawtooth National Forest near Sun Valley, Idaho, where Donald Oman stood up to the ranchers, more than a third of the grazing land was found to be in poor condition. Streambanks had been denuded, leading to erosion, and fish and other wildlife had suffered.

Sheep do their share of damage as well. In the Pine Nut mountains of Nevada some years ago, a team from Stanford's Center for Conservation Biology found the site of a butterfly colony they had been studying completely stripped of herbaceous vegetation, the soil churned into dust and covered with sheep tracks and droppings, and the butterfly colony exterminated.

Reducing livestock-induced desertification in the western United States would not be difficult. Grazing fees on most public lands should be determined by market forces (that is, comparable to those paid for use of private land), and stocking rates (number of animals per square mile) reduced to levels of minimal damage. In the most sensitive areas, cattle should be removed entirely and kept off either permanently or until the land has a chance to recover—which may take centuries. Encouragement should be given to the increasing number of ranchers who maintain the range in good condition or are attempting to regenerate it.

Cattle ranching is a western tradition that can and should be preserved in well-watered areas and even, with care, in some drier regions. It's worth noting that much of the original cowboy and Indian action took place in the Great Plains, just east of the Rockies, which is still a productive cattle-producing area. But the notion that the best use of most ecosystems in the semiarid West is as range for cattle—even range kept in good condition—is an idea that needs to be reexamined.

If the new guidelines were honestly observed and enforced, many of the marginal operations (which generally cause the most damage) would simply go out of business. Some form of compensation might be made available to cushion the financial shock to ranchers during a transition to the new regulations; over the long term it would be a bargain for taxpayers. In appropriate areas, some operators might be temporarily subsidized to convert to game ranching.[18] This, too, could be a good investment, eventually repaid many times over in availability of water for municipalities, recreational amenities, a return of wildlife, and a healthier, productive rangeland.

Times are changing in the West.[19] New Yorkers still may not care how many animal units are grazed for what fee on which parcel of BLM land, but more and more people in western states do. Political power is slipping away from the livestock industry in states like Arizona, and the industry and its lobby would be wise to greet the twenty-first century with some forward-looking plans for ecologically sound operations that will preserve a great western tradition, rather than the mixture of hysteria, stone-walling, and threats that have characterized much of the industry's reaction to necessary change so far.

▬▬ Protecting Land Resources in Poor Countries

Obviously, politics plays a crucial role in degradation of land in a rich nation like the United States. Politics is equally important in poor nations, where people may wish to protect their own environment and resources, but often lack the power to do so. Despite all the fanfare that accompanied the launching in 1977 of the United Nations program to combat desertification, dishearteningly little has been accomplished. At least one expert has attributed the lack of success to a failure to analyze and deal with the social and economic factors that lead to abuse of land, especially in the vulnerable Sahel region of Africa (the band of semiarid land bordering the Sahara desert on the south).[20]

The key to halting the degradation of land clearly lies in local control of resources.[21] Villagers and nomadic herders are not ignorant of the consequences of their actions; more often they lack the means or the authority to take appropriate preventive steps. Where the movements of African desert nomads have not been limited by political restrictions, their traditional life-style has permitted survival even through the droughts of the 1970s and 1980s.[22]

Land degradation seems typically to have followed disruptions of tradition, usually by external forces beyond the control of local people. Land-ownership patterns also play a role, as when poor farmers are squeezed into marginal areas or lose title to their land, as has happened in parts of Latin America, the Philippines, and other areas. In Africa, women do most of the farming but have no access to farm credit or other assistance.[23]

In parts of Asia and Africa, local control was lost during the colonial period. In India, the first step occurred when British colonial administrators removed village common lands from local control and turned them over to the government. When India gained freedom from the British, the new national government inherited the colonial laws and has foolishly failed to revise them.[24]

Anil Agarwal, head of the Centre for Science and Environment in New Delhi, reported an example of how these laws promote environmental degradation.[25] People in the Rajasthani village of Gopalpura were legally prevented from protecting the village watershed. The villagers planted trees on a ridge at the top of the watershed in 1987 and built a wall around the saplings to prevent animals from killing them. Even though the land was in the village, the government immediately served legal notice that the trees had been planted illegally. While a court case dragged on for two years, successive droughts killed the saplings. Because the area was protected from grazers, however, other trees began to sprout naturally. Nevertheless, in 1989 the government imposed a fine for the illegal planting and ordered the wall destroyed.

In spite of such bureaucratic stupidity, the villagers of Gopalpura agreed among themselves on a series of environmental protection measures. Fines are imposed on anyone plucking leaves, cutting branches, or grazing animals in a protected area. There are even fines for observing but failing to report violations.

Another Rajasthani village, Seed, may be the only Indian village with a self-imposed land-use plan. The state of Rajasthan has a unique law, the Gramdan Act of 1971, that allows the adult population of the village to manage resources within its boundary. The Seed village council forbids grazing on some common lands and protects trees on all of them. It has the power to impose fines for violations of land-use restrictions by local people. In the severe drought of 1987, Seed harvested

80 bullock-cartloads of grass, which was divided equally among village households.

Impressive success stories are seen wherever villagers and local grassroots groups have maintained, asserted, or regained control over their local resources and environments. The Indian government recently rewarded a desert community known as the Bishnoi, whose sustainable way of life blends spiritual and ecological practices in an extremely harsh environment. These traditions allowed them to survive handily during a record drought.[26]

The Chinese have succeeded in mobilizing larger groups of people for conservation and restoration efforts, with help from United Nations agencies. The people have terraced some eroding slopes of marginal lands and converted others to tree crops and pasture for livestock over a large area of China's Loess Plateau.[27] After the cropland area was reduced by half and the rest was put to other uses, crop production more than doubled and was augmented by the tree crops and animal products. In this and other development issues, China may be different in succeeding with centrally planned programs, even though this program was implemented and the benefits were reaped locally.

In most developing countries, the problem too often has been focusing too much on the government-initiated large-scale projects and neglecting the well-being of people at the local level. Future success will hinge on changing this pattern to one of improving patterns of land use by giving local people more responsibility for the treatment of the lands they depend upon.

▬▬ Land Use Competition

As mentioned at the beginning of this chapter, one reason for a shrinking agricultural land base is competition for other uses of the land. A fairly obvious, but often underrated competition is between use of land for agriculture and for homes, parking lots, highways, factories, and other urban infrastructure. This competition is clearly being lost by agriculture (rarely are shopping centers torn down to build farms), but is mostly ignored for two reasons. One is the insidious nature of the process; each individual loss of farmland is usually small, especially in poor countries. And land is not lost only at the edges of expanding urban complexes; the interaction may be much more subtle.

NASA's Shuttle Imaging Radar-B mission in 1984 produced images of the land surface of Bangladesh during the monsoon, a time when approximately 75 percent of the countryside is flooded for rice cultivation. The images permitted NASA scientist Marc Imhoff and his colleagues to analyze the competition between living space and agriculture.[28] People must live on bulwarks raised above the water level of the paddies, constructed of soil dug out of the center of the paddies during the dry season. As the rural population grows, the area of bulwark grows at the expense of paddy area, and the water in the dug-out paddies gets too deep to accommodate the highest-yielding rice strains. The result is a marked decline in yields and production.

Imhoff is convinced that similar losses in paddy rice area due to competition with infrastructure are widespread in South and East Asia. He believes the effects are not detected in aggregated data on rice production because economic factors and weather changes can cause considerable fluctuation in production, while improved agricultural technologies and intensified cultivation can compensate somewhat for losses of cropland.

It is estimated that about 36,000 square miles of land (the equivalent of Indiana) will be built on between 1980 and 2000 in poor nations. Much of that will be former farmland. In China, four out of five people live in farming areas, and construction of homes and other facilities needed for the nation's growing population inevitably subtracts from badly needed arable land. India has similar problems. New Delhi's growth consumed 54 square miles between 1941 and 1971, and expansion of the city into farmland continues today.[29]

The situation in rich countries is no different. In California, suburbs now sprawl over large areas of land that once supported truck gardens and orchards. When we arrived at Stanford University in 1959, there were still apricot orchards in the Santa Clara valley, where houses, shopping malls, and electronics industries have since proliferated. Farther south, the sprawling suburbs of Los Angeles have gobbled up many of the state's productive orange groves; citrus production has been forced to move into the Central Valley, where it is more dependent on irrigation and more vulnerable to winter freezes.

For a time in the 1970s, the United States was estimated to be losing 3 million acres a year, a third of it prime farmland, to urban development and highways.[30] During the 1970s, farmers abandoned soil conservation practices adopted after the 1930s dust-bowl disaster and planted "fence to fence" in response to government incentives intended to boost grain exports. The additional cropland gained more than offset the development losses, but largely inferior land was brought into production and additional pressure put on good land. In short, the gains were short-term ones, and accelerated rates of soil erosion and productivity losses were the long-term result.[31]

The second reason that the importance of losing agricultural land to urbanization may be underrated is that generally the *best* agricultural land is lost first. When given a choice, people have always settled first in well-watered areas with rich soils. That's where most settlements have sprung up, and that's why urban growth has spread over good soils. Be it the paddy fields of Bangladesh or the rich farmland of Lancaster County, Pennsylvania, more people needing more infrastructure build it on fertile soils. So, although the world may lose only about 1 percent of its farmland to urbanization per decade, the loss is of first-class land. The United Nations Food and Agriculture Organization has estimated that between 1980 and 2000 some 5,400 square miles of land are being lost each year to urbanization; more than 1,900 square miles in East Asia alone.[32]

The demand for land not only squeezes out farmers; it causes problems for urban dwellers as well. Land to accommodate more people is scarce near big cities, making it difficult for people to live close to their jobs and thus requiring long, time- and energy-wasting commutes. People who work in Los Angeles, for instance, are now settling in Palmdale or Riverside and facing daily round-trip commutes of up to 160 miles.

Highway proliferation itself is a serious problem, gobbling up enormous amounts of land. An automobile-centered society thus not only causes tens of thousands of accidental deaths each year, creates severe air pollution, and contributes substantially to acid precipitation and global warming; it competes with agriculture for good land—and again agriculture generally loses.

Even land for disposal of garbage and other wastes is so hard to find that the "garbage crisis" has even been dramatized on television by footage of a stinking trash-heaped scow sailing the seas in search of a place to dump. As we saw in the last chapter, many cities are predicted to "run out" of space for landfills in the next few decades; more than a few, from New York to Berkeley, have already run out and are dumping on their neighbors—or paying them to take it away.[33]

Of course, people must be housed and provided with workplaces, and waste disposal is necessary. But much urban development is unnecessarily consumptive of land. In the United States, this probably is a holdover of the "endless frontier" mentality—the deep-rooted belief that the nation has boundless land resources. The frontier closed a century ago, but many Americans don't believe it yet. The notion of boundless land is reinforced by land prices; even prime farmland is worth only a fraction as much per acre as land purchased for development. Many farmers near cities count on selling out to developers eventually and retiring on the proceeds, and there is little to prevent them from doing so, even though it may not be in the long-term national interest.

Most European nations long ago realized their land was limited and established strict zoning laws; suburban sprawl is far less prevalent in Europe. Simply employing sensible zoning regulations could prevent much unnecessary loss of good farmland to urban development. Another approach is to undertake a positive program to protect good farmland, as the Farmland Trust, a private group, has attempted to do in the United States.

Our urbanized society has partly lost the sense of appreciation for land, especially land that produces food, that our forebears (who probably grew up on farms) had. As the world moves into an era of less secure food supplies, we may wish we had taken our rich farmland less for granted.

▬ Deforestation

The decimation of Earth's forests is one of the most serious global environmental problems. Deforestation contributes to the flux of CO_2 and other greenhouse gases into the atmosphere, destroys Earth's richest storehouses of precious biodiversity, wreaks havoc on other ecosystem functions, and reduces a critical part of humanity's resource base. The worldwide shrinkage of forests constitutes a vast experiment by humanity to change the basic nature of our planet, on which forests have been the dominant terrestrial vegetation form for several hundred million years.[34]

Since agriculture was invented some 10,000 years ago, perhaps a third of Earth's original forest cover has been removed, especially over the last few centuries, and now no more than 40 percent of the ice-free land is covered by forests and woodland patches.[35] Forests are disappearing virtually everywhere, although new forest growth in Europe and North America has partly offset losses there. Despite some replacement, the United States today has only about two-thirds as much forest area as existed before European colonization. Western European and Mediterranean nations were stripped of forests centuries earlier.

Forests are under assault today for timber, pulp (for paper), and fuelwood, and by clearing for farms, pastures, highways, and suburbs. They are also threatened by air pollution (especially acid deposition) and, ironically, by the climatic change that their own destruction is helping to cause.[36] Regrowth of forest in the temperate zone is occurring, but often as intensively managed stands of even-aged trees of one species—tree farms in essence. Second-growth forests contain neither the rich biodiversity nor the valuable centuries-old huge trees of virgin forests.

The most alarming component of deforestation, however, is the accelerating destruction of moist tropical forests. Between 1979 and 1989, it has been estimated that the rate of tropical deforestation jumped from about 1 percent per year to 1.8 percent.[37] In 1989 alone, forests in the humid tropics disappeared from an estimated 55,000 square miles, an area the size of Iowa. In several nations, including Madagascar and Thailand, forests were shrinking up to five times as fast as the average rate: 8 percent per year. In Nigeria and the Ivory Coast, they were vanishing in the late 1980s at more than 14 percent per year, at which rate they would be gone in less than a decade. Losses were recorded during the 1980s in all of the 34 nations that possess significant tropical forest resources.

If the acceleration of deforestation continues, only half the original tropical forests will remain by 1993. Unless dramatic efforts are launched immediately to preserve what remains, it will be too late. If unchecked, deforestation will destroy most of today's large forest tracts by 2010, and only fragments of those forests and the irreplaceable wealth of biodiversity they now contain will be left by 2025.

Tropical moist forests are being destroyed primarily for timber and for agricultural land. In Brazil and several other nations, the attack is led by farmers who have moved (or been moved by government inducements) into the forest. Traditional shifting agriculture, practiced by indigenous groups, is essentially sustainable, as long as population density is low enough that farmers are not forced to return to old clearings before regeneration is complete. The displaced farmers, however, are often inexperienced in jungle farming and carry on an unsustainable and ultimately destructive practice; forest ecologist Norman Myers calls them "shifted cultivators."[38]

Because the fragile forest soils are generally incapable of supporting continuous cropping, farms are soon abandoned and new land must constantly be opened. The same is true for pasturing livestock, which may temporarily follow the failed attempt at cultivation. The soil's fragility and vulnerability to erosion and leaching of nutrients, along with the marked change in microclimate that follows clearing of more than a few acres, makes regeneration of the forest a very slow process at best. In general, the larger the area cleared and the longer it remains treeless, the less capable of regeneration it is.[39] Even in favorable circumstances, regeneration of smaller patches surrounded by undisturbed forest may take upward of a century.

Some ecologists believe that, if the bulk of a large forest area such as Brazil's Amazon basin is cleared, the resultant regional climate change will prevent regeneration for millennia, if not forever.[40] Certainly the decimation of species and populations would prevent the reappearance of a forest with even a fraction of the original biodiversity. For practical purposes, then, tropical moist forests are a nonrenewable resource, and deforestation on a large scale is irreversible.[41]

Government subsidies have supported both farming and ranching in tropical forest areas in nations like Brazil (although Brazil has recently terminated that policy). International development agencies have also financed many large-scale projects such as ranching, agricultural developments, lumbering and mining operations, and dam projects that cause large-scale deforestation.

Timber harvesting is the second greatest cause of tropical forest destruction. Between 1950 and 2000, annual worldwide consumption of wood for nonfuel purposes is expected to have more than quadrupled, and by 2025 will have increased some six times.[42] Much wood harvesting in tropical nations is instigated from outside by developed nations, especially Japan.[43] Too often, governments of poor nations have been persuaded to sell timbering rights to foreign companies at prices far below the real worth of the timber.

Even "high-grading," selecting only individual high-quality trees to cut, causes enormous damage in a tropical forest because trees grow close together and are interlaced with a network of strong vines. It is almost impossible to take out one tree without damaging or killing its neighbors. Typically, high-grading reduces the canopy by about half, and the resultant drying makes the area susceptible to fire. One such fire in Indonesia in 1982–1983 burned for eight months and destroyed over 14,000 square miles of forest.[44] And the roads cut to reach the trees open an avenue for invasion by settlers.

Deforestation is a major problem outside the humid tropics as well. Seasonal tropical forests—those with one or two pronounced dry seasons each year—are easier to exploit, often have better soils than rain forests, and have been under assault for centuries. The richness of their biotic diversity is not far below that of rain forests. Conversion to agriculture, as well as logging for timber and fuelwood, have caused the demise of seasonal forests in many nations. Much of Central America's forest is of this type, a region that has lost more than four-fifths of all its original forests. While very little remains of the Central American seasonal dry forests, only about 2 percent on the Pacific side, some appear to be capable of rehabilitation, as long as a significant fragment of the original forest and a reasonable sample of its biodiversity remains.[45]

One rehabilitation project, headed by University of Pennsylvania biologist Daniel Janzen, is under way in northwestern Costa Rica. In the recently established Guanacaste National Park, a small, fragmented dry seasonal lowland forest, rich in species and microhabitat, is being restored and relinked to rain forest on adjacent mountain slopes. The project is emphasizing cooperation and collaboration with local people, including schoolchildren who are being taught about forest ecology and trained to help in restoration activities and to be guides for visitors.[46]

In drier areas such as savannas and desert edges, where tree cover is sparse, losses of woodland are mainly due to demand for fuelwood by resident populations. Since 1975, the use of wood for energy has surpassed other uses, and a severe and worsening fuelwood shortage has appeared in many developing nations.[47] In India, the need for firewood and fodder is about 133 million tons annually, yet only some 36 million tons of wood are available in state-run stores. As a result, Indians illegally fell trees or cut off their leaves and branches. One prominent Indian environmentalist envisions "India becoming the biggest desert in the world."[48]

Similar stories can be told of many developing nations, including large parts of China, Pakistan, the Himalayan plateau, and most of the Sahel, where women and children commonly spend many hours each day in search of wood for cooking. The denudation of woody plant life from semiarid and arid areas contributes greatly to desertification. When mountain watersheds are stripped of tree cover, downstream areas are subject to floods, droughts, and irregular river flows. Thus deforestation in the Himalayas, mainly for fuelwood, has probably contributed to catastrophic floods in India and Bangladesh.

Fuelwood is the main or only energy resource for over 2.5 billion people—half the world's population. The total energy provided is estimated as equal to the 21 million barrels of oil produced daily in 1990 by the OPEC nations.[49] In 1980, the United Nations Food and Agriculture Organization (FAO) estimated that well over a billion people in poor countries were cutting fuelwood faster than it could be replaced by new growth, and that number is expected to double by 2000.[50] Attempts to plant trees and woodlots for fuel have not begun to meet the need; by some estimates, planting is sufficient for only about a fifth of the need.

Many large-scale tree-planting projects have failed because local people had no control over the resource or responsibility for managing them.[51] In a Chinese project, trees were planted and villagers were expected to care for them, but the wood was to be harvested for use elsewhere! The failure was hardly surprising. Giving local people control over forest resources may lead to their preservation in many circumstances and a regime of sustainable use, just as local control over land resources could. The Chipko and other movements in India[52] and the resistance movements against logging in Malaysia and Indonesia certainly support this observation.[53]

The destruction of tropical forests constitutes an extreme global emergency that is just beginning to be appreciated. Rising concern about tropical forests in the rich countries, however, has caused resentment in the tropical nations, whose governments and business communities are outraged by what they view as interference in their internal affairs. They also correctly point out that the industrialized nations, especially the United States, have destroyed most of their own virgin forests and are continuing to do so.

Three important differences between temperate and tropical forests, however, make this point less telling. The enormously greater wealth of biodiversity, the potential of tropical deforestation to influence climate on a global scale, and the relative irreversibility of large-scale removal of tropical forests make their importance as an irreplaceable global resource clear and their preservation far more urgent.[54] But these differences are not necessarily well understood by most government officials in developing countries, who simply see their forests as natural resources to exploit, just as rich countries see their own.

Indeed, the United States government is subsidizing the cutting and selling of much of its own remaining virgin forests very much as developing nations are doing. Taxpayers are subsidizing unsustainable logging on more than 100 million acres of national forest land to the tune of $100 million per year.[55] While some two-thirds of the nation's original forest area is still forest, more than 90 percent of it is second growth; the remaining virgin forest is a tiny fraction of the original.[56]

Cases in point are the precious remnants of old-growth forest in the Pacific Northwest. Those remnants help preserve the diversity of organisms that make up a natural forest and the genetic diversity that can give a forest long-term resilience in the face of environmental change.[57] If the multiplicity of species making up the original forest completely disappears, there is no guarantee that lumber can be produced indefinitely. The sustainability of tree farms is untested. One forest scientist described such misplaced faith as "playing genetic roulette."[58]

Even the productivity of heavily managed forests may

prove difficult to maintain in the face of various adverse trends such as increased soil erosion and the mining of soil nutrients. Diseases may gain the upper hand in even-aged monocultures (stands of single species), and rising stresses from pollution, acidification, and climate changes may compound the problems. Each of these trends or some combination of them could gradually make conditions generally unsuitable for tree growth in many situations.

Since there is no "endangered habitat" act, protection of the northern spotted owl, an endangered species, has been used as a legal device by the environmental community to protect the fast-disappearing forests. In the long run, protecting the owl and its forest will not only preserve that productive ecosystem, it will save jobs in the timber industry, if that industry can be converted to a sustainable one. This would entail a shift away from mowing down the old-growth forests, which might last for another decade or so, to a system in which forests could be exploited indefinitely without degrading the region's capacity to maintain them. That means preserving today's old-growth forest remnants and, over many centuries, expanding the area occupied by undisturbed ancient forests.[59] Sustainable logging regimes should be concentrated in the extensive second-growth forests of the region.

Yet in 1990, the old-growth forests were swiftly being destroyed for the financial benefit of a few, putting in jeopardy an entire way of life for thousands. Unhappily, the immediate loss of some jobs was the focus of protests by loggers, who understandably were concerned with keeping food on their family tables this year, not on the viability of their profession in a decade or two.[60] This might seem to contradict the principle discussed above, of local control over resources, but important differences between the Northwest loggers and groups like the Chipko of India are those of scale and true local control. Absentee owners in the eastern United States control the harvest rates and jobs in Oregon and Washington, and the loggers depend on logging for their livelihood, not on the intact forest. The Indian villagers are protecting trees to preserve their watersheds, the fertility of their farms, and the supply of fuelwood.

Logging accounts for only 2 percent of the jobs in Washington state and 6 percent in Oregon. Saving the last of the ancient timber would not have even a significant short-term impact on the nation's housing prices, although, of course, it would make hcmes cheaper in the long run if the timber could continue being produced and harvested in perpetuity.[61]

Plant ecologist Peter Raven described the situation very well: "By treating 500- to 1000-year-old forests as if they were a renewable resource, we are acting out a fiction, and thereby making a grave mistake. Forests are indeed renewable, but once they have been removed from a particular area, the ancient forests . . . will never appear again, *given the nature of human activities in the contemporary world and their consequences.*"[62]

It would be much more in the nation's interest to help retrain some loggers for other work than to lose both a precious heritage and a sustainable logging industry. Fortunately, some foresters are learning that forests are not just collections of trees planted in straight rows to be clear-cut at the right moment as determined by short-term economics. Under the leadership of forest ecologists like Jerry Franklin of the University of Washington and his counterparts in western Canada, foresters are beginning to develop harvest plans that are much less damaging than those usually practiced today.[63]

The guiding philosophy of the new approach is to preserve enough ecological integrity that the forest can regenerate itself more or less naturally and maintain productivity over the long term. The principal strategy, which still must be tested, is to cut the majority of trees (80 to 90 percent) in a tract, leaving snags, rotting logs, and debris in place, then allow the area to regenerate. Maintaining species and structural diversity (or creating it in existing forest plantations) is a major goal.

The bad old days have been typified not just by the rape of forests in Washington and Oregon (and similar plundering across the border in British Columbia), but by the destruction of Alaska's vast Tongass rain forest. That operation has been heavily subsidized by the American taxpayer, with the original intent of supporting an active timbering industry and an economic presence in Alaska (so close to Siberia). The result has been a windfall of cheap timber for Japan (which carefully protects its own forests). The Tongass, with 16.7 million acres, is our nation's largest national forest; from 1980 to 1990, the United States Forest Service *lost* more than $350 million on sales of its timber.[64] The destruction of the Tongass forest was opposed by the fishing industry, fearful of its impact on the spawning streams of Alaska's $400 million-per-year salmon catch, and by Native Americans who depend on the forest for food and materials. It also has been opposed by all who value natural beauty and biodiversity.

Beneficiaries of the Tongass rainforest logging included two pulp mills (run by the Louisiana-Pacific Corporation and the Japanese-owned Alaska Pulp Company). The Forest Service persuaded the two companies to build their Alaskan mills in the mid-1950s by giving them fifty-year contracts at sweetheart prices.[65] It was another example of the Forest Service's "land of many abuses" policies. Under a law drafted by the timber industry and pushed through Congress in 1980 by Senator Ted Stevens of Alaska, the two mills were promised 450 million board feet a year from the Tongass, and American taxpayers were required to give the Forest Service $40 million annually, or "as much as the Secretary of Agriculture finds is necessary" to supply the mills with timber. Through the 1980s, for every dollar of taxpayers' money invested in harvesting the Tongass, only a few cents has returned to the Treasury. Despite this enormous subsidy, employment in the local timber industry declined by almost 50 percent between 1980 and 1987 to about 1,800 workers, partly because the Japanese preferred to buy whole logs, not milled timber or pulp.

More than half the high-value timber was gone from the forest by 1987, and the Forest Service planned to harvest the rest quickly. It persisted in this plan against all economic and ecological reality because of the bureaucratic perks that accompanied Stevens's gigantic subsidy.

In 1990, the United States Congress, under the leadership of Senator Timothy Wirth of Colorado and Representative George Miller of California, finally passed a law to protect the Tongass. It was about time. As Wirth, who has pioneered in bringing environmental issues and the problem of the loss of tropical rain forests before his Senate colleagues, said: "We are no longer embarrassed. Last year I was in Brazil and met with President José Sarney, and the first thing he said was, 'What about the Tongass?' "[66]

In the United States, the federal government could take a giant step forward simply by banning all logging of the pitifully small remnant of old-growth forests on federal lands. Reeducated foresters might help convert tree farms and other heavily managed forests to more sustainable systems that include a variety of tree species and trees of different ages, as well as soil-holding underbrush (now commonly suppressed with herbicides). A more natural forest might be somewhat less convenient to harvest, but it might be maintained as productive more or less in perpetuity while helping to conserve biodiversity. Achieving this revolution in the timber industry will be politically difficult, though, as the protracted battles over the northern spotted owl and the Tongass National Forest have shown.

Saving the world's remaining forests, especially in developing regions, will not be easy, considering the growing human population, its need for more food, and its rising demand for wood products. The destruction of forests might well become much worse if a shift away from fossil-fuel use leads to greater reliance on biomass for energy, thus intensifying the pressures on forests and woodlands. Even if the pressure could be diverted to "tree farms," which already supply large quantities of pulp, and to carefully managed forests, maintaining productivity could prove difficult, given competing land-use pressures and the consequences of the high discount rates now prevailing.

Individuals, firms, and societies naturally discount the future. It is a function of the uncertainty of human life. Almost everyone would rather be given $1,000 today than in ten years. If nothing else, the money could be put in a money-market fund at an interest rate that would roughly double its value over that period.[67] Because of discounting, $1,000 promised a decade in the future is worth only about half as much as $1,000 today (if the interest rate is around 7 percent, it is discounted to $500). Similarly, $1,000 worth of timber that could be harvested from a mature forest plot in thirty years would be worth only $125 today. If the plot's owner could sell the saplings today as fuelwood for $150 dollars, it would make good economic sense to do so, for in a bank at 7 percent compound interest, the $150 would be worth about $1,200 in thirty years.

High discount rates encourage the destruction of natural capital such as soils and forests (when they are privately owned), rather than their sustainable use. Overexploitation for immediate gain is also caused by the "tragedy of the commons."[68] Common resources not owned by individuals who use them (whales, fish stocks, timber in national forests, grass in communally grazed pastures) often are overexploited. Those who consume the most gain the most; there is no incentive to conserve. The tragedy of the commons arises when property rights are not properly allocated, a situation that leads the market to generate externalities needlessly: extermination of whales and fish stocks, timber cutting on public lands, deforestation and desertification of lands held as a commons, whether in the western United States or African savannas. In such circumstances, individual capitalists, making decisions sensibly in their own self-interest, create disasters for their descendants because the long-term interests of society are not included in the accounting. Only when recognized authorities are charged with controlling access to a resource so as to preserve it (as they often are in traditional societies), can the tragedy be avoided.[69]

Protecting tropical forests, in both theory and practice, is much more difficult than saving the Tongass or curbing the denuding of the Northwest—although mechanisms such as debt-for-nature swaps (see below) and a growing awareness of the true value of intact forests, which could help protect them from the consequences of high discount rates and misallocated property rights, hold out some promise.[70] Indeed, one must ask what price would be worth paying just to reduce substantially the odds of a CO_2-induced climatic catastrophe that might cause the starvation of hundreds of millions of people, create tens of millions of environmental refugees, and place enormous financial burdens on rich and poor nations alike.

If civilization commits itself to avoiding such an outcome by scaling back the human enterprise and especially by preserving biodiversity, a massive attempt to restore natural forests over much of the globe will necessarily be an essential element of the endeavor. Programs are badly needed to rehabilitate degraded forest lands where feasible, but at the moment they are blocked by lack of resources and, especially for tropical rain forests, lack of knowledge.[71] Reforestation and rehabilitation of other types of degraded forest ecosystems will also be difficult and expensive to accomplish successfully, but in the long run would yield great economic and esthetic payoffs.[72]

Rehabilitation and restoration ecology are becoming serious endeavors in many regions, but much more needs to be learned in terms of both techniques and limitations before we can be sure whether and how completely particular natural ecosystems can be regenerated. No restored ecosystem, if severely damaged over a large area, will be the same as before, even though it might someday contain as many species and seem equally productive. But preserving a minimum (itself unknown) of the working parts is essential for restoration to work at all. Above all, the potential for restoration ecology's success should never be a justification for failing to preserve as much as possible of Earth's most productive and species-rich ecosystems.

▬▬ Birds as Indicators of Land Degradation

Since the days when coal miners took canaries into the pits, birds have been serving as early-warning systems for humanity. A clear indication of the destruction of biodiversity inevitably caused by human abuse of land is the decline of bird populations, now evident in many areas of the world. In the Americas, increasing attention is being paid to reductions in numbers of various Latin American species that migrate to the forests of the eastern United States each year to breed.[73] The scientific consensus is that their populations have fallen roughly by half over the last few decades and that the decline is likely to continue. There are thought to be two basic reasons for this. The first is the fragmentation of North American forests by freeways, farms, and sprawling suburbs. This has created a great deal of "edge" habitat congenial to an assortment of deadly predators on the nests of songbirds, from brown-headed cowbirds and bluejays to raccoons and small boys. The second reason is the destruction of the winter habitats, especially tropical moist forests, of many migrating bird species.

Behind both trends, of course, are the three elements of I = PAT that are promoting the extinction of birds and other organisms everywhere: the explosive growth of the human population, expansion of per-capita consumption, and, too often, the use of technologies to support that consumption that are unnecessarily destructive of the environment. To accommodate more people, many areas simply have been preempted that would be valuable in preserving biodiversity and maintaining our life-support systems. Individuals whose goal is a maximum number of Americans must understand that that goal could be better achieved by having a *sustainable* population for thousands or millions of years than by trying to cram in as many people as possible over a few decades until the nation collapses into chaos.

Expansion of the United States population, coupled with a lamentable lack of attention to land-use planning—or to control of toxic substances, protection of groundwater, and other environmental issues—is tightly tied to the forest fragmentation and other habitat destruction that makes survival difficult for many bird species, as well as other wildlife.

Drought and drainage for agriculture of pothole habitat in the northern Great Plains has reduced duck populations in North America by about 25 percent in the last three decades. Urban growth and conversion of meadowlands, hayfields, and pastures to row crops in the Midwest has hurt populations of the barn owl, that most widespread of all birds. Loss of vole-rich habitat to the needs of growing human populations has reduced the barn owls' reproductive success. Destruction of old-growth forests in the Northwest threatens the spotted owl and the marbled murrelet.

In the western United States, in wetlands areas such as California's Kesterson National Wildlife Refuge and Nevada's Stillwater Wildlife Management Area, there has been an epidemic of deformed birds. Birds with protruding brains, corkscrew beaks, and no eyes seem to be testifying to the gradual buildup of toxic substances from irrigation water. Children, pregnant women, and nursing mothers have been cautioned not to eat American coots from the Tulare Basin north of Bakersfield, and selenium levels in San Francisco Bay were so high in 1988 that warnings were issued against eating certain wild ducks.

The situation in Europe is similar, where intensification of farming and pollution effects have extirpated many bird populations. Around the world, bird populations are being nickeled and dimed to death as the expanding human population occupies their habitats and consumes or degrades their resources. The number of people has more than doubled since 1950, and their impact on Earth's landscapes has roughly quadrupled in that time.

So it is not surprising that a 1988 report of the International Council for Bird Preservation estimated that over 11 percent of the world's bird species were endangered—more than three times as many as they had estimated a decade earlier.[74] Some of the increase was due to better information, but some reflected the accelerating deterioration of the global environment, especially of tropical forests. Africa's endangered bird species rose in the same period from 65 to 170, and Brazil's from 29 to 121 (64 of them in the rapidly dwindling coastal rain forest north of Rio de Janeiro). Indonesia's endangered bird list rocketed from 14 to 126.

In southern and eastern Asia, the vast majority of natural wetland ecosystems are under threat, largely because of the need to feed 55 million *more* people annually—the population equivalent of a new Thailand every year. Coastal wetlands are being drained for conversion to farming; mangroves and swamp forests are being cleared for aquaculture ponds and timber. As a result, wetlands are in trouble. All five threatened stork species are Asian, as are four of the six threatened ibises and spoonbills, and four of seven herons. The situation in the United States is not much better, as population pressures have led to the destruction of 40 percent of the nation's wetlands and to serious declines in shorebird populations.

Birds, of course, are just one set of indicators of what is happening to biodiversity as habitats planetwide are subjected to assaults launched by *Homo sapiens*.[75] Virtually all flora and fauna are suffering dramatic declines, and the majority of losses are among "the little things that run the world" that few people are aware of.[76] Thousands of species of insects and other terrestrial arthropods are almost certainly vanishing unheralded as tropical moist forests are destroyed. Dozens of plant species were lost forever when the forest on a single ridge in Ecuador was cut down.[77] Populations of frogs and salamanders are in decline over much of the globe, for reasons not yet understood (although acid rain is a suspect in many cases). Biodiversity is in retreat, from the United States Pacific Northwest, the English countryside, and suburban New Jersey to the rainforests of Zaire, Brazil, Madagascar, and New Guinea, and even the coral reefs of the Caribbean. And our own life-support systems are being weakened as a consequence.

Respect for the Land and Its Resources

One of the tragic losses in industrialized civilization is that of respect for the land. It can be seen in the carelessness (or total absence) of restoration after strip mining or other surface-disturbing activities. It can be seen in the insane notion that driving off-road vehicles over open countryside is a legitimate "sport." But modern industrial societies are nevertheless inescapably sustained by the land, and they lay waste to it at their great peril.

The solutions to problems such as deforestation, desertification, and general habitat destruction in a rich country like the United States are relatively straightforward: stop subsidizing the destruction of forests and overgrazing; find ways to overcome the effects of high discount rates and misallocation of property rights; encourage investment by individuals and firms in the long-term health of the environment and society; install mechanisms such as zoning laws and tax disincentives to prevent any more sprawling development. The time has come to stop placing homes, factories, malls, highways, or other development in relatively undisturbed areas, be they grasslands, woodlands, chaparral, wetlands, or whatever. More than enough American land has already been severely disturbed that the needs of our population (assuming its size is controlled *soon*) can be met by redevelopment, rather than further invasion of prime agricultural land or the few remaining semi-intact ecosystems. In crowded urban areas, new housing should be largely restricted to high-density or high-rise buildings, designed with ample areas of native vegetation between them.

Considerable progress could be achieved in making even badly disturbed areas more hospitable to biodiversity without enormous cost or effort. Discouraging the planting of nonnative plants anywhere would be a useful start. If all the lawns in southern California were replaced over the next decade with native plants adapted to desert conditions, not only would a great deal of precious water be saved, but local biodiversity would be much enhanced. In San Diego, for example, careful control of backyard plantings could supply the corridors necessary to connect the small remaining fragments of habitat in undeveloped canyons and help maintain attractive elements of the local fauna such as roadrunners, wrentits, and coyotes (which help control rats and other pests).[78]

Meeting the need for habitat preservation in poor countries with exploding populations is a much more difficult problem. One cannot view the problem of preserving tropical biodiversity outside the context of the debt and poverty in developing nations. Developing countries are heavily burdened with external debt; the loans owed to banks of other nations in 1990 amounted altogether to some $1.3 trillion. Brazil paid about $40 billion servicing its foreign debt during the 1980s but has been unable to reduce the principle of its $120 billion debt at all.[79]

About half of Brazil's 150 million people live in poverty, perhaps one in four or five is malnourished, and by 2020 there may be over 230 million Brazilians. Most poor people have a very high discount rate for obvious reasons, so it is no surprise when they are forced to "mine" their capital—to destroy forests and biodiversity for immediate meager gain. Taking the long view of one's resources is a luxury available to people who know where their next meal is coming from.

As civilization struggles toward Holdren's optimistic energy-use scenario, the rich must help the poor to protect as much land containing relatively undisturbed habitat as possible. One important tool for doing this, the "debt-for-nature swap," was invented in 1984 by Thomas Lovejoy, when he was vice-president for science of the United States World Wildlife Fund.

Conservation efforts in developing nations are often stymied by lack of funds to acquire land for reserves and pay for wardens and fencing, in part because of their huge foreign debt burdens. The debt swaps are a mechanism for using the debt to help protect their biological capital. Banks commonly sell the IOUs in the secondary debt market at prices reflecting the market's estimate of the chances that the debt will be redeemed. In May 1989, a dollar's worth of IOUs from Colombia was worth 55 cents, from the Philippines 46 cents, from Nigeria 20 cents, from Ecuador 12 cents, and from Peru 3 cents.[80]

In making a debt-for-nature swap, a conservation organization or other agency purchases a quantity of debt on the secondary market, and then spends the debt's value in local currency for the protection of biodiversity. In the first such swap, the United States—based organization Conservation International paid $100,000 for $650,000 worth of Bolivian debt in 1988. The local currency was then donated toward the management of the Beni Biosphere Reserve and several adjacent areas. Subsequently, other swaps have been performed, most notably with Costa Rican debt. About 5 percent of that nation's commercial debt has been purchased and exchanged for local currency bonds to use for conservation projects.[81]

The process is somewhat more complicated than indicated here,[82] but it is thought that roughly $200 million of local currency could be generated by $50 million of United States funds each year—*if* United States tax laws were changed to encourage it.[83] At the moment, selling debt for cash is financially more advantageous to a bank than donating it. Provision of a tax credit for the market price of donated debt would correct that; the cost to American taxpayers would be minor compared to the long-term benefits in conserving global biodiversity.[84]

As Aldo Leopold wrote almost two generations ago, we need to restore a "land ethic." His instructions were to the point: "Quit thinking about decent land-use as solely an economic problem. Examine each question in terms of what is ethically and esthetically right, as well as what is economically expedient. A thing is right when it tends to preserve the integrity, stability, and beauty of the biotic community. It is wrong when it tends otherwise."[85] Leopold's focus on the beauty of ecosystems may seem odd, but (apart from the utility of that beauty for recreational values) the complexity and harmony of an intact ecosystem is beautiful to most human beholders, so beauty is probably a good indicator of a system's health and capacity to provide services.

▬▬ Water

Many years ago, Justus von Liebig announced a principle now known as *Liebig's law of the minimum*. It stated that the amount of living matter in a given environment would be limited by the stock of whatever requisite of life was in shortest supply. Fresh water may well turn out to be that requisite for many human populations; water shortage could be the factor that ultimately limits the scale of the human enterprise; it is already serving as a limiting factor in some regions.

Unfortunately, the public and decision makers have largely ignored the impending water crunch, except to grope for "supply side" solutions such as the preposterous North American Water and Power Alliance (NAWAPA). NAWAPA was a giant water project proposed by American interests in the 1960s to grab Canadian water to help quench the thirst of the arid Southwest. The costs would have been colossal, the ecological disruption severe, and the "fix" good only until around the turn of the century.[86] The Canadians were unenthusiastic to say the least, and the project fortunately died, although people keep trying to revive it. American greed for water has also caused tensions with Mexico to rise as Colorado river water delivered to the border has become increasingly salty and polluted; in some years virtually no water was left.

In the Middle East, international relations are not so amicable, and the 1967 Arab-Israeli war was fought in part over the waters of the Jordan, Yarmuk, and Litani rivers.[87] The situation seems destined to go critical again soon. The nation of Jordan is on the verge of a severe water crisis.[88] In 1990, a Jordanian government official despaired of any possibility of supplying sufficient water to the exploding population beyond 1996. Even in 1990, water was supplied to homes only two days a week, and costs hovered around a dollar for a cubic yard— a rather high price for a week's minimal supply of water per person in a country with an average annual per-capita income of $1,500. The Jordanian government considered building solar desalinization plants a strategic need.[89] Tensions have risen over upstream threats to water supplies between Egypt and Ethiopia (the Nile) and Syria and Turkey (the Euphrates). Water management in the Middle East, with its rapidly growing populations and rising industry, is an urgent need, but reaching agreement on a plan among several contentious nations will not be easy.

That people might actually kill each other over water seems irrational, since there are vast amounts of fresh water on the planet. Each year enough rain falls on the continents to cover them, on average, 29 inches deep. Unfortunately, though, that rainfall is delivered very unequally, and much of it falls on sparsely inhabited areas. Substantial amounts of water from northern Canada and Siberia flow into the seas in large rivers such as the Mackenzie, Ob, and Lena. The largest river in the world, the Amazon, drains a vast area containing only a few million people, with a volume of flow equivalent to one-fifth of all the planet's rainfall (equivalent to twelve Mississippis). Equatorial Africa and Southeast Asia also have abundant rainfall in areas of sparse population, as do several temperate rainforest regions.

The uneven distribution of fresh water is one of two major reasons why there are water-rich and water-poor nations. Iceland and New Zealand have more than a cubic yard of runoff per year for each square yard of land surface, while the United States (also fairly water-rich), gets an average of about three cubic feet. Australia receives only about half a cubic foot and Peru somewhat less, while Egypt and Saudi Arabia each get barely a cubic inch of runoff per square yard of surface.

The other major reason for there being water-rich and water-poor nations, of course, is the uneven distribution of people. Iceland, with its abundant rainfall and tiny population, has over 800,000 cubic yards of runoff per person per year. The United States, with moderately abundant rainfall and a population ten times as dense as Iceland's, has only about 12,000 cubic yards of annual runoff for each citizen. China, with slightly more rainfall than the water-rich United States, becomes water-poor because of its huge population.[90] It has a per-capita runoff of about 3,000 cubic yards. Singapore gets almost three times as much rainfall per square yard as the United States, but with a population 170 times as dense, that boils down to an annual runoff of only 275 cubic yards for each citizen of Singapore. Egypt, with little rain and many people, has only a pathetic 20 cubic yards per person, explaining Egypt's dependence on the Nile to bring runoff from upstream countries.[91]

The distribution of water is not only uneven between nations; it is also uneven within nations. The western half of the United States has only about a third of the nation's runoff; nearly all of Australia's rain falls near the coasts. Far more serious is the unevenness of runoff in *time*. Over much of the United States, 40 to 70 percent of the runoff occurs during a flood period of two or three months each year.[92] In many tropical and subtropical regions, rainfall is even more seasonally restricted. Moreover, precipitation, and thus runoff, varies not just from month to month but from year to year and on longer time scales by up to 50 percent or even more. California's drought, 1990–1991 being the fifth consecutive year with annual precipitation substantially below average, is a case in point. So, in the other direction, are the occasional "100-year floods"— excess water of no use to anyone.

Finally, much of the water that falls in inhabited areas quickly becomes unavailable for human use. Almost 60 percent of it is lost to evaporation; much runs into the oceans by routes from which it cannot be practically retrieved; and a great deal is made unusable by pollution. In the United States, more than fifty times more fresh water is stored underground in soil and aquifers than exists at any given moment in lakes, rivers, and streams,[93] but only a small fraction of that vast groundwater store is economically accessible. Most water in soil is soon lost through evaporation or transpiration by growing plants (including crops) and must be continually replenished by rain. Some soaks downward, very slowly recharging aquifers. Despite the uneven distribution of rainfall, fresh water from surface sources is by far the easiest to obtain in most areas.

Human beings attempt to deal with the vagaries of surface water availability in a variety of ways. Dams can help even out the annual flow of rivers, but at a cost in construction, increased evaporation, flooding of land, and disturbance of riverine ecosystems. Dams moreover are temporary structures, since they eventually silt up, making them useless for either storing water or providing hydroelectric power. In many poor nations with denuded watersheds, silt often fills the reservoir a scant few decades after the dam's completion.

Wells to extract groundwater can be expensive to drill and maintain. They also may be temporary in an era when many aquifers are being pumped out far faster than they can be naturally recharged.[94] Pollution can be cleaned up, sewage can be treated, and, in general, water can be processed for reuse. But each of these processes can be costly, technically difficult, or both. In poor nations, the funds and technical capability are often lacking altogether. Local authorities even in rich nations sometimes have trouble financing adequate treatment.

Vast differences in levels of water use exist between societies also: the average American uses about 2,500 cubic yards of water each year, four times as much as the average Swiss and 70 times as much as the average Ghanian. The Swiss

are no poorer than Americans; both nations are highly industrialized, which accounts for much of their high water use, but
Swiss agriculture is not dependent on irrigation. In Ghana,
neither irrigation nor industry is a significant factor; nearly all
water use is by households, and that, too, is much lower than
in rich countries.

Of course, water can be conserved. American farmers
in some parts of the high plains have reduced evaporative water
wastage in irrigation from around 40 percent to as little as 2
percent. Israel has worked miracles of water-use efficiency in
agriculture. Many California homeowners have cut their water
use by 40 or 50 percent, even placing bricks in their toilet
tanks to reduce the amount of water used for each flush. But
85 percent of the water used in the state is consumed by agriculture, which had done relatively little in any organized way
to curb its consumption, even after five years of drought.

The Los Angeles basin graphically illustrates the collision between population growth and limited water resources.
In a region where air is in short supply, its huge and growing
population is supported by water imported from relatively water-
rich northern California.[95] The price of turning the Los Angeles
area into a smog-choked megalopolis has been the gross alteration of many riverine ecosystems elsewhere, the drainage of
Mono Lake and disruption of its unique ecosystem, and destruction of the agricultural productivity of the Owens Valley.

Southern California, still adding a million new residents
every three years, is headed for severe water shortages soon,
regardless of the current drought, unless other sources can be
found.[96] Arizona fought and won a larger share of the Colorado
River's water (so it too could continue its high rate of population
growth), and court decisions have compelled Los Angeles to cut
back its diversion of water from Mono Lake.

Conservation is one solution; so is enlarging the already
ecologically disruptive California Water Project, which brings
enormous amounts of water from north to south, but this is
opposed by northern Californians. Stemming the influx of people
would help, but seems unlikely to happen immediately. The
cheapest and most available source of additional water is the
huge amount of heavily subsidized project water allocated to
Central Valley farmers. If the farmers reduced their water use

only 10 percent, enough could be made available not only to support decades of continued urban growth but also to restore many streams and fisheries.

We personally favor ending population growth in southern California for a plethora of environmental and social reasons, and generally oppose cities taking water from agriculture. But California agriculture is so grossly oversubsidized—to the point that irrigating pastures and alfalfa fields is profitable—that an overhaul of the state's irrational water system is badly needed. Until 1990, it seemed politically and legally impossible, but persisting drought may break the logjam.

A similar disaster may occur in Colorado if Denver and its booming suburbs succeed in their attempts to seize and redirect water from the western slope of the Rockies, thereby depleting the water resources not only of the western half of Colorado, but all the other areas from the Utah border to Mexico and Los Angeles that depend on Colorado river water. Here is another case of misplaced priorities governing the allocation of resources.[97] Colorado water rights are basically allocated by historical accident—whoever staked the first claim—with little regard for need or societal values. But the rights can be sold, traded, or leased. As is true in most states, cities and developers can afford to pay much more for water than can farmers and ranchers, who in turn can outbid natural ecosystems (which until recently had no standing at all in the market).[98]

Decades of rapid population growth and proflilgate water use in many regions around the world have brought the inevitable collision in water-short regions. A 1990 story in *Time* vividly described problems in various nations: the Indian white-collar worker in Madras who must stand in line at a public tap at 3:30 A.M. for his daily five buckets of water; over 8,000 Indian villages with no water at all; Beijing's increasingly desperate water-supply situation, with one-third of the municipal wells dry and the water table falling as much as six feet per year.[99]

Water-supply problems as they relate to agriculture are very serious. People can survive without flush toilets, but they cannot go without food, and food cannot be produced without water. Irrigation is the largest single *consumptive* use of water, accounting for roughly 85 percent of consumption in the United States. Agriculture accounts for only about 40 percent of water *withdrawals* in the United States, but unlike domestic and industrial users, it does not return significant amounts of water to rivers or streams.

Worldwide, more than two-thirds of freshwater use is for agriculture, and about 17 percent of the world's cropland is irrigated.[100] In many regions, the increasing demand for water for agriculture is an important component of the trend toward shortages and rising competition over supplies of fresh water. Water planners and developers have yet to consider seriously how global warming and climate change might affect future water supplies in areas where shortages are already a problem, at least sometimes. But many water-short regions, including California and the Great Plains states, may find their scant rainfall further diminished by shifts in atmospheric circulation.[101]

▬ Environmental Impacts of Water Use

Shortages and competing needs are not the only water problems facing humanity; there are two major environmental consequences of human use of fresh water. One is the familiar problem of "water pollution"—contamination by human beings of both fresh and salt waters, discussed in Chapter 5. The other is that human consumption and diversion of water have led to widespread degradation of ecosystems and ecosystem services. As with air pollution, the consequences for society of the effects on ecosystems are likely to be the most damaging and long-lasting.

The most serious environmental impacts of human mobilization of fresh water have arisen from allowing populations to overshoot long-term carrying capacities of entire regions, and those impacts often undermine the carrying capacities. A classic example is the Sahel region of Africa, where tubewells drilled to provide water allowed nomadic societies to expand their herds of cattle and goats beyond the carrying capacities of the fragile grasslands. The resulting overgrazing and desertification, when the droughts of the 1970s and 1980s came, led to an intensification of the drought and widespread famine. In 1990, the once-proud Tuareg tribe had lost their herds to drought and had been reduced to destitution and dependence on foreign aid, living in shanties around Timbuctu in Mali. [102]

Giant dams, long favorite projects of development agencies, often prove severely disruptive not only to natural ecosystems but to human settlements and local agriculture, especially in developing nations. [103] In tropical regions, such dams can create massive disturbance of regional hydrologic systems and ecosystems. [104] A billion-dollar project in Brazil, built to provide electricity for expanding Manaus, destroyed a half-million acres of forest, massacred wildlife, displaced indigenous peoples, created a vast shallow eutrophic lake, and brought hunger and disease to river dwellers. Similar untoward effects have followed large-scale dam projects in many developing nations from Egypt to central and southern Africa. But the big dam developers are undeterred; a huge project being planned in China will displace more than a million people, flood 140,000 acres of land, and cause untold damage to local wildlife; another in India is arousing much local opposition. [105] In both cases, opponents of the dams have pointed out that several smaller dams built higher in the watershed could provide the water storage for flood control and electricity production at considerably less financial, social, agricultural, and environmental cost.

The greatest havoc on ecosystems has been wrought by the use of water for irrigation. Irrigation has been a major component of the expansion of humanity's feeding base since mid-century. [106] Agriculture in southern California and the Aral Sea region of the Soviet Union is utterly dependent on irrigation. Without it, those regions could not be supplying the majority of their nation's fruits and vegetables. Without irrigation, yields of grain in the high plains of the United States, northern China, and northwestern India, all critical to world food production, would drop by a third to a half. Without irrigation, Egypt would have to import *all* its food instead of just two-thirds. Nations as diverse as Japan, Pakistan, Peru, Israel, and Indonesia depend upon irrigation for more than half their food production.

So irrigation is good; it supplies humanity with a major portion of its food. That's one way to look at it. But in another way, irrigation has been disastrous. In most circumstances, it is a temporary game, leading to the slow destruction of land, sometimes through waterlogging, more often through a gradual buildup of salts (salinization) in soils. As the land slowly deteriorates, productivity falls over decades to centuries, depending on the circumstances. Salts also accumulate in reused irrigation water, accelerating the salinization process in irrigated areas downstream.[107] The temporary nature of most systems creates a perpetual need to bring new land under irrigation. Since the supply of arable and irrigable land is finite, irrigation can be viewed as just one more way humanity has found to live on its capital.

To put it another way, irrigation has often been said to have "made the desert bloom." Deserts, except for the most extreme ones, always bloom with their own unique and interesting plant diversity—plants that support a fascinating set of unusual animals. Irrigation destroys that biodiversity permanently to allow a temporary bloom of human population. Charles Galton Darwin commented starkly four decades ago on the Sukkur barrage on the Indus River, a dam that diverted water to irrigate 6 million acres of India: "After a few years the effect was only to have a large number of people on the verge of starvation instead of a smaller number."[108]

Numerous societies in history have declined or failed because of dependence on irrigation systems that eventually became unproductive. The destruction of the ancient hydraulic civilization in the Tigris and Euphrates valleys is a classic example of the ultimate unsustainability of civilizations built on irrigation.[109] It remains to be seen whether damming the Nile and redistributing its water in irrigation systems will lead to a collapse of Egypt's 6,000-year-old civilization, long sustained by Nile water and the continual refertilization and creation of new land from Nile silt. All that now seems to be coming to an end as unchecked population growth and the Aswan Dam (by reducing the Nile's essential nutrient and silt flow dramatically) have reduced Egypt to a client state of the United States, teetering on the edge of catastrophe. Diversion of Nile waters by rapidly growing upstream nations, silting of the dam, and the possible inundation of the precious delta lands by rising sea levels may well push Egypt over the edge.

Irrigation schemes can be made more enduring, however, through careful planning, maintenance of the system, and use of appropriate drainage techniques.[110] The water needed in irrigation systems can be reduced by 20 to 30 percent at reasonable cost simply by plugging leaks in piping and lining the bottoms of canals to prevent seepage.[111] Such precautions should be standard in developed nations where the consequences of not carrying them out are clear; unfortunately, lack of incentive and oversubsidization of water often prevent their implementation. In developing nations, the knowledge, materials, and financial support may be missing.

Sophisticated methods such as drip irrigation can accomplish both goals by using only a fraction of the water required by conventional systems, so salt accumulates much more slowly and waterlogging is less of a problem. These systems, though, are considerably more expensive and difficult to install and maintain, and substantial amounts of energy are needed to make the plastic tubing to carry water to the plants. In addition, the water used must be exceedingly clean, or the tubes become clogged. All these factors make drip irrigation systems unattractive to farmers who would reap the benefits of such systems relatively far in the future, especially when government subsidies now provide them with abundant cheap water. Drip systems until recently were mainly restricted to areas like Israel, where water was in extremely short supply and people had the resources and expertise to build and maintain them.

Much irrigation water is pumped from underground aquifers, rather than taken from rivers. In dry regions around the world, aquifers are being pumped out at rates far above those of natural recharge—a highly unsustainable strategy. In some cases, aquifers have already been so depleted that further pumping is uneconomic, including parts of the United States.[112] The American grain production increases of the 1960s and 1970s were partly thanks to water taken from the giant Ogallala aquifer underlying the plains states from Nebraska to Texas; now as wells run dry, more and more of that land is being returned to less productive and less dependable dryland farming.[113] In the more arid West, urban population growth, along with agriculture, has provided the impetus for tapping groundwater. In other areas such as northern India, both population expansion and the need to increase agricultural production have led to overdrafts. This is still another example of living on capital, emulated around the world as aquifers have been so drained that it is no longer economically feasible to pump from them.

Worldwide, problems with irrigation now seem to be

coming to a head. Huge irrigation projects are a recent phenomenon; three-fifths of today's irrigated acreage has been added
in the past 40 years, and now almost a million square miles
(about 2 percent of Earth's ice-free land surface) are irrigated.
That requires the diversion of about 800 cubic miles of water
annually from lakes, rivers, and stores of groundwater, the rough
equivalent of the flow of six Mississippi Rivers. But the rush to
irrigate of recent decades is slackening; most of the attractive
areas have already been put to use. On a per-capita basis, the
worldwide acreage of irrigated farmland has been falling since
1978, as more and more damaged land is being taken out of
production.[114]

The potential environmental impact of massive diversions of water is epitomized by the fate of the Aral Sea. The
Aral Sea, Asia's second largest lake (after the gigantic Caspian
Sea, which lies to the west), covered 26,000 square miles in
1960. Now 11,000 square miles of that area are exposed,
parched seabed, dotted with stranded boats and isolated
docks.[115] The disaster is the result of a 1918 decision to divert
the flow of the Amu Darya and Syr Darya (the two large rivers
that fed the lake), principally to irrigate fields of cotton, but
also fruits, vegetables, and rice, in Uzbekistan (where the sea
is located) and Turkmenia (to the south). Now the bed of the
Amu Darya, the Oxus of classical times, is dry where it reaches
the sea.

Before the rivers were diverted, the Aral supported
10,000 fishermen in the town of Muynak. In 1965 they sailed
from that seashore town to harvest bream, pike, and perch,
supplying the Soviet Union with 3 percent of its fisheries catch.
Now all 24 native species of Aral fish are extinct, and the former
fishing port is more than 20 miles from the receding shore. A
local cannery is kept alive by the government, which ships in
frozen fish from the Baltic and the Barents seas, each 1,700
miles away. The climate of the now-landlocked town has also
changed for the worse, with temperatures exceeding 100° F for
extended periods in the summer.

The loss of the lake's fauna is not the only ecological
disaster caused by the salting up and shrinking of the Aral.
Huge "salt storms" are generated as vast clouds of salty dust,
more than 40 million tons annually, are swept by fierce north

winds from the dry lake bed. Enormous quantities of this dust are deposited on farmland hundreds of miles away, and some has been traced as far away as the Arctic coast and Soviet Georgia.

The process, if continued, will ruin croplands and degrade natural ecosystems over a vast area. Meanwhile the cotton fields, for which the whole exercise was originally started, are beginning to suffer from the curse of irrigated land over much of the planet—salinization, caused by the accumulating salt load left on the fields when the river water evaporates. Cotton production, not surprisingly, is declining.[116]

As is also typical of cotton cultivation, pesticides and defoliants have been vastly overused in the Aral region.[117] In places the fields are not only encrusted with salt but with pesticide residues—and those residues are swept up in the salt-dust storms and find their way into water supplies. High levels of agricultural chemicals and salt in air, rivers, and groundwater are blamed for the region's soaring rates of throat cancer, stomach and liver disease, respiratory and eye ailments, birth defects, and infant mortality.

The Aral story illustrates, as an extreme case, the environmental consequences of ill-advised diversions of water for irrigation. A temporary gain, in this case a few decades of self-sufficiency in cotton production for the Soviet Union, has been bought at the cost of an ecological and public-health disaster.

There is considerable debate in the Soviet Union about how to save what remains of the Aral Sea and prevent an even more widespread ecocatastrophe. One school, infected with the NAWAPA mentality, would solve the problem with another vast water project, which probably would cause further ecological devastation. This project would divert waters now flowing into the Arctic Ocean from the Irtysh and Ob rivers through a canal 1,500 miles long to feed into the Aral. More sensible plans call for cutting back the amount of diverted water by one-fourth and improving the efficiency of irrigation, allowing more water from the rivers to reach the lake and stabilize it. Even reducing the cotton production quota has been discussed. But no plan, however grandiose, can ever restore the vanished native fishes; and some damaged farmlands may not recover for centuries, even with strenuous efforts at restoration.

▬ Wetlands

Wetlands are another source of surface fresh water that has suffered considerable abuse. These biotically rich and extraordinarily productive systems[118] have been drained or filled the world around, largely to expand farmland. The consequences include disruption of local hydrologic systems; wetland areas serve as natural reservoirs and as such reduce floods and maintain stores of water through droughts. They also help maintain air and water quality, support a diversity of economically valuable wildlife, especially waterfowl and fishes, and provide timber from their associated forests. All these services are lost when wetlands disappear.

The process of loss is usually the cumulative result of many independent decisions by landowners and others—another case of environmental destruction by nickel-and-diming.[119] In the lower 48 United States, more than half the original wetland endowment has been lost; in some states over 90 percent has vanished. The impact on wildlife is measurable by substantial reductions in populations of ducks and other birds dependent on wetlands. The vast majority of conversions have been for agriculture; until a few decades ago these were actively encouraged by the government, which in the nineteenth century regarded swamps as threats to human health and progress.[120]

The United States has belatedly begun to try to preserve the remaining fraction of the nation's original wetland heritage (apart from established wildlife refuges), starting with the 1972 Clean Water Act, which put the Army Corps of Engineers in charge of issuing permits for filling in wetlands. Despite accusations that this amounted to appointing the fox to guard the henhouse, at least some members of the corps seem to be making a serious effort to carry out its charge responsibly.[121] The 1985 farm legislation included disincentives for farmers who converted wetlands to crops and provided subsidies for those who preserved them. Since then efforts have been made to establish a government policy of "no net loss," to encourage restoration or artificial replacement (a less satisfactory option than preservation) of lost wetlands.

The "no net loss" policy has run into considerable opposition, not only from farmers who make some (short-term at least) financial sacrifice to preserve or restore wetlands only to see developers on the next property turn them into highly profitable houses and commercial buildings. Developers and industry are also resisting the policy, seeing desirable land suddenly being put out of bounds. The loss of wetlands had not yet stopped in 1990.[122]

The United States is by no means alone in its destruction of wetlands. Much of western Europe, including Britain, was originally marshy. The Netherlands survives by keeping the water out of its low-lying land. But the wetland losses have been paralleled by decimations and extinctions of wetland fauna— waterfowl and other birds, even butterflies dependent on wetland plants.

In developing nations, similar losses have occurred. Modernizing China's agriculture since 1960 has included converting thousands of small wetlands and ponds to cropland, often to the detriment of local crop production. Protective coastal mangrove swamps have vanished along tropical and subtropical shores around the world, resulting in far greater damage from tropical storms and high tides to coastal settlements. Some societies have seen the light and begun restoring the mangroves— a trend that deserves serious encouragement by development agencies and governments. Such restoration will prove even more valuable when and if global warming raises sea levels.

Fresh water, like forests, productive land, and soils, is a precious and ultimately limited resource that if treated wisely is perpetually renewable. Sadly, though, too much of humanity's use of water is far from wise and is rendering much of the resource for practical purposes nonrenewable, as aquifers are overdrawn, rivers diverted, lakes and swamps drained, and their ecosystems destroyed, all for temporary gains of water or land. Much land, too, is being degraded as a result of irrigation. The implications of all this for future agricultural production are less than rosy, as the next chapter will explain.

Humanity, nevertheless, is not about to run out of water; much of the difficulty is that water is so often mismanaged or allowed to become unusable through pollution. Irrigation systems can be made much more sustainable than most are today, with appropriate investment and precautions. Allocations of water among sectors of the economy can be rationalized and priced according to their true value to society. And the most neglected consideration of all, the need for water to preserve the integrity of natural ecosystems, as well as the sustainability of agricultural systems, must be incorporated into planning and management.[123] Water's place in environmental and social planning should reflect its importance as a vital resource.

In the crucial decades ahead, the renewable resources of land and water—capital now being squandered and made nonrenewable—will be needed more than ever before to sustain humanity through its population outbreak until rapid growth can be reversed. Both land and water are particularly indispensable in supporting agriculture, and we now turn to this vital and complicated subject.

Unsustainable Agriculture

"Some people are so poor that God can appear to them only in the form of bread." So observed Mahatma Gandhi many decades ago while struggling successfully to liberate India from British rule. But the battle to feed the Indian population, now more than twice its size at independence and divided into three separate nations, has scarcely advanced at all. The same is true for many poor nations. Humanity's struggle both to feed the poor and to overfeed the rich constitutes one of the principal causes of environmental degradation, one that perhaps will be the most difficult to correct. But it must be corrected if Earth is to be healed. Because of the size of the human population, the nature of many diets, and the way most agricultural systems are run, eating is one of the most ecologically destructive of all human activities.

The subcontinent of South Asia still includes the greatest number of hungry people in the world.[1] Assessments vary considerably, according to how the availability of food in societies is measured, but the World Bank estimated around 1988 that almost 500 million, about half the population, were undernourished in South Asia.[2] More conservatively, a United Nations committee on nutrition estimated in 1987 that "only" about 170 million South Asians were hungry.[3] Chronic hunger also plagues sub-Saharan Africa. There a quarter to almost half of the population, some 100 to 200 million people, have diets that supply too few calories for an active working life. Another 100 to 300 million people in other developing regions also get too little to eat.[4] Overall, it appears likely that almost a fifth of the world's population, more than a billion people, were unable to obtain sufficient energy for normal activity in 1990.[5]

Even the lower estimates by the United Nations Food and Agriculture Organization (FAO) represent a human tragedy of gigantic proportions, particularly when it is remembered that hunger is most prevalent among small children and infants.[6] It also should be remembered that early malnutrition directly threatens the future health and productivity of the children who survive. The FAO even noted that the absolute number of hungry people was climbing, although the proportion of them in the total population had been declining (at least until 1987).[7]

Until the mid-1980s, efforts to keep world food production ahead of population growth was fairly successful, although the rich continued to gorge themselves while the poor often went hungry. Regular reports of global grain gluts in the early 1980s lulled much of the public in western nations into believing there was no world food problem.[8] Disastrous recurrent droughts and famines in remote and desperately poor nations such as Ethiopia and the Sudan were explained away as extreme cases of maldistribution of food supplies or malign political repression against rebellious provinces.

Enough truth was behind these explanations, especially the existence of "food wars" (in which people are deliberately starved to achieve political ends or starvation is a prominent side effect of conflict),[9] to persuade many observers that the widespread hunger in developing nations was simply a matter of poverty and inequitable distribution of food supplies. Food production has been increasing more rapidly than the population, they maintain, and if food were equitably distributed, there would be no world food problem.[10]

Between 1950 and 1984, global food production did outpace population growth. Cereal grain harvests (the principal feeding base of humanity), expanded 2.6-fold, while the population slightly more than doubled.[11] Unfortunately, though, the dietary improvements made possible by the bigger harvests were mostly seen in developed nations and some middle-income nations, where people were eating more meat and dairy foods produced with grain-fed livestock. So the remarkable gains in global food production in the face of very rapid population growth were far from universally shared.

But the situation is not so simple—or so easy to solve.

It can best be placed in perspective by considering how many people could be supported by the 1985 world food harvest (which was a world record, though not much below the later 1989 harvest) if the primary food supply—crops and range-fed animals—were more fairly distributed.[12] That supply would feed 6 billion people a principally vegetarian diet, 4 billion a diet with about 15 percent of its calories derived from animal products, and 2.5 billion people a "full but healthy diet" (perhaps 30 percent of calories from animal sources).[13] These numbers are not reassuring in a world whose population passed the 5.4 billion mark in 1991, and is destined to reach 6 billion in 1997.

The complacent picture usually presented of global food security up to 1987, moreover, left out several disturbing underlying trends and circumstances, such as accumulating environmental problems and the serious land degradation also taking place that may threaten future food production.[14] The earlier strides in increasing food production were not maintained in the 1980s. Even though more food was being produced per person than ever before, the average annual rate of expansion in global food production had declined from a high of 3.5 percent between 1966 and 1976 to about 2.2 percent between 1980 and 1990—only slightly above the rate of population growth.[15]

The slackening of food production increase in the 1980s did not occur uniformly. Instead, harvests were lagging significantly behind population growth in two major regions, tropical Africa and Latin America, and the trend in South Asia was none too encouraging.[16] Even the celebrated surge in China's grain production fell back,[17] as farmers turned to producing other, more profitable crops and as prime farmland was converted to housing, factories, and other uses.

A half-century ago, most regions of the world were self-sufficient in food production, and many exported grain. The exception was Europe, which imported much of the grain it consumed. Since then, for various reasons, the pattern has changed dramatically; more than 100 nations now regularly import grain from the United States, Canada, western Europe, Australia, and a few other surplus producers. Roughly half the world's grain is produced in the developed world (which has barely a fifth of the population). Nearly all developing nations have become dependent on grain imports to keep their populations adequately fed,[18] although many of them produce cash crops for export while neglecting the nutritional requirements of their own poor. This growing worldwide dependency on a mere handful of suppliers for basic foodstuffs could itself spell trouble for global food security. However reliable present-day suppliers may have been in the past, such concentration of production is risky in the face of global climate change, which could make the system even more vulnerable to economic and political disruptions than it is now.

In the late 1980s, grain production received several setbacks, and global harvests have failed to keep pace with population growth on a worldwide basis since 1984. Adverse weather was a major cause of poor harvests in 1987 and 1988.[19] A record drought in 1988 struck three of the world's most important breadbaskets—North America, the Soviet Union, and China—and jolted the world at least briefly into realizing that the food situation was not nearly as secure as had been believed. Whether the 1988 drought was a harbinger of global warming may never be known, but climatologists have warned that such weather events could be expected more frequently as the greenhouse gases keep building up in the atmosphere.[20] Regardless of the cause, grain production suffered an absolute decline between 1986 and 1988 of about 10 percent (while the population grew by over 3.5 percent).[21] Better harvests in 1989 and 1990 barely returned per-capita production to the 1984 level.[22]

■■■ The End of Expansion

Some analysts now think that the era of rapid food expansion is essentially over and the real battle to feed the growing human population is just beginning.[23] The 1970–1990 slowdown in the rate of increase in global food production may portend deepening problems. As the human population, some 5.4 billion in 1991, continues to expand by 1.8 percent annually,[24] some 95 million more people must be fed each year with 24 billion fewer tons of topsoil,[25] trillions fewer gallons of groundwater, and declining ecosystem services. Humanity is squandering its inherited capital of agricultural soils, groundwater, and biotic diversity in the effort to keep food production rising—resources that are both finite and nonrenewable on any practical time scale.

In addition to the steady loss of vital resources underpinning agriculture, other constraints on the expansion of food production have begun to appear. These may limit future production regardless of problems that may arise from adverse changes in weather and climate caused by global warming.[26] One is the limited amount of land suitable for cultivation. To varying extents, people already live on or use some two-thirds of the world's land (90 percent or more if severe deserts and land underlain by permafrost are excluded).[27] The 11 percent of Earth's land now planted in crops is almost all that is suitable for growing them; most of the unexploited tillable land remaining is of marginal quality.[28]

That is why since mid-century the opening of new cropland has accounted for only a small fraction of the rise in food production, and why its importance has steadily dwindled. In recent decades, more and more degraded land has been removed from production, offsetting areas newly opened for agriculture. The great surge in global grain production since 1950 has come principally from increased yields (production per acre). The acreage of grain cropland *per person* worldwide will be only half as much in 2000 as in 1950 because the human population will be 2.5 times as large.[29]

The factors that have led to a doubling or tripling of grain yields (depending on the crop species and the setting) over four decades are the familiar technologies of the green revolution: high-yield strains of cereal crops and increased inputs of inorganic fertilizers, pumped irrigation water, and synthetic pesticides. Where enough of these inputs (all tied in various ways to fossil fuels) have been provided,[30] dramatic increases in yields of the major grain crops, especially corn and wheat, have often been attained. But yield increases from green-revolution technology may now be playing out, for various reasons.

In many regions, fertilizer applications have passed the point of diminishing returns, and fertilizers can boost yields only so far even under ideal conditions (at high levels they can even be toxic to crops). Sooner or later, the cost of adding more fertilizer cannot be compensated by the earnings from any additional crop produced. So substantial further rises in grain yields are unlikely to be gained from green-revolution technologies in industrialized nations, where farmers already apply large amounts of fertilizers and other inputs.

Moreover, the scope for further yield increases in developing countries under prevailing economic conditions is rapidly shrinking. Green-revolution technology has been adopted by those farmers in a position to do so in almost all areas with suitable climates and soils.[31] Most of the remaining opportunities for further spread of green-revolution technology are among poorer farmers in some developing nations who have neither been able to pay for the needed inputs nor been supplied with supporting services.[32]

Green-revolution farming involves much more than access to supplies of the seed, fertilizer, water, and pesticides: an infrastructure that provides information and advice on using the technology, adequate farm credit to purchase inputs, and access to markets that can absorb the additional crop produced. Subsistence farmers are often faced with "inadequate demand," even in areas where people may be grossly underfed. "Inadequate demand" simply means that poor people lack funds to buy the food they need; hence economists sometimes talk of "the need to boost demand" in hungry areas. If increased local production results in dropping prices, people may buy a little more food, but the farmers often cannot cover their costs. In these situations, the green revolution is no bonanza, and the farmers are likely to give it up, unless a wider market is available or the government helps by subsidizing the inputs.

The use of synthetic fertilizers carries environmental penalties as well.[33] Runoff from farm fields can overfertilize streams, lakes, and estuaries, leading to serious eutrophication with disruption of aquatic ecosystems and fish kills. Nitrates (from nitrogen fertilizers) often seep into groundwater, reappearing in wells providing drinking water, and threatening human health directly.[34] In the era of global warming, a more ominous environmental threat has been discovered: the role of nitrogen fertilizers in producing nitrous oxide, an important, long-lived greenhouse gas.[35] Too little is known as yet about whether and how emissions of nitrous oxide might be curbed through different fertilizing techniques, but the possibility that this problem may constrain fertilizer use in the future must be considered.

Fertilizers are highly dependent on fossil fuels for both their manufacture and (for nitrogen fertilizers) as raw materials.[36] Eventually, depletion of fossil-fuel reserves will affect fertilizer production and supplies.[37] Even now, when petroleum prices rise, so do fertilizer prices, putting them beyond the reach of poor farmers and usually resulting in a drop in grain harvests. Future supplies of potassium and phosphorus fertilizers also pose potential problems.[38]

Increased fertilizer applications have little effect on yields unless plenty of water also is available. The expansion of irrigated land has therefore been a major feature of the green revolution.[39] Between 1950 and 1980, the land area under irrigation more than doubled, although the rate of expansion has slowed considerably since then. The 17 percent of the world's cropland that is irrigated produces a third of the food—an indication of irrigation's importance in almost tripling the global harvest over the last four decades.[40]

But the gradual deterioration of land and water resources that commonly accompanies irrigation may be catching up with us. Around the world today, as aquifers are drained and stream-fed systems deteriorate and accumulate salts, the amount of irrigated land being taken out of production is rising. Because the easiest and cheapest irrigation projects were naturally built first, the costs of building and maintaining new irrigation systems are also increasing. Consequently, the amount of new land being put under irrigation each year is diminishing, and irrigated acreage per person worldwide has been falling since 1978.[41] Within a few decades, the amount of land under irrigation may well cease expanding and even begin shrinking.

The third essential element in the green revolution technological package is pest control—mostly through the use of chemical pesticides. Since insects are among humanity's principal competitors for food (and also transmit many important human diseases),[42] problems of insect control are likely to climb higher on the human agenda and require more and more attention from agricultural scientists. At the same time, environmental problems caused by widespread, and often excessive or inappropriate, use of these toxic chemicals are also rising.

The current state of the art in agricultural pest control contains both good news and bad news. A recent study by the National Research Council suggests that the role of pesticides in generating yield increases, while important, has been exaggerated.[43] The global insect-control picture is still dominated by chemical control programs that involve spraying insecticides repeatedly on a schedule, whether or not pest populations pose any visible threat. Many programs are designed to keep produce cosmetically perfect rather than simply to avoid significant nutritional losses. This pattern persists even though decades of experience have shown that heavy and repeated use of pesticides militates against successful control in the long run.[44]

Heavy insecticide use promotes the evolution of insecticide resistance and impairs the natural pest-control functions of ecosystems, often leading to the creation of new pests. It may also reduce residual pest populations to levels so low that they can no longer support the populations of predators and pathogens that normally help to control them.[45] The predators die out, and the pests can resurge to record levels, free of their enemies. Resistance management programs are now an important element in developing successful systems of pest management and increasing food production.[46]

In the long run, farmers have no chance of winning the coevolutionary race between crops and pests by attempting to eradicate pests through the broadcast use of deadly chemicals. Pesticide use carries a short-term economic advantage; about four dollars are returned to American farmers for every dollar spent on pesticides, but that figure does not reflect the hidden costs to the farmers and to the public.[47] Those include direct risks to the health of people who apply the poisons, work in poisoned fields, or consume residual poisons with their food and water, as well as indirect health risks arising from the assault mounted by pesticides on the services supplied by natural ecosystems. A more serious hidden cost of locking farmers on a "pesticide treadmill" is that abuse of pesticides, like abuse of narcotics, leads to continual demand for higher doses and great difficulties in kicking the habit when the effect wears off.

Pesticide misuse is a global problem, and its consequences have had severe impacts on agricultural and economic systems, as well as natural ecosystems. The external costs connected with pesticide use are becoming too high; they must be recognized and internalized. More to the point, new pest-control techniques must be developed and adopted.

Potential problems in maintaining the flow of inputs needed for green-revolution agriculture (not to mention increasing them to keep harvests up with population growth) are a serious enough threat to future food production. What about the most important resources of all for growing food—good quality farmland and soil itself?

Deterioration of farmland and grazing lands is a worldwide problem, most dramatically revealed in estimates of global annual topsoil losses from erosion. The most extreme environmental problems facing agriculture and livestock production are erosion, desertification, and the deterioration of irrigated land, which were discussed in the last chapter.[48] All three processes have led to abandonment of tens of millions of acres of farmland that is no longer productive, in rich and poor nations alike, at a rising rate in the last two decades.[49]

▬ Beyond the Green Revolution

As the remaining scope for expanding food supplies through green-revolution technology shrinks, no promising new technology appears on the horizon that could carry the process farther on a global scale. Despite the hype that has attended recent developments in molecular biotechnology, no breakthroughs have yet occurred that could substantially increase the yields of major crops beyond the green revolution's achievements. The goals of current genetic engineering research consist of such improvements as genetically based resistance to pests and diseases or incorporation of the ability to fix nitrogen in grains. These improvements mostly would help lower farmers' costs and reduce the need for inputs. These are worthy goals, of course, but the effect on overall crop yields would be slight at best. And each of these projects poses environmental risks and trade-offs.[50] Sometimes the "improvement" may have unfortunate economic consequences. In one case, bacteria were given the ability to synthesize cocoa butter, so it could be produced in biotech "factories." That triumph of biotechnology threatens the livelihood of farmers in Madagascar and other poor countries who grow cacao beans, the traditional source of cocoa butter, as a cash crop.

Even if a biotechnological breakthrough in a major food

crop materialized, at least ten years would be needed for full field testing and widespread adoption of the new variety. Some improvements will doubtless be seen in various crops, farming practices, and pest control, but even numerous such small changes are unlikely to match the global impact of the green revolution, through which yields of major crops doubled, tripled, and even quadrupled in a decade or two. A big jump in agricultural productivity resulting from advances in biotechnology during the critical few decades just ahead is not foreseen.

Humanity now may be approaching physical and biological limits that ultimately will constrain potential world food production. The energy fixed in photosynthesis and made available to animal life, net primary production (NPP) is the primary resource for human life as well: the basis of all of civilization's food and much of its energy and material wealth. Thus potential NPP comprises the ultimate constraint on human population growth.[51] Of the 40 percent of NPP on land diverted into human-directed systems or suppressed as a result of land-use changes, most can be ascribed to agriculture and livestock production. To the extent that the human takeover of land and NPP means displacement of natural ecosystems and the organisms that comprise them, other elements of the human life-support system are also threatened.[52] In future decades, the impoverishment of biotic systems may play an increasingly important role in determining the limits to food production.[53]

Lester Brown, an agricultural economist who is president of Worldwatch Institute, has concluded that the convergence of several constraining trends—the lack of good new land available for farming, diminishing returns and rising environmental costs from green-revolution technologies, and the widespread deterioration of agricultural lands—is causing a slowdown in the historical expansion of global grain production.[54] In the 1990s, he estimates, the world's farmers may have difficulty maintaining grain production increases of more than an average of 0.9 percent per year—only half the current average rate of world population growth. If Brown's assessment is generally right, the outlook for continuing to expand global food supplies fast enough to keep pace with population growth is not very bright—even without considering such massive potential disruptions as global warming.

Global Warming:
A Wild Card

Even if the world's climates were locked into today's relatively favorable mode, those involved in the battle to feed humanity clearly would face enormous difficulties. But global warming could greatly exacerbate all the problems of increasing food production,[55] accelerating and intensifying a global food crunch that was already in the cards for humanity.

A simple computer model of the world food situation was developed at Stanford University to test the possible effects of global climate change on world food security.[56] Optimistic baseline scenarios were devised in which, in the absence of unusual climate events, food production increases kept pace with population growth. When the model simulations imposed weather-caused harvest failures at various frequencies, hunger-related deaths rose, in some cases doubling over the levels of recent years. Over the past two decades, at least 200 million people have died of hunger or hunger-related disease.[57] The model indicated that even small drops in the global grain harvest—or 5 percent below the general trend—especially if they occurred two or three years in a row, could significantly raise death rates above recent levels.

For the decade or two immediately ahead, though, a more likely eventuality appears to be a progressive failure of food production to keep pace with population growth, as resource constraints tighten and farmland continues to deteriorate. Model scenarios where annual food production fails to keep pace with population growth (grain harvests rise by 0.9 percent and the population by 1.7 percent per year)[58] indicated that, *even without climate changes*, hunger-related deaths could increase fivefold to nearly a billion over two decades.[59]

The results of this model may signal a fundamental change in the global food situation. For the first time since 1950, absolute food deficits, compounded by inequities in the food production and distribution system, may occur in the near future and cause widespread famine. The model also highlighted the role of rapid population growth in generating future food shortages. Clearly, a concerted effort to reduce population growth would pay off in easing a future population-food crunch, as well as in mitigating any adverse effects of global warming on agricultural systems.

Elements of Unsustainability

Present-day "conventional" agriculture, based on green-revolution technology, is in many ways an inherently unsustainable system.[60] Given rising oil prices and an expectation of tightening supplies after 1990, agriculture's heavy dependence on a fossil-fuel-intensive, highly mechanized farming technology is one component of unsustainability. Prices of fertilizers (and to a lesser degree pesticides and the costs of running farm machinery) are dependent on supplies and prices of petroleum and natural gas. Farmers in rich nations were squeezed financially by high oil prices in the mid-1970s and early 1980s; in poor countries, many were priced out of participation in green-revolution farming or at least forced to cut back on input purchases—with a consequent drop in production.[61] Furthermore, the need to reduce consumption of fossil fuels to abate greenhouse gas emissions in coming decades will inevitably clash with the equally compelling need to keep raising food production, unless population growth is substantially slowed.

Another unsustainable aspect of modern agriculture is its emphasis on monocultures—the planting of a single crop strain over vast areas, as is commonly done in the American Midwest.[62] For instance, in 1969 an average of 5 million acres was planted in each of nine strains of wheat, and six strains of corn averaged 11 million acres each.[63] Planting large monocultures is necessary for mechanized farming and is very efficient in the short term, especially when growing the same crop year after year in the same field is reinforced by the structure of farm subsidies. But monocultures, particularly when repeated for many years without rotation, are highly susceptible to pests and crop diseases and require heavy pesticide applications for protection. Continuous monoculture planting also leads to depletion of soil nutrients, only partially compensated by fertilizers, which deliver only two or three of some twenty essential minerals. The entire monoculture system also contributes to high rates of erosion in vulnerable soils, particularly when erosion-prone row crops such as cotton, corn, and soybeans are planted continuously.[64]

Another problem related to the planting of huge monocultures is that of the genetic uniformity of crops.[65] The global shift to high-yielding crops has led to dramatic declines in the diversity of crop strains grown in every region. Fewer strains, in turn, means that each crop species contains less genetic diversity than previously. But genetic diversity is the raw material that plant breeders use in their struggle to keep crops resistant to pests and pathogens and to maintain adaptability to different climates. Substantial portions of harvests of both Indonesian rice and American corn were lost in the early 1970s: the rice because the widely planted strain was too susceptible to a virus, and the corn because it was attacked by a new genetic strain of a fungus causing corn-leaf blight. If other varieties had been planted too, some would have been naturally resistant and losses would have been smaller. Genetic resources luckily were still available to permit new resistant strains of the crops to be produced in subsequent years.

The importance of preserving the genetic diversity of crops looms even greater today than it did in the 1970s as the possibility of rapid climate change grows imminent. Continued reliance on a narrow selection of today's high-yield varieties may carry the high cost of mortgaging the ability of plant geneticists to develop new strains adaptable to tomorrow's climatic conditions, unless extensive precautions are taken. Fortunately, the urgent need to preserve genetic variability in crops has been widely recognized, and programs were launched to remedy the problem under the International Board for Plant Genetic Resources, founded in 1974.[66] There are still major technical problems to be surmounted, but at least an effort is being made.

High-yield agricultural technology, including the unsustainable aspects of irrigation described earlier, has already contributed to the gradual degradation of cropland and losses of productivity in the north temperate regions where the technology was first developed and applied.[67] While the deterioration of farmland in many areas is often recognized, it is a slow process, and short-term economic pressures operate against deployment of measures that will pay off in sustained or enhanced productivity over the long term.[68] These pressures can be

effectively counteracted only by wiser agricultural policies that
provide incentives for farmers to adopt soil-conserving practices,
such as the United States Farm Acts of 1985 and 1990, which
provided special payments for conservation set-asides of vulnerable land.[69] The 1985 law resulted in a sharp reduction of
United States soil losses within a very few years;[70] the effects
of the expanded 1990 program have yet to be demonstrated.

Tropical Agriculture

Of course, the greatest immediate need to increase agricultural productivity is in tropical nations, especially in sub-
Saharan Africa, South Asia, and Latin America. But the inherent
instability of "modern" agriculture becomes obvious much more
quickly in the tropics—and is a major reason that food production has increasingly lagged behind population growth in
those regions. Especially in the humid tropics, as we noted in
the discussion of global warming, year-round favorable growing
conditions intensify pest and crop-disease problems. Most Temperate Zone insect populations are subjected to an annual control
regime by winter; in contrast, tropical insect populations appear
to be controlled mainly by predators and the coevolved resistance mechanisms of the plants themselves. Unfortunately, populations of predators, be they birds, lizards, or other insects,
often are more susceptible to the disturbances created by agriculture than are populations of herbivorous pests.

Soils, too, are more fragile in many tropical regions
(although there are important exceptions, such as parts of Indonesia). In moist tropical forests, a large portion of the system's
nutrients are contained in the vegetation, which often grows on
thin, poor soils that are subject to erosion, leaching, and com-
paction when cleared. Conventional agriculture as practiced in
temperate regions has been introduced repeatedly in the humid
tropics, frequently with disastrous results.[71]

If the soil of cleared tropical moist forest can support
cultivation at all, it is typically only for a few years. Much
cleared tropical land is used as pasture, but even this often is
a temporary game, good for five to eight years. As the tropical
rains erode the soil and leach out the nutrients, the land rapidly
becomes a wasteland and is abandoned. The farmers or herders

move elsewhere, but the forest is unable to reclaim large swaths of land so depleted. Land-use figures on the Amazon basin and similar tropical areas often show little change in the amount of land cultivated or grazed from year to year, simply because of constant turnover. But the forest area is shrinking rapidly.[72]

The dream of turning the Amazon basin and most other moist tropical regions into productive breadbaskets on the Iowa model is an illusion.[73] Continuous agriculture may successfully be practiced in a few areas blessed with unusually good soils (such as Java), but these areas make up a small fraction of the total land in equatorial regions. For most such areas, agricultural success will depend on protecting large portions of the forest and indeed on using it as part of the food- and materials-providing sector of the economy, essentially as the indigenous people have done for centuries.[74]

One promising type of agricultural production in forested areas is agroforestry, in which trees and tree crops are grown together with a variety of shade-tolerant cereals, tubers, and vegetables.[75] Where the forest has already vanished or must be sacrificed, success most likely will hang on creating new agro-ecosystems that resemble the forest—not on planting vast mono-cultures of row crops.[76]

Building a Sustainable Agriculture

Making agriculture sustainable in the world's bread-baskets is a relatively straightforward task, especially in North America. It is important to deploy better techniques to reduce soil erosion and dependence on farm chemicals. These might include intercropping or polyculture (planting two or more crops together in a field, simultaneously or overlapping in time). "No-till" cultivation, in which plowing is avoided (thus reducing erosion) can be advantageous, but the current approach involves too much use of herbicides. Similarly, better techniques to supplement nutrients without overusing inorganic fertilizers (which do an incomplete job and cause serious problems of water pollution) need to be developed. Ways should be sought, including the use of subsidies if necessary, to encourage farmers to plant a variety of strains of each crop to provide flexibility in the face of possible rapid climate change; if some strains don't produce well under some set of adverse conditions (which cannot be predicted), others might. And the transition toward alternative methods of pest control already under way in the United States and Britain should be accelerated (see below).

A few pioneering agroecologists are experimenting with new kinds of agricultural systems (sometimes based on traditional systems when they can be recovered) that more closely resemble natural ecosystems than Iowa cornfields.[77] In essence, these systems are characterized by complexity, with several crops grown together (polycultures). Some plant scientists are trying to develop perennial crops, which would develop more extensive root systems and require no plowing and replanting between crops, thereby reducing soil damage and erosion.[78]

Farmers who frequently rotate crops and intensively intercrop can benefit from decreased erosion, soil enrichment, and reduced pest problems (and reduced use of expensive and dangerous pesticides). Benefits can be attained even by tolerating "weeds" that do not seriously compete with the crops for sunlight or nutrients, but help hold soil in place. For example, when wild heliotrope was planted with leguminous crops, weed populations were reduced by almost three-fourths, and several insect pests were reduced below levels at which they caused economic damage.[79]

Intercropping may involve planting together in one field, say, a tall crop like corn, to be harvested first, with one or two shorter, shade-tolerant crops such as beans, rice, squash, or sweet potato that can be harvested a month or so later. The corn stalks can provide support for vines of beans or peas, which in turn contribute nitrogen to the soil. The combination of several crops helps hold soil in place and discourages weeds and insect pests. The crops must be chosen for compatibility with each other and each particular locality, and planting timed to minimize competition as each crop matures.[80]

Agroforestry is an intercropping system based on tree crops; classic examples are the planting of coffee, tea, or cacao beneath shade trees, which themselves are often nitrogen-fixing legumes that supply valuable wood products. This sytem is particularly well suited to many tropical and subtropical areas where conventional grain farming does not thrive.

Making agriculture both more sustainable and more flexible will demand substantial departures from the "factory farming" practices now common in rich nations and cash-crop production areas of poor ones. There are problems, of course. A more flexible and experimental, less chemical-intensive farming practice also would require provision of extra education and advice for farmers, presumably by knowledgeable agricultural agents. At the least, a different sort of mechanization will be required to facilitate the new techniques if adopted in developed countries; many intercropping systems do not lend themselves to standard cultivating and harvesting devices. Are farm machinery companies thinking about this?

Moreover, such a farming approach is considerably more labor-intensive than the highly mechanized practice today. The labor problem raises a sticky question for rich nations; who will do the work? Must farm labor be imported from poor countries, through a higher immigration rate, to fill the need? Of course, the need for more labor could prove a solution to a pressing problem in poor countries—if productivity can be raised sufficiently to make more employment on the farm economically sensible. Most poor countries have large numbers of unemployed and underemployed rural workers who could greatly benefit. Ecologically oriented agricultural systems may well prove much more productive and more sustainable in tropical and subtropical settings than conventional agriculture, and it is there that rapid production increases are urgently needed.

Unfortunately, tropical agriculture has long been neglected by the agricultural development establishment. The green revolution was based almost exclusively on high-yield strains of the "big three": wheat, rice, and corn (maize). Wheat is largely a Temperate Zone crop and an inferior performer in many tropical developing countries. Rice is a traditional crop in Asia but not elsewhere, and its use in Latin America and Africa is limited so far (but could be expanded). Maize is fairly widely grown in Latin America and Africa, but not high-yielding hybrid corn (which requires special cultivation techniques and expensive inputs). Innumerable traditional crops have been forgotten or overlooked by plant breeders; some may have been lost forever as they were replaced by the dominant big three, regardless of relative suitability in many tropical settings.

An obvious possibility for increasing food supplies,

though, is to find and develop new food sources—or to recover old ones. Some 75,000 species of plants have edible parts, but humanity today relies heavily on only about two dozen species.[81] Even so, when analyzed nation by nation, the feeding base was recently found to be more diverse than is usually thought. Some 103 plant species make significant contributions to the food base of at least one country.[82]

The need to preserve traditional, genetically diverse varieties of the major crops is now widely acknowledged (if not perfectly carried out). Less clearly recognized is a related need to preserve minor crops and traditional farming systems.[83] And the preservation of edible wild plants in natural ecosystems, especially in the tropics, is essential. Wild plants not only can supply genetic variability that could be transferred to close relatives under cultivation, they also are a largely untapped reservoir of potential new crops, many of which could surely be of great benefit in feeding humanity in the future.

One example of proven potential is a group of plants in the cockscomb family known as grain amaranths. These plants were once widely cultivated in Latin America, but after the Spanish conquest they were mostly replaced by corn. The Aztecs had ascribed religious significance to one amaranth species, which led to its suppression by Catholic missionaries. Seeds of some amaranth species have very high quality protein, and leaves also tend to be protein-rich. Some 800 species of amaranths, many of them tropical, are a still largely unexplored source of new crops for feeding hungry human populations,[84] even though some amaranth products have appeared in American health-food stores. Another example is wild species of rice, of which twenty are known.[85]

Even widely used traditional staple crops of the humid tropics such as cassava are only beginning to attract the attention of agronomists.[86] Yield improvements probably will follow, but implementing them among extremely poor (and often illiterate) farmers will be a big task.[87] Much remains to be done to improve agricultural productivity in tropical nations, especially in Africa, where poverty is severe, land resources are meager, and populations are growing rapidly. The toughest part will be to expand and upgrade agriculture without further degrading the land.

Meanwhile, the human population is still increasing by

nearly 2 percent per year (a current doubling rate of about 40 years); in many poor nations, the growth rate is 3 percent or more (doubling in 17 to 23 years). Global food production therefore needs to be increased by nearly 2 percent each year just to stay even; and to improve diets in poor countries, a higher rate would be desirable (considering that the rich are unlikely to reduce their consumption very much).

The principal strategy for producing that additional food obviously is to increase yields on land already in production. But, if the green revolution has little more to offer on a global scale, how might that be accomplished—particularly in the crucial decade or so just ahead? Some time must be bought for population programs to lower birthrates and for redesigned, better supported agricultural systems to yield results, even if a serious concerted effort were put into both programs immediately. Are there strategies that could boost food supplies in the short term without jeopardizing future production?

▬▬ Reducing Food Losses

One very good way to increase food supplies quickly is to reduce losses to pests, and to do so without injuring ourselves or the ecosystems we depend on. No one knows exactly how much the world's agricultural production is lowered by weeds, or what fraction is lost to insects, rodents, diseases, and spoilage before and after harvest, but a range of 25 to 50 percent seems a reasonable estimate.[88] Suppose the average loss were 40 percent and that a coordinated effort toward more effective crop protection could cut that by half. Then the human food supply would be increased by roughly 20 percent, enough to feed a billion more people. That would buy us a decade to slow population growth and increase agricultural productivity by other means.

One of the easiest and most effective ways to cut losses is to build or improve crop storage facilities, especially in poor nations where enormous losses occur from spoilage alone. Many possibilities exist; the main problem often is getting facilities where they are needed. Simple rat-proof metal sheds could reduce spoilage enormously in many areas. Tight containers that exclude insects and rodents could be made more available for use in homes by poor families. Techniques are being developed to reduce losses in large grain elevators. In one case, an ecological approach was devised to control insect pests that attack stored grains.[89] Supplying simple, durable solar-powered refrigerators to Third World villagers could help cut pest losses and spoilage further.

In the field, the basic task of protecting crops is to keep pest populations over broad areas from expanding to the point where they seriously reduce yields. The drawbacks of reliance on synthetic pesticides are making it clearer every day that better, less environmentally damaging methods are needed, especially in developing regions. Farmers must substitute other methods for pure chemical control as quickly as possible while simultaneously achieving a greater degree of control over pests.[90]

In rich nations, adjusting the desired level of control could significantly influence the amount of insecticide required. Much insecticide use in rich nations is to keep insect damage far below so-called "economic injury levels," which requires keeping insect populations suppressed to very low levels.[91] The goal is to improve the cosmetic appearance of the crop, not its yield. Thus minor insect damage on a few apples is deemed unacceptable by a public that doesn't realize that an occasional worm is a small price to pay for reducing pesticide residues on fruits, damage to ecosystems, and injury to farm workers in the field.

Among the scandals of the international system is the United States practice of exporting insecticides that are banned for domestic use. An attempt was made to change this in the 1990 Farm Act, but the clause was dropped before passage.[92] The poisons, of course, return to Americans on imported food and cross our southern border borne by wind and water. The issue is complicated by the real need for cheap and effective means to control insect-borne diseases, especially malaria, in the developing world. DDT is the method of choice today. Perhaps this conundrum could be solved by banning exports of the chemicals for farm use but licensing DDT for targeted use against disease-bearing insects. Another approach would be to step up educational programs in developing nations on the hazards of pesticides and other powerful chemicals.

The good news in industrialized nations is that "integrated pest management" (IPM) techniques are increasingly being adopted by farmers, and their experience indicates that ecologically sound insect control is economically feasible. IPM relies heavily on such traditional tactics as using plants bred for resistance to pathogens and insects, using biological control agents (such as predacious beetles to control aphids and scale insects), and some cultural controls (crop rotation, plowing under crop stubble in which pests overwinter). IPM systems do not totally avoid using pesticides but reserve them for use only when necessary to maintain control.

IPM systems have often proven to be very economical. A major effort was made in the United States by the Consortium for Integrated Pest Management (CIPM), an interdisciplinary effort by 17 universities. Its objective was to improve IPM systems on four major crops—alfalfa, apples, cotton, and soybeans—and it was a rousing success.[93] Its major results were "an increase in acreage of major crops that are now under IPM, a reduction in the use of pesticides (principally insecticides) for protecting major crops, and a *reduction in production costs.*"[94] IPM is a technology whose time has come.

Taking the IPM approach a step farther, organic farmers in several industrial countries, including the United States, avoid the use of synthetic chemicals—pesticides and fertilizers—altogether.[95] They use biological and cultural controls against pests and rely entirely on natural fertilizers (crop residues, manure, and composts). In general, they depend less on monocultures, produce a greater variety of crops and livestock per farm, and are more experimental in their approach. While in some cases, their yields may be slightly lower than those of farmers on similar land using conventional methods, their costs are also lower and profits roughly the same. Perhaps more important, their land shows little or no sign of deterioration, and erosion losses are lower.

Although standard IPM is a more moderate step than organic farming toward controlling pests without causing undue ecological havoc, attempts to implement IPM programs in developing countries have encountered numerous problems.[96] Implementing IPM successfully requires considerably more sophistication and effort on the part of farmers than simply operating spray-rigs. To integrate IPM into a nation's agricultural system, a substantial commitment to support it must be made as well by society—financial, legislative, and educational. In many developing countries, a concerted effort, supported by assistance from rich nations, toward making IPM or chemical-free farming techniques succeed will be necessary.

━━━ A New Emphasis on Agriculture

This effort could be a part of a renewed commitment on the part of developing nations to the agricultural sectors of their economies. In the era of "development" since World War II, many poor nations have emphasized the economic development of the urban and manufacturing sectors, to the detriment of agriculture. It is no accident that the most successful developing nations today (especially in East Asia) are mainly those that made an effort to secure their agricultural sectors first. The neglect of food production (other than cash crops for export) is now exacerbating poverty, hunger, and land degradation, but it can also hinder development in other sectors and lead to social unrest as well.

It is not too late to redirect the world's concern to agriculture. The looming crisis of global warming could lead to a serious (and long overdue) reevaluation of the entire global food system, from cultivation methods to the grain trade. Scrutiny is especially needed of policies that encourage or discourage ecologically sound, sustainable production and more equitable food distribution, and appropriate changes should be made. Global warming, by causing unpredictable weather changes, may force farmers to become much more flexible in their ways: to plant a variety of crops each year with different timing, temperature, and moisture needs so even if some fail, others may flourish. Plant breeders also need to develop crop strains that produce well under a wider range of weather conditions than most do today.

One simple and useful step in nations like the United States would be to educate the public to the increasing seriousness of the human predicament in general and of the need to establish and support ecologically sound agricultural systems in particular. Many Americans seem to believe that their food materializes magically in supermarkets overnight; few have any concept of the complexities or the fragility of the agricultural enterprise. Indeed, most of us take completely for granted that food will always be available. Children need to appreciate the agricultural system also; does your local school system give appropriate emphasis to the most important human endeavor— growing food?

Our civilization is now entering dangerous times; all of us can help alleviate the dangers by demanding government policies that will promote a more sustainable agricultural system, one that is robust enough to see us through the population outbreak, even if severe disruptions occur because of global warming.

Risks, Costs, and Benefits

CHAPTER Eight

Humanity clearly must make many decisions about the risks involved in present and future alterations of Earth's life support systems. What, if anything, should be done now to slow global warming? Are the ozone protocols tough enough, and what enforcement measures are justified? Is the damage done to natural ecosystems by acid precipitation sufficiently serious to justify the expenditure of billions of dollars in an attempt to reduce it by half? Would it be worthwhile for rich countries all to spend one half of one percent of their GNPs in attempts to foster ecologically sound agricultural development in poor nations? All such questions relating to the healing of Earth are difficult to answer, since they involve many uncertainties. Nonetheless, rational people and societies are (or ought to be) interested in comparing honestly and systematically the costs and benefits of actions they might take.

Throughout human history, people have had impacts on natural systems, and for millennia those impacts have generally increased Earth's long-term carrying capacity for people. The agricultural revolution some 10,000 years ago, when people first began to grow most of their food rather than hunt and gather it, was a prime example: it led to the transformation of huge tracts of grassland and forest into farmland. That eventually made it possible to support billions of people on Earth rather than millions.

216 In this century, humanity has become a truly planetary ecological force, its effects going far beyond the transformation of the landscape to include altering the composition of the at-

mosphere globally, interfering with planetary nutrient cycles, modifying climate, and exterminating other life forms. In historical perspective, the switch—from a modest to an overwhelming scale of impact, and from a positive to a negative impact on carrying capacity—was almost instantaneous and monumental: human activities now appear to be lowering the long-term carrying capacity and incurring risks on a scale unimaginable less than a lifetime ago.

All human activities have unavoidable effects on ecosystems. In order to remain alive, people must take in oxygen from the environment and release carbon dioxide to it. We must acquire energy directly or indirectly from the chemical bonds that plants manufacture in the process of photosynthesis, and we must return some of that energy to the environment in the form of waste heat. Every move you make changes your environment—sometimes the changes are not perceptible, but they happen. So there is no question as to whether or not we should affect our environment—the questions all center on the *scale and consequences* of our influence. These questions generally boil down to assessments of risks, costs, and benefits, as a basis for making decisions in the face of uncertainty. The penalties of making environmental mistakes plagued our ancestors at least since the first fire set to drive game toward hunters got out of control and killed someone. But until very recently, the penalties were borne by a few or a few million people, locally or regionally. Now they can be borne by billions of people globally.

In the modern era, people are constantly faced with decisions involving risks, costs, and benefits; and more and more of them are related to environmental issues. Faced with a plethora of environmental problems, societies must make all manner of decisions about which pose the greatest risks, which are most important to deal with first, how much effort (if any) to allocate to each, and so on. All such decisions involve trade-offs; for instance, money spent on scrubbers for factory smokestacks will not be available to improve the quality of the factory's waste-water effluents; scientists designing sophisticated weapons cannot spend their time designing energy-efficient automobiles. A society that subsidizes organ transplants and medical care for the aged will have fewer resources to spend on maternal and child health or family planning clinics.

Many decisions, unfortunately, are made without being

informed by careful risk assessments or cost-benefit analyses. Of course, such evaluations add to the costs of decision making, and may be unnecessary when the choices are fairly clear. But every politician knows of cases where hindsight has revealed too late a need for careful study in advance. In the context here, though, familiarity with the nature of these evaluations could be particularly helpful in thinking about ways to heal the Earth.

"Risk" is sometimes used as a synonym for "probability" in insurance policies—the risk of a loss. In analyzing issues like those discussed in this chapter, "risk" means the consequences of an event multiplied by its probability (or frequency) of occurrence—or, to put it another way, risk is a probability times a bad consequence.[1] For example, if the collapse of a particular dam would cause 5,000 deaths and the probability of such a collapse is 1 in 1,000 per year, then the associated risk is 5,000 deaths times 1/(1,000 years) = 5 deaths per year. Similarly, one measure of the risk of driving automobiles in the United States is 50,000 deaths per year. You can think of that risk as the average number of deaths per automobile accident multiplied by the probability or frequency of auto accidents (accidents per year).

If an activity can have more than one kind of adverse consequence, its risk can be measured in more than one way. For instance, the risk of automobile driving can be measured, in addition to deaths per year, by injuries per year or by property damage per year. Different risk measures also result from different formulations of the probability or frequency involved. One might be interested not in the total number of automobile deaths in the United States, but in the number per year in a particular state, or the number per million vehicle miles, or the number per hour of driving.

In comparing risks of different activities, one must take care that the measures of risk are really comparable with respect to both the consequences and the type of probability measure used. As an example of consequences that are not directly comparable, consider two different types of cancer. Suppose exposure to chemical A gives one chance in 10,000 of causing pancreatic cancer and chemical B one chance in 10,000 of causing the commonest form of skin cancer.[2] The former almost always causes a rapid and painful death; the latter is very rarely fatal. Then the risks of the two exposures would be considered equal if expressed as cancer cases per year; but since the two cancer types are not equivalent in their implications, a more informative measure would be fatal cancers per year.[3]

Very often the environmental and safety risks that one wishes to evaluate cannot easily be expressed in monetary terms—risks of loss of many human lives, of the extinction of a million species, or of increasing the speed of climatic change. In such cases, the risks and benefits of given decisions are compared in a "risk-benefit" analysis.[4] If the risks can be expressed in monetary terms, then those are included with other costs in a more comprehensive "cost-benefit" analysis.[5]

An important feature of risk-benefit and cost-benefit analyses is that often neither the risks nor the benefits of a particular decision are known with assurance. Therefore, while risk is a probability times a bad consequence, benefit is a probability times a good consequence.[6] For instance, deciding whether or not to purchase a piece of land to make a reserve to protect a given species involves an assumed risk: the probability of the species going extinct if the reserve is not created times the seriousness of its loss to society. Costs would include the price of the land,[7] the expenses of fencing and policing the reserve, and so forth. The benefits, too, must be calculated: the increased chance of the species' survival provided by the reserve (that is, the increased chance that the risk will be avoided) times the value to society of preserving the species (including esthetic values, possible direct economic values, possible values in helping to supply ecosystem services, and so forth).[8] If the costs are less than the benefits, then creating the reserve would be a good decision economically.

Of course, quantifying the various factors in such an analysis can be difficult, but the exercise nevertheless can be extremely useful. Often, just analyzing the factors that can be readily quantified will provide the basis for a sound decision; many an environmentally unsound development project has been stopped by a "straightforward" economic cost-benefit analysis that neglected important considerations because they were difficult to quantify. But often such limited procedures can result in serious errors.

The trap of ignoring what can't easily be expressed in dollar costs has been described by economist Herman Daly as concentrating on things that are countable rather than things that count.[9] Focusing on what can be quantified rather than what is important has become a major problem in the social sciences.[10] It is essential to take care always to consider the qualitative (and ethical and moral) contexts in which risks, costs, and benefits are assessed. For instance, as Daly has pointed out, a key role in avoiding ethical responsibilities in quantitative analysis is played by the use of randomness. Somehow the deaths in auto accidents of 50,000 random, unknown people in the United States each year are generally acceptable. But if those people were predetermined and named, you can bet that the pressures for more auto safety would be enormously greater.[11]

Neither Daly nor we would argue that such moral considerations should displace quantitative analysis—only that they must be explicitly included along with the results of the analyses when courses of action are determined. We can't say what is "acceptable"; that is a decision for society as a whole. We do personally believe that society would be better off if more effort were made to personalize the 50,000 American accident victims or the millions who die of malnutrition in poor countries each year. In schools and in the media, more attention to selected individual stories of those who are now just faceless statistics would help.

The other important problem with quantitative analyses is what energy analyst John Holdren calls "the tyranny of illusory precision"—using quantitative risk estimates, divorced from the underlying assumptions and the uncertainties associated with the estimates, to announce firm conclusions that are unjustified.[12] A classic example was the "Rasmussen Report" on the safety of nuclear power plants.[13] In that study 0.024 excess deaths per reactor-year of operation was publicized as a meaningful estimate of the risk of reactor accidents. Submerged in the fine print were the uncertainties associated with that estimate—according to the authors of the report, the risk could be twentyfold larger or smaller (and according to competent critics it could have been 100-fold larger or smaller). Also lost in the shuffle was any acknowledgment that critical components of risk, such as earthquakes and sabotage, were not included in the analysis at all.

A more recent occasion that presented elements of illusory precision was the pompous announcement of the conclusions of the 1990 National Acid Precipitation Assessment

Program (NAPAP) by one of the key participants on the CBS television program *Sixty Minutes*.[14] A precise (and small) number of lakes were stated to be at risk from acidification in the northeastern United States. The assessment itself, while containing some good science, was summarized so as to produce an illegitimately low evaluation of the threat by using data selectively and ignoring important dimensions of the problem.[15]

Pancreatic cancer illustrates one of the most difficult aspects of risk analysis, a so-called zero-infinity problem. The probability that you will get the cancer is clearly very small—almost zero. The consequences, should that unlikely event occur, from the viewpoint of the affected individual, can be thought of as infinitely bad. But many voluntary activities involve trade-offs with the risk of a painful death. If people were unwilling to consider those trade-offs, every car would be a Sherman tank and the speed limit would be 15 miles per hour. People are unwilling to pay for a tank or to drive that slowly, so they accept what they perceive as a small risk of the infinitely bad.

In many people's judgment, the risk of a nuclear war is now moving into the zero-infinity area; the chances of such an event are dwindling, but the deaths of hundreds of millions of people are nonetheless likely if the event occurs. Other environmental risks seem to be moving toward what we would call a "likely-infinity" situation—one where the probabilities may be 20 to 50 percent or more, with hundreds of millions or even billions of lives hanging in the balance. Rapid climate change due to global warming is in this category, as ozone depletion will be if the recently agreed protocol to end the use of CFCs and halons is not complied with, or if some still-unknown factor in atmospheric chemistry (similar to the unexpected ice-cloud chemistry that created the Antarctic ozone hole) makes that protocol inadequate.

There are two other important issues related to the zero-infinity problem. One is that few people would judge as equal two risks measured as "one death per year" if in one case that risk were generated by a 1 in 10,000 chance of a disaster killing 10,000 people and in the other a 1 in 2 chance of killing 2 people per year. There is a soundly based aversion to running the risk of big, concentrated disasters (which are likely to have psychological and social consequences very different from risks causing gradual attrition).[16] Most people consider some disasters too big to entertain the risk no matter how small the probabilities. If a nuclear reactor accident could kill everyone on Earth, few would deem it tolerable to generate electricity with nuclear power even if the annual accident probability were only one in a trillion. We suspect that if people were well informed on the consequences of very rapid climatic change or the extermination of most of Earth's life forms, they would find the activities promoting them intolerable as well.

The second issue is that people often find the infinity more believable than the zero. People understand that experts are more successful at estimating the consequences of big disasters than in estimating their probability. There was no question about the consequences of the *Titanic* sinking; it was the probability of its sinking that the experts miscalculated. So the public doesn't really believe that the risk stays constant as the consequences rise and the probability falls.[17]

▬ Evaluating Risk

At the personal level, risk-benefit decisions often involve the question of whether to avoid substances that may be harmful to health. Is the flavor of a steak well marbled with fat worth a possibly increased probability of dying of a heart attack? Is the relaxation, pleasure, and possible ease of weight control that accompanies smoking a sufficient benefit to counterbalance the substantially increased possibilities of dying young of lung cancer or circulatory disease? Do the benefits of using aerosol underarm deodorants or hair sprays compensate for possible health effects, which are certainly small (and may be zero), of inhaling them? People informally weigh such risks all the time, always in the face of uncertainty and often even without access to the basic information about the risks that society possesses.

Even when considerable information is available, and the power to act is in your hands, decisions may not be easy. Suppose it were announced that your town's drinking water contained that chemical that gives you a one in 10 million chance of dying of liver cancer if you drink that water for the rest of your life. Would you spend $10 a year on a filter to remove the material? $100 a year? $1,000? Your answer would obviously depend on many things, including your age and financial situation.

Consider a more familiar example. There are undeniable benefits to driving a private automobile—convenience being paramount among them. But when everybody is driving a private automobile and commuters spend hours daily in near gridlock, the convenience factor is reduced and the direct (accident) and indirect (air pollution) risks escalate. Even now the benefits of driving are difficult to balance against the escalating risks. How does one calculate into the conveniences and inconveniences of automobile commuting the risks of additional exposure to airborne carcinogens and heart-threatening carbon monoxide or the contribution that the automobile makes toward global warming? Is the residual convenience worth a month-shorter life expectancy? A year? Two years? Is it worth a 1-percent chance of subjecting your grandchildren to food shortages by contributing to future crop failures caused by global warming? A 2-percent chance? A 5-percent chance?

Both the risk of diminished life expectancy and the risk of starving your grandchildren are especially difficult to analyze because they contain both personal and social components. If you give up driving or riding in cars, the risk to your personal life expectancy from the use of automobiles will be reduced, but not to zero. You could still be run down as a pedestrian, or die of lung cancer caused by smog. Only if a substantial portion of society gave up driving would that component of risk decrease very much. The risk to your grandchildren from global warming is even more difficult to reduce by personal effort, so one is tempted to take no action because "it won't make any difference." On the other hand, one can always try to set an example, or even try to persuade society as a whole to make the "right" decision (as civil rights, peace, and environmental activists have done).

Probabilistic thinking, dealing with the odds rather than certainties, is both a pain and unavoidable in analyzing risks. Suppose (to pick a random example) you were a biologist at Stanford University. How should you react to the announcement that there is a 50 percent chance of a large earthquake on the San Francisco peninsula in the next 30 years? If you were 60 years old, you certainly might react differently than if you were 30. In practice, you presumably would want to estimate (that is, make an educated guess at) the probability of being hurt or killed in such an earthquake. That would involve estimating the strength of the building where you work, the degree of earth-quake resistance of your house, the proportion of time you will spend in each place, and how much of the time you are likely to be out of town.

Then you might want to consider the new hazards you might expose yourself to should you move: tornadoes in Kansas, hurricanes in Florida, running out of water in Tucson, a smaller chance of a more catastrophic quake in St. Louis, a greater chance of being robbed or murdered in New York City, and so on. Of course, virtually no one actually sits down and tries to make this kind of calculation, but people nonetheless weigh these kinds of odds in a more casual way every day. And while the public perception of risks does correlate somewhat with the calculations of experts, a combination of lack of information and a much richer variety of elements entering the evaluation leads to some striking differences. Nuclear power, for instance, is generally considered much more risky by the public than by risk analysts. The former often weigh such factors as the po-tential for catastrophes or threats to future generations more heavily than the estimates of annual fatalities preferred by ex-perts. In addition, people tend to view the world as already unacceptably risky.[18]

The problems experts face in measuring both costs and benefits are even more difficult because costs and benefits are usually distributed so unevenly among individuals, because in-dividuals differ in their responses to risk, and because people judge risks on criteria other than simple probability times con-sequences. For example, the statistical risks incurred by smok-ing are large—a smoker runs a yearly risk of around 1 in 300 of dying because of the habit—but many people still choose to smoke, and some heavy smokers enjoy the habit thoroughly and die of accidents at a ripe old age.[19] Some individuals are prob-ably genetically resistant to some of the deleterious conse-quences of smoking, and many individuals are virtually immune to propaganda attempting to persuade them to give it up.

Comparatively, the chance of getting liver cancer from a trace of a toxic chemical in the water supply may be one in a million annually, but that risk, unlike that of smoking, is taken involuntarily. Most people evaluate involuntarily assumed risks as much more serious than those assumed willingly.[20] This difference in attitude about voluntarily and involuntarily assumed risks led to the rapid passage of antismoking legislation in many places when the hazards for nonsmokers from sidestream smoke became clear, even though that danger is small compared with the hazard to smokers themselves. No legislation has been proposed to prevent people from smoking in private.

There also may be serious problems of scale in assessing social risks; a behavior that may not be harmful for a small group of people may become extremely dangerous if practiced by a large group. For example, a town of 1,000 people piping its sewage a mile out to sea may not overwhelm the ocean's waste-disposal capability, and beaches may remain clean and safe. A city of one million doing the same may cause a high incidence of infection in swimmers and surfers who frequent its beaches.

Social risk-benefit choices are always made on the basis of incomplete information. Decisions to use novel chemicals or deploy new technologies are generally made to capture benefits already identified—killing insects that compete for our food, generating cheap energy, providing fast, convenient transportation, and so on. But the sizes of the benefits may be quite uncertain. Some risks can be identified early, but others may become manifest only after deployment of the technology. Thalidomide is a drug that was originally used widely for the benefits it provided as a sleeping pill and sedative, and it has proven helpful in the treatment of leprosy. It was readily available for a few years in Canada and some European countries and was taken by many pregnant women. We now know that it caused about 7,000 children to be born with unformed limbs.

━━━ Scientific Disputes over Environmental Risks

The difficulty society has in evaluating scientific risk-benefit issues related to the environment makes our collective head hurt, and the problem is exacerbated by a common misperception. People unfamiliar with science often assume that when there are two opposing views on a scientific issue, the truth lies "somewhere in between"—an assumption that may sometimes work for political or social issues. It is far from the rule in science, however. We call this misconception the "ether breeze" hypothesis. In the last century, some scientists believed that outer space was filled with a rarefied element called "ether," and that Earth moving through the ether imparted motion to the ether near it, just as a ship sailing through the sea imparts motion to the water near it. Others doubted the existence of ether or of an "ether wind" caused by Earth's passage. In 1887, A.A. Michelson and E.W. Morley performed one of the most famous of all scientific experiments, testing to see whether the velocity of light would be different if it were measured in the direction of Earth's motion or against it. Their result, since confirmed many times, was that there was no directional difference caused by an ether wind, as predicted on the basis of the ether theory. The answer was that one group of scientists was right, and the other was wrong—there was no "ether breeze," the answer did not lie in between.

Similarly, Darwin was right and his many opponents were dead wrong; organisms were not created once in a fixed form; evolution is a central fact of biology.[21] Copernicus was right that Earth rotated on its axis and circled the sun; those who thought it was the center of the universe, although backed by the common wisdom of the time, were also dead wrong. Louis Pasteur was right in claiming that life cannot be spontaneously generated (worms springing from mud, maggots from garbage, and so on), and people who thought that life was continuously arising in those forms were completely wrong. Continents do drift around on Earth's surface, as Alfred Wegener said; they are not fixed, as Wegener's many scientific detractors contended.

Accordingly, people should not automatically assume that a "compromise" will provide the answer to scientific disputes in matters relating to the environment—and indeed, intermediate views often have not carried the day. Some scientists believed, for example, that CFCs did not pose a serious threat to the ozone layer; Molina and Rowland thought they did. Molina and Rowland were right. Some biologists warned in the 1940s and 1950s that overuse of synthetic organic pesticides such as DDT would lead to the evolution of resistance and damage to ecosystems. They were considered alarmists by others, but time has proven them correct. In the late 1960s, some scientists were greatly concerned about the problems of feeding 3.5 billion people; others were sure that many times that number could be fed. Since then some 200 million people have died of hunger and hunger-related diseases, and at least a fifth of today's population is not being adequately fed, despite irreversible depletion of stocks of soil, groundwater, and biodiversity in the attempt.

Other scientific disputes related to the environment will be solved one way or another, we expect, in the next few decades. The question of the seriousness of global warming is the most prominent. It will be settled by the gigantic experiment humanity is running—and we fear that the scientists who judged the threat to be severe (rather than nonexistent or minor) will be proven correct. Much the same can be said for debates over such questions as the potential damage from acid precipitation (will forests over huge areas be killed?) and ocean pollution (can the seas be pushed over a threshold that, among other things, will cause even more widespread collapses of fisheries?).

In sum, when one group of scientists says that an environmental problem is potentially very serious and another makes light of it, it is not safe to assume that the "truth" is that some difficulties may be encountered, but the problem will be manageable. That may indeed sometimes be the case, but frequently it will not be. In the case of DDT, concerns about the evolution of resistance and ecosystem effects proved correct; but scientists' worries that high loads of DDT in mothers' milk might cause a cancer epidemic have, so far at least, proven incorrect. But signs of more subtle problems are now appearing,[22] and our children may thank us later for banning DDT in the United States. To date, the record indicates that in environmental issues, the scientists who warned of hazards have been correct more often than those taking a more Panglossian perspective. The growing prominence of environmental problems, the continuing rise in the absolute numbers of hungry people, the failure of technological miracles (with the arguable exception of the green revolution) to materialize, all generally support that conclusion.

One should be cautious not to confuse true scientific disputes on issues, such as the relative dangers of natural and synthetic carcinogens or whether global warming has been detected, with the essentially political debates in the media on scientific matters. The prototype of a media debate on a scientific matter that does not reflect a scientific dispute is the evolution-creation imbroglio.

There is no dispute among qualified biologists about whether life evolved, just as there is none among physicists about the impossibility of perpetual-motion machines. Unfortunately, evolutionists, ecologists, and conservation biologists have been much less successful than physical scientists in educating the public about their disciplines.

Workers in the media are expected to present a balanced picture, to show "both sides of a controversy." This approach is appropriate for many social issues on which people may disagree (although even there, fringe views are often weeded out), but not for many scientific questions. Journalists generally recognize the limits to reasonable debate in the physical and medical sciences. Reporters and editors are well enough educated that stories of space-shuttle launches are not "balanced" with presentations by groups who think that the whole space program is a hoax created in TV studios. Images of Earth transmitted from the shuttle are not paired with equal time given to representatives of the Flat Earth Society. Astronomers and astrologers do not appear together on talk shows. Stories describing new surgical procedures do not require equal time for faith healers.

But when Ben Wattenberg, a journalist, states that continued population growth in the United States is good because it will strengthen the country militarily, that's "news," mainly because of its novelty.[23] A little analysis shows it to be not only novel, but nonsensical.[24] If Julian Simon, a specialist in mail-order marketing, announces he is unpersuaded that a serious extinction crisis is under way, that's considered news too—again because of its novelty. Unfortunately, such views are often presented as if they came from a substantial segment of the scientific community, because many journalists (and many other well-educated people) lack the background to evaluate them. After all, how many people were taught either in high school or college the relative population sizes of victorious and defeated nations in either world war? How many students learn, even today, about the relationship of a habitat's quality and area to the extinction of species living in it? The views of Wattenberg and Simon are taken seriously by a segment of the public, even though to a scientist they are in the same class as the idea that Jack Frost is responsible for ice-crystal patterns on a cold window. And those views are embraced by much of the business press (*Forbes*, the editorial pages of *The Wall Street Journal*). As a result, the business community, which needs accurate information on environmental issues, is being poorly served.

Almost all people in the media have enough knowledge of the physical world to filter out many outrageous notions. The notion that the world is flat is not in the spectrum of ideas that must be included in the news for balance. Few reporters or editors, however, have had the basic education that would provide a similar filtering capacity for statements on environmental issues. As a result, the public gets the impression that there is far more uncertainty about many issues than actually exists among the scientists studying them—as has recently happened over the issue of global warming.[25] The doubts of the tiny group of scientists who are very skeptical about the global warming projections of climatologists have been exaggerated and continuously quoted by nonscientists and a handful of other scientists who fear the economic impacts of attempting to abate the warming.[26] The resulting confused reporting places an even greater burden on citizens and decision makers who are attempting to evaluate an issue already made difficult by very incomplete knowledge.

Even when personal preferences are clear, and individuals are well informed and aware of the uncertainties, risk-benefit analyses are not simple. In a society of millions, the complexity multiplies enormously, especially when both risks and benefits would be distributed differently among individuals. Science can provide estimates of the probabilities of certain events and of their consequences, but it often gives little or no guidance on actions to take in response to those estimated risks. While science can outline steps that could be taken to slow global warming, it cannot decide for us how much "insurance" against catastrophic climate change it is worthwhile to pay for.

Science also can't deal easily with the statements of antievolutionists, perpetual-growth advocates, racists, or other elements of the lunatic fringe who simply make assertions rather than carry out careful investigations and analyses. Science is, to a large degree, a carefully orchestrated process in which the actors operate by mutually agreed-upon rules enforced by group sanctions. When science interacts with public policy, the scientific process can be helpful, but only in providing background for nonscientific social decisions.

▬ Our Values and Personal Risk Assessment

Science alone clearly cannot provide answers to the many difficult questions of risk assessment raised in this book. Of necessity, therefore, we as authors do make many nonscientific value judgments on issues of risk, using a set of broad guidelines that we think would be reasonable for society to adopt. These are the personal conclusions of two people who have attempted to inform themselves on the issues, not answers arrived at by the standard research procedures of biology, physics, or chemistry. So before we go further in discussing cures for Earth's environmental ailments, we would like to explain our general approach to environmental risks and benefits—the values we bring to our analyses.

Our basic value judgment in the area of risk is that both personal and social risks should be handled *conservatively*. How to be conservative is *relatively* easy to see when considering risks in one's personal environment. Statistically, the most important things to do to protect your own life and health are not to smoke (or chew) tobacco and to wear a seat belt when you drive. In those cases, you are avoiding well-known, well-understood risks—although as ex-smokers we are also aware of the psychic pain involved in avoiding the horrendous risks of smoking. Avoiding overexposure to sunlight by wearing hats, long-sleeved shirts, and sunscreen when outdoors for long periods (especially in summer) is a way to minimize relatively painlessly another known and increasingly serious risk.

But there are many potential personal risks that are not so easily dealt with. What can one do about the repeated warnings of cancer threats in food, water, personal care products, and so on? How about cholesterol, saturated fats, sugar, and all that putatively bad stuff? Those are tough questions even for scientists to answer. It is often very difficult to assess tests of the safety of products or the health effects of various foods. Much work is done by looking for correlations—for example, finding out whether people in different cultures with different intakes of fats differ in their rates of heart disease. They do, and people who consume less saturated fat tend to have less occluded coronary arteries and fewer heart attacks.

Direct testing is more difficult. In most cases, it would be unethical to conduct experiments on human beings. For instance, it would be unethical to dose a group of volunteers with high-cholesterol foods to see if they had more heart attacks, or to feed them high doses of the fungal poison aflatoxin to determine whether they suffered more cases of liver cancer. So other animals, often mice or rats, are substituted for people in tests.[27]

Usually, a very low probability of harm is considered sufficient reason not to employ a product, but the whole process of determining the probability and the nature of the damage is very complex.[28] Given that a product can never be proven absolutely safe and that one is confronted with thousands of products, what strategy is both conservative and reasonable? How should one deal with these complexities and the absolutely unavoidable uncertainties? If you give up everything that might possibly be bad for you, as the old saying goes, you may live longer, or at least it will *seem* longer.

Some risks of cancer are doubtless present in substances you use or eat or find in your drinking water, but generally the evidence is they are small compared to those of smoking or chewing tobacco. If you are an American or European, the noncarcinogenic components of your diet probably are a greater threat to you than carcinogens in it, since in those cultures heart and circulatory disease are the number-one killers. Both the carcinogens to which we are exposed and the potentially harmful elements in our diets, of course, should be controlled within reason by such public policies as policing standards of water purity, testing foods for pesticide residues, testing food additives, educating the public, and so on. The question of what is "within reason" now is, and doubtless will continue to be, a matter of debate, although "zero tolerance" (permitting no detectable quantity of contaminants in food) is not reasonable. The technology of detection is continually being improved, so that ever tinier and tinier amounts of chemicals can be identified, and many deleterious substances are likely to be detected sooner or later in minuscule quantities in many or most foods.

The risks associated with what is ingested can also be partially controlled by private behavior, and in that sphere defining "within reason" is up to you, although public education and full-disclosure labeling should be available to inform your choices. We personally try to eat a sensible, balanced, relatively low-fat and high-fiber diet, but we don't make a fetish of it and don't often forgo anything we really enjoy.

▬ Our Values and Risk-Benefit Decisions by Society

We may be conservative in dealing with risks personally, but we are extremely conservative—that is, risk-averse—when considering the health of Earth's ecosystems. We believe great caution is called for when making potentially irreversible changes in humanity's only home or in taking chances with hundreds of millions or even billions of lives, including those of our own grandchildren. We don't agree with political scientist Aaron Wildavsky, who wrote in 1989 that "the secret of safety lies in danger."[29] He sees the free market as an ideal mechanism for balancing risks and benefits. If an activity is too risky, he believes people will decide not to pay the costs and the activity will fade away. Wildavsky is very concerned about the possible risks of overestimating the dangers posed by new technologies (forgoing some or all of their potential benefits).[30] He greatly downplays the risks that may be associated with the harm such technologies may do to people or ecosystems, basically claiming that rich nations can buy their way out of any unforeseen difficulties.[31]

Wildavsky's predilections are widely shared in business and government, and even in some segments of the scientific community.[32] It normally appears in the form of statements such as "If you can't *prove* that doing X will do serious harm to the environment, then we should go ahead and reap the vast benefits of doing X." Like the accused in our criminal justice system, technologies are innocent until proven guilty. In other words, wait until the existence of negative externalities is proven before taking action that would internalize them. That was the defense of the chemical industry, which managed to delay badly needed action on the ozone front for a decade and a half.

This type of reasoning applied to the pharmaceutical industry, for example, would change the standard for drug approval to one where the drug could be sold without testing until someone "proved" that it was deadly. It is not clear that we should have different standards for the health of the ecosphere than for human health. In either case, one needs to weigh the expected costs and benefits of any action, not simply apply either "zero tolerance" rules or laissez-faire.[33]

Under a laissez-faire philosophy, it would have been folly to oppose the conversion of the United States to a nation dependent on the automobile for commuting. The technology provided the imperative; the market demanded it. Of course, automobile and tire manufacturers and oil companies lent the market a hand,[34] but that is a detail. We have reaped the benefits in mobility, jobs, economic growth, and privacy on the way to work. It wasn't until the 1960s that some of the costs became apparent in the form of smog, loss of community, and traffic jams. While the potential of carbon dioxide for causing global warming had been known in the nineteenth century, the automobiles' contribution to the problem was not widely discussed until the 1980s. Now it is clear that allowing the automobile to dominate the structure of American society entailed many unforeseen costs, and the United States is faced with a potentially traumatic and extremely costly task of reversing its dependence on individual automotive transport.

The risk of global warming was anticipated but long ignored. One can imagine many ways to improve society's performance in dealing with such recognized risks. But what about risks that remain unknown long after the technologies generating them have been deployed? The time it takes for costs to appear after decisions are made can be impressive, as the history of CFCs clearly illustrates. Even when the risks are anticipated by a few people, they often remain unappreciated by decision makers for a long time, as in the cases of the AIDS epidemic and global warming.

Social Risk
Assessments

"Societies" can't make the kinds of conscious risk assessments that informed individuals do. Instead, those decisions are made by small subsets of the society, with most individuals (including many leaders) unaware that they are being made. Furthermore, although questions of intergenerational equity do play a role in individual risk assessment (women will, for example, avoid consuming substances in pregnancy they think may later harm their children), we believe that it should play a much larger role in social risk-taking. Social systems have adapted to the consequences of inevitable individual deaths. But should our generation's "decision" to wipe out much of Earth's biodiversity be carried out, the consequences could plague our descendants for a hundred thousand generations or more.[35]

The consequences of mistakes in social risk-taking can be therefore much more widespread and long-lasting than those of individual risk-taking. As has often been said about the extinction crisis, we are not just causing the deaths of many other organisms, we are causing an end to their births. To most people, the latter is a much grimmer consequence, be it for rhinos or for humanity. Since the probability of the extinction of many organisms is very high and the consequences are deemed very serious, the risk being run by not dealing effectively with the extinction crisis is gigantic.

Social risk-benefit decisions differ from personal ones in another important way. The kinds of risks one faces day to day tend to be familiar and easily perceived, as are the benefits. Most of us have seen many automobile accidents, if not in person, then on the evening news. The risks one runs in driving are part of everyday life, but so is the convenience of personal transportation. Virtually everyone over fifty has seen a friend or relative die miserably of cancer and has lost someone close from a heart attack. Costs that are thought to lie far in the future may be heavily discounted—the "getting lung cancer in my seventies is a small price to pay for the joy of smoking until then" syndrome. But easily perceived risks are often ignored if one values the benefits highly enough. Skydivers are in no doubt about the potential costs of the chute's not opening but know the exhilaration of the drop. Many regularly confront the zero-infinity dilemma of jumping out of an airplane for the fun of it and accept the risk.

In contrast, few people recognize the benefits of sharing Earth with millions of other organisms or think much about the importance of a reasonably stable climate. Furthermore, the most critical environmental risks are difficult to perceive. With the possible exception of a zoo keeper or someone else observing the death of Martha, the last passenger pigeon (or some other sole survivor) in captivity, no one has ever watched one species go extinct, let alone thousands. No one has ever observed the results of heating the planet by a half-dozen degrees over fifty years or so, either. The costs associated with individual risks are usually paid in minutes or months; those associated with environmental risks may be paid over millennia or more. The costs may be virtually infinite, but they are distant in time and amorphous, unlike those of hitting the ground at 100 miles an hour after your parachute has failed to open.

Environmental risks often do not present themselves in forms that are easily perceived by people.[36] A car swerving toward our car or a snarling dog lunging at us are the sorts of stimuli that millions of years of biological and cultural evolution have prepared us to recognize as threats. But the gradually climbing zigzag line on Keeling's chart of the CO_2 concentration above Mauna Loa or statistics on habitat destruction are not. For most of human history, people could do nothing about environmental changes occurring over decades. People didn't cause them, and people couldn't stop them. Consequently, human beings evolved perceptual systems that were not good at focusing on gradual change in the environmental backdrop against which their lives were played out—in fact, they actually suppress detection of such change. With the background held constant, the sudden appearance of a hungry tiger or a tasty gazelle (or a speeding car) was all the more readily perceived and dealt with.[37]

Despite all these difficulties in perception, society must constantly weigh risks and benefits. Suppose, as the Stanford computer study suggested, that very rapid climate change in the next century could result in hundreds of millions of premature human deaths? What costs should be incurred to slow that change if its probability were 1 percent? 5 percent? 25 percent? 50 percent? These are not easy questions to answer, and there is no *scientific* basis for answering them. Since most of the measures required to slow the warming, from greatly increasing energy efficiency to land reform in poor nations, would carry other great benefits, and since the probability of unprecedented change appears to be in the vicinity of 50 percent, a very large social effort, one incurring substantial costs, seems to us not simply justified but virtually mandatory. In our view, the delay in dealing with chemicals that attack the ozone layer was inexcusable when the principal costs to societies would have been those of a transition to substitutes, a move that could have brought out the best in industry's competitiveness and innovation.

The debate over the risks, costs, and benefits associated with society's possible responses to global warming has hardly begun, as we saw in the discussion of the Nordhaus analysis. It is clear that the social process will be fraught with disputes. For instance, Nordhaus approved of the Bush administration's inaction, saying it "should be commended for following a reasoned approach and avoiding measures that would lock the economy onto a path that could not respond flexibly to new information or emerging technologies." We, of course, strongly disagree and find the administration's behavior execrable. We also wonder how research and development of new technologies for alternative energy sources could make society less flexible!

We do not disagree with Nordhaus because we think slowing global warming will be cost-free. Very substantial costs will inevitably be incurred in solving the suite of environmental dilemmas now facing civilization. The monetary amount of those costs and the monetary benefits will be difficult to calculate and doubtless always partly in dispute. But even now rough orders of magnitude can be discussed. Suppose the economists who claim that it would cost a trillion dollars ($10 billion per year) over the next century for the United States to do its part in slowing global warming are correct. Suppose that slowing global warming would save 1 billion lives over that period (10 million per year).[38] In that case, the American contribution to saving each life would be $1,000. If other nations contributed in proportion to their GNPs, the rough price paid to save each life would be about $3,000. Expressing the value of human life in dollars is difficult. By most of the criteria that can be used, however, from our own subjective judgments or jury awards for wrongful death to evaluations of productivity lost by premature demise or the amounts society is willing to pay to save lives, $3,000 is a few percent of the monetary value of a human life— which should probably be valued at least at $100,000.[39] Now let's suppose that the chances of global warming costing 1 billion lives over the next century were 50-50. Clearly, inaction in the face of the threat of global warming would be difficult to defend, since we would be avoiding an "expected" loss of 50 trillion dollars (1 billion lives × 0.50 probability × [say] $100,000 per life) for an investment of 1 trillion dollars—a great bargain in cost-benefit terms.

Of course, the steps envisioned in the trillion-dollar estimate to abate global warming would not deal with all other environmental problems—although they *would* help ameliorate many (toxic air and water pollution, for example). To be conservative, let's suppose that to assure our environmental security comprehensively and effectively would cost ten times as much as just tackling the warming. In that case, the cost would be $100 billion annually. Could that be afforded? We think so; it represents about a third of that spent on military security, an expense that almost everyone believes should decline in the future. And, of course, just as military expenditures provide subsidiary benefits such as jobs and the development of technologies that also have nonmilitary uses, so one would expect subsidiary benefits from expenditures on environmental security. They would include having a more pleasant (not just more secure) environment, better health, and lower health care costs.[40] In short, even without being able to calculate risks, costs, and benefits exactly, rough estimates indicate that *we can afford to insure ourselves against environmental catastrophe.* Not to do so would be bizarre for a society that routinely insures lives, limbs, homes, automobiles, and businesses against risks whose probabilities of occurrence are much smaller than those, say, of unprecedented climate change due to global warming.

We believe it is important to take out that insurance for another reason. Not all risk decisions are as straightforward as those involved in abating ozone destruction or even global warming. Yet in both cases, the risk-benefit question and decisions are coming much too late—when the problems are entering the "probable-infinity" range and may have so much inertia that avoiding serious consequences is impossible. The lesson of the unexpected depletion of Antarctic ozone due to previously unknown chemistry in stratospheric ice clouds is an important one. Insurance should be taken out *before* engaging in risk-taking. Once the parachute has failed to open, it's too late to call your insurance broker.

So we are convinced that a more conservative approach is needed in the environmental risk philosophy of our society. A primary component of that philosophy should be to *increase humanity's margin for error*. And the basic way to accomplish that is to reduce the scale of human activities. Lacking crystal balls, analysts cannot accurately judge all the possible hazards of our activities, much as they may try. But we could move to a situation where even serious technological errors (like the manufacture of CFCs) would be unlikely to overwhelm the capacity of the planet to heal itself before they are detected and corrected. So once again, we see that reduction of scale is the principal challenge for the next stage of human evolution.

Chapter Nine

Escaping the Human Predicament

As long as human populations were relatively small, and the number of "energy slaves" per person was small also, even severe effects of bad environmental policy were not critical to humanity's future. People were capable at most of creating local environmental disasters. The disappearance of the irrigated civilization that once thrived in the valleys of the Tigris and Euphrates, the destruction of the classic Mayans, the degradation of the once fertile and forested Mediterranean basin, and the desertification of the arid western United States all testify to the power of local destruction that *Homo sapiens* has long possessed.

But the situation today is utterly different; human impacts are degrading the entire planet. Not only has our large-scale global civilization exceeded the capacity of its environment to support it on income, but it is committed, by the momentum of population growth, to substantial *further* expansion.[1] Even the most optimistic demographic projections indicate virtually a doubling of the present population (to over 9 billion) before growth can be humanely stopped and reversed. Economic growth too is expected to multiply severalfold globally. This explosive further enlargement of the human enterprise and (inevitably) its impacts is the essence of the human predicament. Can the resources be mobilized to support 9 to 14 billion people, even briefly, without permanently degrading the resource base for subsequent generations?

It should be obvious that the global predicament cannot be resolved through purely local measures; important though they are for meeting local needs, such efforts are doomed to failure unless people in localities everywhere also take appropriate action. No matter how carefully Rajasthani villagers in Gopalpura protect their watershed, if citizens in Hamburg, Falmouth, and Kansas City don't stop burning huge quantities of fossil fuels, the climate may change so that there is not enough water to protect. Likewise, local efforts will fail if they are designed to deal only with symptoms. The 1987 harvest of 80 cartloads of grass may have been sufficient, divided equally among households, to see the villagers of Seed through the drought. But in 2017, if those cartloads had to be divided among twice as many people, they might prove utterly inadequate.

Locally or globally, people cannot escape the grip of the I = PAT relationship. No foreseeable reduction of per-capita consumption or improvement in the environmental characteristics of technologies can keep impacts declining in the face of continued population growth. Suppose, through heroic efforts, humanity managed over the next eighty years to reduce both the average consumption per person and the average environmental impact of technologies globally to half of today's level. If the current growth rate continued, expansion of the human population would overwhelm those advances and leave humanity's total environmental impact unchanged in the 2070s. This, of course, is just a hypothetical example, since even today's level of impact clearly could not be sustained for eighty more years, nor do demographers anticipate continued population growth at the current rate, which would produce a population of over 21 billion in 2070.

By now, we should have convinced you of the ubiquity and seriousness of environmental problems. They are not problems that can be solved with slogans about "sustainable development," "clean air," "agroecology," or even "stop at two." They won't be solved by the United States doubling its foreign aid from 0.2 to 0.4 percent of its gross national product.[2] They won't be solved by just putting smog controls on automobiles, recycling, or using energy-efficient light bulbs. All these activities can—must!—be part of the solution, but by themselves or even in combination, they will be inadequate, especially in the long run.

The dilemma facing human societies, rich and poor alike, is how to manage the unavoidable population and economic expansion ahead while reducing the scale of total impact on the environment, and designing a new system in which humanity can live in dignity on its income and even maintain a comfortable cushion against future environmental disaster. Civilization looks forward to perhaps a century of limited resources, an increasingly impoverished environment and disrupted ecosystem services, and largely unpredictable consequences from global warming. Beyond that difficult period lies unknown territory. Whether it will prove to be catastrophic or an era of human recovery and the rebuilding of a new, sustainable civilization depends very largely on what happens in the near future—the next decade or two.

Humanity can decide to take control of its long-term future and mitigate its total impact through all three factors in the equation: population, affluence, and technology. Plenty of opportunities still exist for reducing the contributions from all three, but delaying action will inexorably result in foreclosure of many of them: more future parents are born daily, more forests are cut down, less soil to grow food and less water for crops are left, less time remains to develop alternative energy and transport systems. Escaping the human predicament will be enormously difficult since virtually all environmental problems are connected to one another and to overpopulation.

Controlling population growth is critical. We cannot emphasize too strongly that significant resources must be directed into programs that limit population growth both in the United States and abroad. Because of the built-in time lags, unless the surge in human numbers is halted soon and a gradual population *shrinkage* begun, there is no hope of solving the problems discussed in this volume.[3]

Yet no less important as well are efforts to keep each person's impact as low as possible through reduced consumption by the rich and the provision of everyone's needs through more environmentally benign technologies. The Holdren scenario (described in Chapter Two) for limiting the energy use of the rich while increasing it for developing nations over several decades can provide a key framework for controlling per-capita impacts during the transitional century. The scenario offers no details on the technologies to be used other than the existing opportunity of greater efficiency, but turning to less destructive energy sources than today's will clearly be essential.

The human predicament is also intimately connected to *Homo sapiens'* severe and persistent social problems, especially the inequitable distribution of wealth and resources, racism, sexism, religious prejudice, and xenophobia. Unless these divisive traits can be managed, they could cripple attempts to develop the global cooperation needed to create a new civilization.

All these problems making up the human predicament are also bound together by the difficulties of finding and intelligently deploying the human and physical resources necessary to solve them. Increasing food production, mobilizing energy, and eliminating absolute poverty for a still-growing population on a finite planet, while reducing environmental impacts and preserving our resource capital, is a tall order for any century.

The transition may take a century, but it must be launched soon. Perhaps a decade remains to begin making the necessary changes that will help us avoid the worst consequences of the human predicament. Of course, only a decade is not remotely enough time to inform the majority of the world's people adequately about the scientific details of what is happening and what needs to be done. Fortunately, that is not necessary. What is required is to make as many people as possible aware of the true costs of their actions, and then let their individual self-interest and concern for their families take over. People almost always respond more readily to economic incentives than to moral suasion.

The world's leaders, however, do need to be educated so they can both promote that process *and* free people to manage their local environments properly. The world's people do not need to understand the details of the nitrogen cycle or how ionizing radiation alters DNA to grasp the fundamentals of environmental problems. They do need leaders to put the big picture together for them, tell them what sorts of local, regional, and national actions are required to prevent a global collapse in decades ahead, and give them the means to take action. Among the most important changes needed is a reorientation of economic thinking.

Today's buzz phrase is "sustainable development"—which means improving the human condition today without damaging the prospects of future generations. Development has long been on the human agenda, but the idea that limits might apply to that development has not fully penetrated the development community. The *Brundtland Report* had the germ of the idea right when it stated that "sustainable development requires that those who are more affluent adopt life-styles within the planet's ecological means—in their use of energy, for example."[4]

The *Brundtland Report* recognized, correctly in our view, that continued economic growth in poor nations and redistribution of income within them is essential. The report was overly optimistic, however, in its depiction of the future of economic growth, especially in rich nations. We believe that growth (in the form it has traditionally taken) will soon come to a halt in the developed world, because markets are fairly saturated and population growth has nearly stopped. The trend for at least two decades has been one of slackening growth in the most industrially advanced economies. We would even contend that, in terms of net benefits to human beings if the external costs were properly counted (such as faltering ecosystem services, mining of soil and groundwater, and higher medical expenses due to pollution), the economies of most rich nations have stopped growing already.

▬▬▬ The Issue of Economic Growth

In many respects, economic growth has been an enormous success. It has catapulted one large segment of humanity to standards of health, education, and general well-being far beyond the dreams of oriental potentates of yore. After all, the beneficiaries of growth have flush toilets and heated homes, and the potentates did not. Given the immeasurable benefits of growth, especially in early stages of modernization, it is not surprising that it has been enshrined as an unvarnished good in the thinking of mainstream economics. This view, however, is based on recent history and betrays a deep misunderstanding of how the world works.

Recent generations of economists unfortunately appear to believe that economic growth can be divorced from the physical world. The clearest symptom of this is the diagram that graces standard economics texts showing the generation of gross national product (GNP).[5] The peculiar thing about the diagram is that it shows no physical inputs or outputs; wealth is generated in a perpetual closed ellipse—by perpetual motion, as it were.[6] When economic thinking congealed on the topic of growth in the 1930s, it basically made sense. Very bright people like John Maynard Keynes were working in a world in the grip of a terrible and persistent depression. Hundreds of millions were out of work; industrial plants stood idle; and only a catastrophic war was finally capable of kick-starting the economy again. The furthest things from the minds of economists then were thoughts of global externalities constraining the scale of the economy.

But now such thoughts should move to the top of the economic agenda.[7] We believe that economists, perhaps even more than ecologists, hold the key to the human future. Economists have developed a system that is very good at determining how to achieve a "Pareto optimal" distribution of goods.[8] But mainstream economics simply does not address itself directly to the most critical question of the day: What is the maximum scale of the economic system that can be *safely* sustained in the long term? How big can the human enterprise—the population and all its accoutrements and activities—be without destabilizing the systems that underpin it? It is as if economists concentrated on distributing the load in a canoe very carefully so that it would not tip, but had no way of determining whether the canoe was sinking on an even keel.

In theory, a message about sinking should come through market mechanisms: as the scale of human activities increases, the depletion of natural resources should lead to higher prices that would reduce that rate of growth, possibly bringing it to a grinding halt. And indeed, these market mechanisms work very well and by and large have kept the canoe afloat (precisely by helping us allocate the weights correctly) for a very long time. As long as the scale of the human enterprise was relatively small compared to the natural systems supporting it, the externalities not considered by the market had relatively little import. The human predicament in large part results from the existence now of massive and mounting externalities not captured by the market—or mistakenly captured as pluses, not minuses, because they swell the GNP. One good example is the $40 billion per year in excess health care costs to Americans due to air pollution. These costs certainly do not contribute in a positive way to public well-being; they serve a corrective role and are paid through a different part of the economy than are the activities that generate the pollution. These rapidly rising externalities are traceable to the incredible expansion of the human enterprise over the past century, during which the population quadrupled while its total impact (as measured by energy use) increased 13-fold.[9]

It has long been clear to ecologists that most economists fail to appreciate either the magnitude or the significance of environmental externalities.[10] Most politicians, businessmen, and the public at large also fail to recognize the importance of those externalities, and therefore the increasingly serious predicament of *Homo sapiens*.

The subdiscipline of environmental economics has recently emerged, in which some promising work is being done.[11] But environmental economics remains fairly low in the professional pecking order. So far it has had only minimal influence on the mainstream of economic thinking. Far from being central in economic education, it is chapter 45 in the standard texts— an appendage, an afterthought.

Nevertheless, environmental economics has been moving in the right direction and picking up speed. There is strong interest in market-oriented policies for environmental protection in Europe, and a tradable permit system was just incorporated into the United States Clean Air Act Amendments of 1990. Carbon taxes are being widely discussed, as are tradable permit systems that would help correct the inferior positions that poor nations occupy in the global economy. All these steps point in the direction of market mechanisms that will "sense" and constrain the scale of the physical economy globally. Progress has been less encouraging in creating national and international institutions that can deal with the necessary monitoring and regulatory tasks, especially those assuring some level of equitable distribution, that cannot be managed adequately through market mechanisms.

In spite of some progress in one corner of the economic profession, most mainstream economists believe that the scale of economic activity can be increased indefinitely, or at least so far into the future that limits to growth need be of no concern today.[12] Their education has left them largely unaware of the externalities embodied in stress on natural systems. Their perception is partly rooted in two related (although not always explicitly recognized) axioms of mainstream neoclassical economics: (1) that externalities are the exception rather than the rule, and (2) that technology can without limit either reduce demands for natural resources or allow the substitution of one natural resource for another, especially when the market forces provide the incentive for such developments.

The first axiom causes economists to be somewhat complacent about the state of the world because they expect market mechanisms to allocate resources appropriately most of the time. They feel justified in this view because historically that has usually been the case. The second axiom also has a historical basis. Every economist is taught that Malthus made dire predictions for humanity based on the inability of arithmetic growth of food production to keep up with the geometric growth of the human population. And every economist knows that Malthus was "wrong" because he did not foresee the tremendous technological innovations that have allowed food production in this century to increase even faster than population growth rates (which themselves would have been unimaginable to Malthus). The problem is that many economists today cannot imagine that this spectacular success might soon end—and might even end in a crash.

The power to improve the human condition that technology has displayed has led many economists to assume that further improvements in technology can be counted on to remedy any unforeseen problems. The noneconomist may ask: How can economists be sure technology will "save" us? The economist will answer, "It always has before," an answer with some justification when applied to the rich nations. It will therefore be no small task to convince economists to reexamine their axioms in light of what ecologists and physical scientists know about biological and physical limits.

Economists nonetheless could play a crucial role in helping humanity move toward a sustainable civilization. If they are to do that, though, the professional training of economists will require substantial revision. Many economists already recognize that graduate education in the discipline focuses too little on important questions of policy and too much on learning to manipulate esoteric mathematical theory based on unlikely assumptions. A survey by two economists showed that many graduate students in economics realize that advancement in the field requires emphasis on such questionable exercises and "that graduate economics education is succeeding in narrowing students' interests."[13] The survey of students' opinions of the importance of reading in other fields to their development as economists did not list ecology or any other biological science among the fields to be scored. The lowest score for relevance given by the students was to physics. Only 2 percent of the students considered it very important, 6 percent important, 27 percent moderately important, and 64 percent unimportant. Small wonder economists have so little conception of physical and biological constraints and so much faith in the power of technology!

With effort, a new ecological-economic paradigm could be constructed that unites (as the common origin of the words *ecology* and *economics* implies) nature's housekeeping and society's housekeeping. A basic rule of the paradigm should be to give first priority to keeping nature's house in order. Unless considerable instruction on the basics of how the physical-bioligial world works is included in the training of all professional economists, they will continue to give faulty advice to politicians, businessmen, and the public. Then many people will continue to see growth of the global economy as the cure rather than a symptom of the disease, and the chances of turning the human predicament into an opportunity to shape a brighter future will be very much diminished.

Of course, resistance to the faulty advice could be much enhanced if every high-school and college student in the nation were required to take at least one course that gave a basic overview of the "state of the planet." At Stanford University, most students (and most faculty) remain ignorant of the basic shape of the human predicament and the background required to understand it—including ecology, demography, evolutionary theory, the laws of thermodynamics, and (to move to economics) the areas in which market signals can be relied upon to prevent or abate environmental deterioration and the areas in which they cannot.[14]

In the United States and most other nations, the educational system's failure to produce citizens equipped to deal with the modern world is a disgrace. Stanford and some other universities are now moving toward upgrading their students' educations, but the conservatism of faculties and administrations remains a major barrier to finding a way out of the human predicament.

Can economists awaken from their complacency and help pull humanity through the crisis decades ahead? Can they begin to interact more with ecologists so that the blissful ignorance of important economic concepts of many ecologists can be corrected? We believe they can and will. Some progress is being made at our own university, where ecologists, economists, and other social scientists are now meeting regularly to discuss the human predicament and starting to outline collaborative research on it. It has been an interesting learning experience for the ecologists, and we think also for the economists. Both groups have found they had much more in common than they had expected.[15]

Some of the issues under discussion include ways to protect society from the consequences of the private decisions of individuals and firms, the role of advertising in creating demand and encouraging consumption, and whether making more ecological information available to individual decision makers would alter their actions and lead to less destructive behavior overall. Economists have expressed their puzzlement over biologists' concerns over biodiversity, and biologists have briefed them on the reasons. Ecologists have had the economic pros and cons of various ways of reducing carbon dioxide emissions explained to them. And Stanford ecologists and economists are both starting to look at ethical and value issues that cannot be solved by their disciplines, but which they must consider personally. Economics cannot determine whether the best use of a gallon of oil is to keep a poor man's home warm for a day or to drive a rich man's yacht an extra half-mile, but most economists would have an opinion. Many ecologists are troubled by questions raised by "deep ecology," such as whether organisms and ecosystems should have values or rights beyond those related to their usefulness to humanity, but they must grapple with these issues nonetheless.

In each case, familiarity with the science may help to inform their value judgments—at the very least preventing the recommendation of courses of action that would be impractical or impossible. Economists know that everyone cannot be a billionaire (in 1991 dollars). Ecologists know that saving all animal species by moving them to zoos is neither feasible nor desirable.

Real progress in the wedding of ecology and economics would be signaled by the following sorts of changes. The President's Council of Economic Advisors would be subsumed in a new "Council of Ecological and Economic Advisors." Mainstream economists would be busy devising market and/or regulatory mechanisms to internalize environmental externalities that were sensitive enough to keep the scale of the economy sustainable—that is, maintaining a scale that would permit it to function permanently (with a substantial safety margin) within environmental constraints. Economic growth would always be discussed in a context of counterbalancing shrinkage and redistribution, as well as consideration of whether the population in question was growing or shrinking. To balance this change, all ecologists and environmentalists would be required to have a course in environmental economics as part of their training. When all this has occurred, mainstream economics as a profession will have become a powerful force for survival, and mainstream ecologists will look increasingly to economists for desperately needed help in creating a sustainable society.

Economists and ecologists will have to struggle to change the conventional wisdom, especially to eliminate two fallacious beliefs. The first is belief in the ability of *current* market mechanisms to allocate resources properly in the face of global externalities, and the second is the belief that technology has a nearly infinite ability to prevent or repair environmental damage. Until these notions are recognized as wrong and are discarded, much of the resistance to the necessary solutions to the human predicament will be based in fallacious economics. Most politicians and businessmen will continue to believe that economic growth in its present form is the optimal path to a healthy society and a decent quality of life.

Ironically, they are undermining in the worst possible way the very goals they seek. Perpetual growth in the scale of the human enterprise is an impossibility; *its growth will cease.* Many who fail to grasp this simple fact (or believe that any limits are so far in the future they can be ignored) make the growth imperative the very basis of their social orientation.

At the extreme, this has led right-wing ideologues to view any move toward environmental sanity as part of a general liberal or communist plot. In 1982 the Republican Study Committee of the House of Representatives warned of the "specter of environmentalism."[16] Ronald Reagan's disastrous choice of Secretary of the Interior, James G. Watt, once said of environmentalists, "They are political activists, a left-wing cult which seeks to bring down the type of government I believe in." Watt's colleague, John B. Crowell, Assistant Secretary of Agriculture during the early part of Reagan's reign, claimed that the National Audubon Society, Sierra Club, and similar organizations had been "infiltrated by people who have very strong ideas about socialism and even communism."[17]

The latter is certainly true today (though in a sense completely opposite from how Crowell meant it!), now that the environmental disasters that accompanied communist rule in eastern Europe have become clear to everyone. Watt's and Crowell's statements simply reflected their fear that environmentalism would hinder the Reagan policy of raping the environment for the short-term further enrichment of his friends and supporters.

Actually, the environmental movement has deeply conservative roots; indeed, conservation has always been, as the name implies, part of the conservative agenda. From George Perkins Marsh and Teddy Roosevelt to Russell Train and Senator John Heinz, conservatives have generally shown more concern for the human environment than have the followers of Marx and Engels.[18]

Marxists, however, are largely correct in emphasizing the importance of the political and economic dimensions of the human predicament. The world is increasingly divided between haves and have-nots—not just rich and poor nations, but rich and poor people within nations, including the United States.[19] People who are well off naturally want to protect their positions; they generally don't want drastic changes. The existing system has put them near the top; the prospect of change is automatically viewed as a threat to their position (as indeed it may be). On the other hand, those near the bottom of the economic ladder often fear that remedying environmental problems would divert resources now maintaining the social safety net (as it might).

Of course, many people, beyond having a general aversion to environmentalism, feel directly threatened by it. They are afraid that changes in the environmental status quo will destroy their livelihoods. Coal miners, highway construction workers, utility managers, makers of fuel-inefficient autos, real-estate developers, pesticide manufacturers, and loggers have legitimate reasons for concern. A poorly planned attempt to create a sustainable society could be disastrous for them.

▬ Paradigm Change

Economics is one of the prime areas in which new thinking is required, in our view, but many others also need revision. Indeed, a compassionate resolution of the human predicament will require a transformation of human thinking comparable to the one that accompanied the agricultural revolution, and in a much, much shorter time. The dominant social paradigm—the collection of norms, beliefs, and values that form the world view most commonly held by western culture—needs to be modified in certain essential respects.[20] In particular, society's conception of what human beings are for and how they should relate to each other and to their environment will have to be reexamined. Widely held attitudes about these relationships, not only in the West but in other cultures, will have to change.

Such a revolution in attitudes will be necessary because human beings of all cultures, as a result of their genetic and cultural evolution, are basically animals with a small-group, parochial, geographically local, short-term outlook.[21] While most cultures include some individuals who have achieved a more global, long-term perspective, relatively few are sufficiently interested or have much opportunity to focus on the "big picture," let alone *do* anything about it. Several billion people are too poor to worry about much beyond the essential tasks of finding work, obtaining firewood with which to cook and warm their abodes, providing water and food for their families, and so on. They haven't time to consider threats to their well-being from global warming or depletion of the ozone layer, even if they are aware of them. They are absorbed by the overwhelming difficulties of day-to-day survival.

Even in the United States, more people now are in economic positions that make being informed and caring about the broad issues facing society and the world an unaffordable luxury. During the 1980s, the gap between rich and poor in our nation became more extreme than in any other major industrial nation.[22] The share of wealth held by the richest 0.5 percent of American households shot upward for the first time after a downward trend of forty years, while the average income of the lowest 20 percent slid downward. To the homeless person sleeping on a sidewalk grate trying to keep warm, the problems associated with the loss of biodiversity seem truly remote.

Unfortunately, the continuing decay of the American educational system has produced an "educated" population that is largely ignorant of science and geography, and innumerate to boot. As a result, in this richest of nations, few citizens have the basic tools that would allow them readily to grasp the "big picture." The American "elite" in politics, business, the media, and education itself is woefully lacking in the background to understand what is happening to their world.

Some other rich nations do better, but all too often vaunted technical educations (such as those obtainable in Japan) are geared to producing clever engineers rather than people with a broad outlook. And in the shambles of the Soviet empire, parochialism is clearly on the rise, although one could hope that with the deadening hand of communism removed, education might rebound (if it is not stultified by even more outdated religious dogmas). Humanity is ill prepared by standard education systems to face the most serious crisis in its history; too often, their education has given people an unfounded faith that "science will solve our problems."

But there clearly are limits to what science can contribute toward resolving the human predicament. Science can lay the groundwork for more environmentally benign technologies, it can provide safer and more convenient contraceptives, and it can help in designing more effective ways to protect natural ecosystems. It could even provide insights into such questions as how to encourage people to have smaller families and how to design economic systems that deliver their benefits without perpetually increasing the human impact on the environment.

Scientific analysis cannot determine whether the United States should plan for a future with 100 million people leading a relatively affluent life-style, or 400 million people living as subsistence farmers and factory workers, although it clearly can inform the decision. Science cannot prescribe the ideal distribution of income or oppportunities between individuals, communities, or nations; it cannot tell us how heavily we should weigh the fates of our grandchildren or our great-grandchildren in our planning. These are moral and ethical choices that must be based elsewhere. Our personal moral and ethical preference, based on what we have been taught by our parents, friends, and society, is that these decisions should be made as democratically as possible by a well-informed public.

▬ Toward a Sustainable Society

In the context of that democracy, our vote is for the United States and other rich nations to begin *now* to move toward a sustainable society. We believe that goal offers the best chance for the long-term persistence of society. Most people think "sustainable development" refers to improving conditions in developing countries, and indeed it does. But what they don't realize is that sustainable development of the rich is equally important, for their behavior most threatens the future of civilization.

Sustainable development of the rich involves far-reaching changes. Not the least of these would be establishing policies designed to begin population shrinkage in countries where it has not yet started, and to accelerate it where it has begun. This means "stop at two" programs, backed where necessary with a system of incentives and penalties carefully designed so that children benefit from both. The poor should continue to receive tax breaks to help them properly rear their children, and improvement of opportunities for the education of all children is essential. But with higher incomes, people should be increasingly taxed for reproduction to cover the extra costs they are imposing on society—something that the 1990 change in American tax laws unintentionally introduced in a small way.

After all, it is not primarily the children of the poor who will soon start driving gas guzzlers, flying in jets, using air conditioners, buying second homes, or eating luxury foods imported from poor countries. All these activities carry high social costs not considered in decisions of production and consumption; the externalities produced by children of the middle classes and the rich are now largely negative and should be internalized.[23] While the children of the affluent often may be more productive than those of the poor, what they produce is mostly not critical to their lives, while what they are using up is. Their contributions generally are not remotely enough to balance the enormous costs their consuming habits levy on the world's life-support systems. VCRs, automobile air conditioning, cellular telephones, junk bond manipulations, and even organ transplants don't improve life nearly as much as global warming, ozone depletion, soil erosion, overdrafts on aquifers, acid precipitation, and toxic pollution threaten it. Indeed, in the 1950s virtually everyone who had a "good life" had it without any of the results of productivity listed above.[24]

Making some of those external costs an element in the decision making of the parents of future consumers would be wise social policy—as long as it was accompanied by universal education about controlling fertility and easy access to the means to do so. Putting taxes on consumption would also help call attention to the costs of population growth.

Poor nations will certainly continue to struggle for old-style industrial development unless the rich nations clearly and dramatically set an example by changing their ways. Unless the rich are widely observed to be reducing all three factors in the I = PAT equation, thereby reducing their pressure on the environment, there is little hope that the poor will cooperate to save Earth's life-support systems.

▬ Helping Poor Nations

While one can fairly easily outline steps that rich nations need to take toward sustainable development, helping poor nations to do so will be much more difficult and complex. Almost all serious analysts agree that the prospects for a bright future hinge on narrowing the global gap between rich and poor. Many agree that the rich have a moral duty to help the poor; they feel their own happiness diminished by knowledge of the vast human misery that afflicts others both at home and abroad. But some people are not troubled by moral questions relating to the poor, or somehow believe the poor deserve their poverty. Even they should wish to narrow the rich-poor gap, though, if only out of pure self-interest, for the poor have the power to destroy civilization. They might do it simply by continuing to cut down forests and burn coal as they struggle for a decent life.

Both rich and poor nations should collaborate in changing the rewards within the system of international trade. Less developed nations that benefit the world by not becoming industrially overdeveloped and by maintaining natural areas and ecosystem services of global significance should be able to reap economic rewards by so doing. In particular, they must be assured the financial resources to gain access to industrial products just as do citizens of rural areas of industrialized nations today. While this could mean higher prices for some foods, higher costs for tourists seeking views of exotic birds and animals, and so on, these higher costs would be compensated by lower environmental costs for everyone.

This, of course, does not mean that every family in poor countries will (or should) immediately have a car, refrigerator, TV, VCR, microwave oven, washing machine, and dryer. Given the current numbers of people lacking these items, the world's environmental systems could not support that, even if the world's industrial capacity could meet that huge demand. But it does mean that a global effort should be made to arrange controls on carbon, CFCs and other emissions, and to provide assistance where needed, so that on average each person should be able to achieve the basics of a decent life: clean water, adequate food and food storage facilities, energy for cooking, basic medical care, schooling, and opportunities to work. Beyond these essentials, at least bicycles, transistor radios, and access to communal TV viewing facilities could be made available to virtually everyone. We believe this minimum level of development could be achieved *if* funds now being wasted on weapons and military forces were reallocated, *if* rich nations underwent sustainable development (or perhaps "redevelopment" would be more accurate), and *if* population control measures were universally successful.

A major problem will be to design measures to help the poor that not only fit their needs but involve their broad participation in planning and execution of programs. Aid should be openly self-interested in the sense of not being a charity operation but based on the common need to create a sustainable global society. Some portion of the "peace dividend," earned by ending the cold war, might be used to increase aid. As an investment in global security it could hardly be better placed. As much as $100 billion annually might be gleaned to assist poor countries through projects such as financing small local factories to produce millions of simple, sturdy bicycles and durable kerosene or solar-powered refrigerators for those who need them—and in the process provide employment for unemployed people in those nations. Developing countries might help finance such activities through international carbon emission offsets (ICEOs) as described in Chapter Three.

Closing the rich-poor gap will be a gargantuan task; no society has ever done it completely or even approximately for very long. Moreover, the argument of many conservatives that even approximately equal distribution squelches investment in development and progress is supported by a lot of evidence from the communist experiments. But the gross disparities that now exist between nations and in many societies can surely be reduced: the destitute can be given access to the basic necessities of life and the most profligate of the rich can be induced to share some of their wealth. Finding ways to accomplish this redistribution on a global basis, though, will require the best minds we can muster.

Somehow poor countries must be steered into paths of development that are also sustainable and not environmentally destructive—paths that do not repeat the sorts of mistakes made by today's developed countries. For instance, if the rich nations fail to help developing countries plan their development based on energy sources other than fossil fuels, all nations will suffer the consequences of huge additional injections of CO_2 into the atmosphere. But the difficulty of planning appropriate ways to aid poor nations pales before the task of dealing with their political and social structures to get that aid properly applied. It will take great care and sensitivity on the part of the leadership of rich nations as well as courage (and luck) to make essential reforms in poor nations.

Changes in Attitudes

It seems extremely unlikely that any permanent solution to the human predicament can be achieved without fundamental changes in the attitudes of people toward nature and toward each other. Environmentalism that seeks only to put patches on the present socioeconomic system seems to us doomed to failure. It is a system in which losses perpetually outstrip gains. A costly and difficult battle might preserve a small area of seminatural habitat—temporarily—while thirty others are plowed under or paved over. As one species is officially added to the endangered lists in the United States, 10,000 genetically distinct populations perish at home and 10,000 species disappear from the tropics. Old-growth forests are logged and "replaced" with tree farms. Gas taxes are raised slightly, or the CAFE standards of automobile efficiency are improved by a few percent, but the population continues to grow, the car population grows even faster,[25] and more freeways are built or expanded to create more and bigger traffic jams. The threat of a toxic substance like asbestos or lead is recognized, and efforts are made to protect people from it, but little or nothing is done about the flow of thousands of other potentially dangerous novel substances into the environment, including hundreds of new ones daily. A protocol to stop the influx of ozone-destroying, greenhouse-enhancing gases into the atmosphere is signed, but it may be too little and too late. Society debates the reality of global warming and whether it represents a threat. Little significant progress is made either in reducing emissions or making adjustments to possible climate changes.

The environmental movement, especially during the 1980s, has been forced to treat symptoms on an emergency basis, rather than addressing the basic disease—the relationship of people to their planet and to each other. As Peter Berg wrote, "Rescuing the environment has become like running a battle field aid station in a war against a killing machine that operates just beyond reach, and that shifts its ground after each seeming defeat."[26]

The sort of philosophy represented by the "deep ecology" movement may point the way—it is eclectic, yet based on a coherent view of humanity's place in the scheme of things. Deep ecology is partly rooted in oriental philosophy; in the language of its founding philosopher, Arne Naess, it is "simple in means, rich in ends."[27] Its main thrust is to make the humanity-environment relationship less human-centered, to show how artificial is the distinction between people and their environments, and to create a more egalitarian world where empathy for other beings, human and nonhuman, is valued as highly as scientific rationality.[28] The movement is quasi-religious, but it recognizes that a successful new philosophy cannot be based on scientific nonsense. Deep ecologists (and "deep economists")[29] do not shy away from value issues, but they attempt to ground their philosophy in a scientific understanding of how the world works. They recognize that the extreme anthropocentrism of our society is not strongly grounded in either science[30] or philosophy,[31] and has been a major factor in generating the human dilemma.

We are aware that ecologists and environmentalists are often suspected of valuing "fuzzy-wuzzies" above human beings. Let us just say that in addition to having a deep appreciation and respect for the other forms of life with which we share the planet, much of our work is devoted to awakening people to how utterly dependent they are on those "fuzzy-wuzzies." The fate of humanity in general, and our grandchildren in particular, depends on a wider recognition of that fact of life.

A Sustainable United States

How one gets the needed change in the dominant social paradigm is, of course, the critical question, especially in the United States, which has so long suffered from a vacuum of leadership at the top. Questions about the purposes of human life and relationships among people and with nature have been the subject of philosophical discourse for millennia. Now, unless such topics quickly rise to the top of the political agenda, there may soon be no arena of civilization in which they can be discussed.

Obviously, no American legislator could survive today trying to start a debate on "humanity's place in nature." There are, however, other issues that can be raised repeatedly and strongly that bear very closely on this. Issues that we think concerned people should push as subjects for greater debate are: What should be the automobile's place in society? What are the real determinants of national security? What is good about economic growth (at least in the form it currently takes)? Is physical growth of the economy necessary (or desirable) for a postindustrial society with a nongrowing population? What population policy should the United States have?

One of the major mistakes of human beings, especially in the United States, clearly has been to design much of society around the needs of automobiles rather than people. The very notion that large numbers of human beings would be willing to spend up to four hours daily isolated in plastic and metal boxes, traveling through a haze of poisons in order to earn a living, exemplifies how people have become subservient to Berg's "killing machine." Even people who enjoy driving rarely enjoy commuting.

At least three of the tools that could help loosen the automobile's stranglehold on society are already being widely discussed: substantial increases in gas taxes, "feebates," and a return to mass transport. A modest increase in the federal gas tax was enacted in 1990, and California boosted its state gas taxes even farther by a ballot initiative. The federal increase was too small, but it may have opened the door to other important changes in the energy sector.

Feebates would be assessed at the time of automobile purchase. Individuals purchasing efficient cars would get a sizable rebate; those preferring gas-guzzlers could still buy them, but they would pay a hefty fee to reflect the social costs of their actions. At the start of the program, the rebate (or fee) could be based on the efficiency difference between the car traded in and the new car—as an incentive to get the worst gas-guzzlers (which would be scrapped and recycled) off the road fast. Feebates could be a stronger incentive to buy an efficient car than gasoline taxes at levels currently entertained, since fuel costs comprise only about a fifth of the expense of owning and operating a car, and efficient cars are now often higher priced than inefficient ones, making the total costs of owning them about equal to a gas-guzzler despite the lower fuel bills.[32] California has recently instituted feebates, and several other states are considering them, so progress is being made.

This is a very hopeful sign, since the feebate principle could be applied to buildings, appliances, indeed any energy-using device; and feebates (like direct energy taxes) would automatically encourage innovation by rewarding manufacturers of the products with highest efficiency that did not suffer significant drops in performance. Feebates also have the potential advantage that they could be designed to be revenue-neutral, with the fees covering the costs of the rebates, so taxpayers would not be burdened.

While feebates could help to facilitate the needed transition away from dependence on automobiles, ultimately a new and efficient system of mass transport will be needed to replace cars and roads. Mass transportation carries far lower environmental burdens for each passenger mile traveled. The switch would also bring economic benefits. According to one estimate, a $3 billion investment in mass transportation would generate almost 200,000 jobs, which could buffer some of the economic costs and personal dislocations associated with the reductions of automobile use.[33]

Many other ideas need to be injected into the debate. One is a nearly absolute ban on the building of new freeways and (except under extraordinary circumstances) roads. All one need do is visit the Los Angeles basin, Japan, or England to view societies being strangled by highways and roads. The rich countries already have too many roads; more can only hasten the deterioration of other parts of the economy, including agriculture, and exacerbate already horrendous traffic problems.

Redesigning towns and cities to eliminate the need for most commuting is an issue that could generate much deliberation on what people are for and what a decent life is. Imagine a society in which almost all people walked to work through at least seminatural settings. The benefits both in personal health and social appreciation of the natural world would be gigantic. For those who really enjoy a drive in the country in a responsive sportscar, think how much more possible and pleasant that would be if they weren't competing for road space with commuters and shoppers.

Making the complete transition would take a great deal of time and financial investment; but the world was turned over to the automobile in about 50 years, and probably could be largely rescued from it in a similar period, perhaps less. The population would have to be relatively stable in size and social makeup to permit planning. Urban centers could be redeveloped with a mix of cleaned-up industries, commercial businesses, high-rise housing, and parks. New patterns of shopping facilities established (perhaps with scattered mini-malls so that most people were close enough to walk to one. The details need not concern us; the problem is to get society to focus on the task of redesigning the human habitat for the health and convenience of a given number of people rather than an ever-growing number of motor vehicles. Getting the job done will not be easy or cheap, especially if population growth is not halted soon.[34]

▬▬ United States Military Conversion

The "peace dividend" expected to be realized from cutting the United States defense budget has already been claimed by many sectors of society to meet neglected needs. But suppose (to take a utopian example) the United States chose to allocate it to creating a more sustainable society and that the nation cut its military budget by half after 1993. Obviously, such a sharp drop in the military budget would be neither possible or wise in the immediate future, although it certainly could be dramatically reduced within a few years. Sizable portions of the financial and human resources still being pumped into fiscal black holes like stealth bombers and "Star Wars" missile defense systems[35]—programs with little or no military purpose, designed primarily to subsidize segments of industry—should be redirected as rapidly as possible. The partial success of Patriot missiles in defending against Iraqi "Scuds" in the Gulf war is no excuse for draining America's resources into an unworkable system of defenses against nuclear-tipped ICBMs. And the proposal in 1991 to produce a brand-new fighter aircraft with virtually no foreseeable military mission was an act of stupidity that deserves a prompt rejection.

The United States could use two-thirds of the savings annually (roughly $100 billion) to improve its environmental security, including funding its own redevelopment.[36] One obvious area into which to redirect financial resources is energy research, which under the aegis of the Reagan administration dropped from almost 6 billion 1990 dollars in 1980 to less than $3 billion in 1990.[37] Another is the building of mass transport systems and the repair and refurbishing of America's infrastructure—roads, bridges, low-cost housing, water-distribution systems, sewage-treatment plants, and so on. Helping to make buildings across the nation more earthquake-resistant might also pass a risk-benefit test. And the nation might recapture from Japan the lead in high-technology commercial products if so much of the American effort in the area weren't directed toward military programs.

Human resources from the military could also be used to the great benefit of the nation. The nation could convert part of its military forces into a "National Service Corps," employing ex-soldiers, -sailors, and -airmen to revive currently languishing public services or to serve in a national conservation corps, analogous to California's successful state program. Both financial and human resources are desperately needed in America's schools, libraries, hospitals, and even employment agencies. Many people with technical specialties could be quickly retrained as aides for science teachers to help American children catch up with their counterparts in Japan, Germany, and other developed nations. Other ex-service personnel could find work as paramedics, research assistants in scientific laboratories, or in providing day care for children or the elderly, work in large-scale reforestation programs, help replan and rebuild cities and provide homes for the homeless, work in a land-stewardship agency to fight desertification and control off-road vehicles. Young Americans could also provide valuable services abroad in the Peace Corps or other assistance agencies. If the Peace Corps were revitalized and its personnel trained in environmental restoration, it might even surpass its previous popularity.

One important assignment for the military in the next several decades will be to clean up the thousands of toxic (and often radioactive) waste sites at military bases and the Department of Energy's nuclear-weapons-manufacturing facilities.[38] Some military personnel already have the technical expertise to deal with these problems. Others could be trained in a military setting as environmental responsibility is incorporated into the military mission.

There would be no shortage of jobs for patriotic Americans wishing to serve their country and the world, although difficult questions would need to be answered before a program could be put in place. Should the sorts of incentives (job training and educational opportunities) now used to attract people to the military be used to encourage enrollment in a national service corps? Or would a draft be necessary? Should the funds be raised by taxes and awarded to private firms to do some jobs (with the requirement that they hire national service personnel)?

Even the Central Intelligence Agency (CIA) might be turned to peaceful missions. In 1990 it was reported that, with the dissolution of the Soviet threat, the CIA "doesn't know who 'the enemy' is any more."[39] Since the agency has specialized in integrating and synthesizing information, environmental scientist Donella Meadows suggested that its talents be adapted to discovering nature's secrets. The CIA could use its spy satellites to track tropical deforestation, gather intelligence on rates of soil erosion, monitor the size of the Greenland and Antarctic ice sheets, detect illegal whaling operations and clandestine polluters, and watch for violations of international agreements on CFC or CO_2 emissions. The latter monitoring assignments would fit in reasonably well with responsibilities for verifying the observance of arms-control and -reduction treaties, which will surely increase during the 1990s.

■ Environmental Security

The time is now ripe for a renewed discussion of national security. The evolving situation with the Soviet Union makes it clear that, whatever the previous chances of East-West combat, those chances have now been reduced. Increasingly, people have started to understand that national—indeed global—security now depends primarily on *environmental security*. Environmental security includes security of food and water supplies, security of essential energy flows, security of coastlines, security from extreme weather events, and security from exposure to cancer-causing ultraviolet radiation, air and water pollution, and toxic substances.[40]

Environmental security, in fact, has been more important than most people realize over the past two decades. More than ten times as many people have died of hunger and hunger- or poverty-related disease during that period than have died in wars.[41] The importance of security from the threat of war cannot be overestimated, but neither can the importance of security from hunger, thirst, disease, and poisoning. Today there is no credible threat (aside from accidental nuclear war) to the basic military security of the United States. There are many credible and growing threats to its environmental security; indeed, enough American citizens are already hungry or ill and denied basic health care to constitute a national disgrace.

The questioning and reduction of swollen military budgets has already begun in both the Soviet Union (under the pressure of economic near-collapse) and in the United States. The conversion of those budgets, and of defense industries and military personnel to addressing problems of environmental security, must proceed rapidly, despite the "lessons" of the Gulf war as propounded by some right-wing pundits.[42]

A failure to grasp the new dimensions of security was made crystal clear by the Bush administration's military approach to solving the problem of western dependence on petroleum from the Persian Gulf. Even before the massive movement of American armed forces to the Gulf, oil from that area was costing an additional hidden amount of roughly $25 per barrel (beyond the market price), when the expense of routine military deployments to secure supplies was considered.[43] After the transfer of massive ground forces to Saudi Arabia, the hidden price jumped something like an *additional* $4 to $25 per barrel, depending on the assumptions made.[44] The price, of course, escalated much further with the start of the war.[45]

Even after the war began, the administration did not propound a domestic "war for energy efficiency"—even though the money spent on military operations, if properly applied to such measures as distributing efficient light bulbs, subsidizing the retirement of gas-guzzling automobiles from the roads, and encouraging other forms of energy efficiency, could substantially reduce the dependence of the United States on imported oil— especially oil from the Persian Gulf region—in a relatively few years. One can hope that an eventual outcome of the Gulf war will be a strengthened United Nations with the power to intervene multilaterally to suppress resource-environment wars before they get started, although there is little sign of such a development. But sound national resource-environment policies and international efforts to reduce the scale of the human enterprise (and thus the competition for resources and the costs of transnational pollution) would greatly reduce the need for such interventions.

One of the most critical challenges in making the world secure will continue to be the development of poor nations without their having to repeat the mistakes of today's rich nations and destroy Earth's habitability in the process. While all nations need to face this challenge for themselves, some are particularly well placed to do so. For instance, among the rich nations, Australia is blessed with a scientific community with considerable expertise in the field of solar cells and with an abundance of sunlight. Australia might strive to be the first nation to deploy solar-hydrogen technology on a large scale to test whether the technology is practical. If it were (as we and many others believe it would be), Australia could then move quickly to offer the new technology to India, China, and other nations, and to help them deploy it.

While Australia has the potential for leading the world into the solar age, Costa Rica, a developing country, is already setting a fine example in several other areas—especially the preservation of biodiversity. Costa Rica abolished its army in 1947 and subsequently has poured its limited resources into, among other things, schooling, health care, and the establishment of national parks. The proportion of Costa Rica's land in parks and reserves is on the order of three times that set aside in the United States. That nation also has at least established a population policy and made a start in controlling its population growth, something the United States government has scarcely even considered.

All societies must soon begin to debate these issues, to grope for satisfactory tactics to use in solving the human predicament. Leading the social debate in the United States will require moral courage on the part of many. Questioning economic growth as an unalloyed benefit today carries risks for American politicians that far exceed those of attacking our dependence on automobiles or reallocating funds now being wasted on the military. But a social discussion badly needs to be generated on the subject. Does ending physical growth of the economy necessarily mean that *all* kinds of economic growth must end? As one analyst put it, "The economics that impels all modern successful societies is based on two interwoven ideas: that the wealth of the society as a whole should increase, and that each individual in it should also grow progressively richer."[46] As long as wealth is defined in terms of material possessions and economic power, both of those dreams in the long run are impossible; and the sooner that is widely recognized, the better the chance of avoiding the collapse of civilization and being able to fashion a transition to a sustainable world.

To provide the changes in the dominant social paradigm that would permit the transition, a wide-ranging discussion of wants, needs, and wealth should be started. The questions are endless, and of a sort familiar to philosophers; now they have great practical importance. Is a well-educated teacher with an annual income of $35,000 richer or poorer than an uneducated football player who makes $700,000? Is someone who spends four hours a day commuting to and from work in a top-of-the-line BMW and works ten hours a day as CEO of a plastics manufacturing company richer or poorer than a carless individual who walks twenty minutes down a wooded lane to work eight hours as a carpenter building fine furniture? Is someone living on an income of $50,000, with a guaranteed "social net" for his or her entire family, richer or poorer than one making $200,000 but faced with the possibility of impoverishment if the company goes bankrupt? Does one become wealthier trying to accumulate gadgets or friends and lovers?

What size should the future American (and world) population be? Population size will play a substantial role in determining what future life-style options are possible. One hundred million Americans clearly could have more energy slaves per person than 400 million; on a planet with one billion inhabitants, each person could be more than ten times as wealthy as if there were 10 billion inhabitants.[47] Some of our own views are, of course, implicit in the questions, but we are hardly immune to the acquisitiveness that is now so imbedded in this culture.

The very complexity of our society, itself partly a function of large scale, will make generating the needed dialogue difficult. But not impossible. One could see its possible beginnings in the 1990 wrangling over the federal budget. Should rich retirees still get all the benefits of Medicare? How progressive should the tax system be? What would be the costs and benefits of a high tax on gasoline? Should there be surtaxes on luxury items? Is it fair to add more taxes to a poor man's beer? The entire debate was carried out against a background of budget deficits and fiscal constraints—as well as looming deadlines. What is now needed is to upgrade and expand such a discussion, placing it in the wider context of resource deficits and environmental constraints as well as fiscal ones.

In short, people who believe that the United States and the rest of the world cannot escape from the downward spiral of resource-wasting growth and environmental deterioration are simply suffering from a failure of imagination (or paralysis encouraged by a brain-dead leadership). At least in democracies with an informed electorate, there is a cure for brain-dead leadership. But just as democracy seems to be undergoing a renaissance abroad, American voters ironically appear to be rejecting it out of cynicism or indifference. A spirit of responsibility for the future and personal commitment leading to political action are needed now, more than ever before. So we now turn to considering how individuals can help generate the leadership needed at all levels of our society to create that better future.

Influencing Policy

CHAPTER Ten

When people consider what they can do personally to help extricate humanity from its predicament, actions like bicycling, recycling, and eating less meat inevitably come to mind. And well they might. After all, those are exactly the sorts of things recommended in the dozens of books published for Earth Day, such as *50 Simple Things You Can Do to Save The Earth*.[1] Yet few of those books mentioned the single most important direct action a person can take: limit oneself to one or two children. The reason for the latter should now be clear to you, and the other direct steps are abundantly covered in other sources. Instead of reviewing them, we will concentrate on what people can do to improve the human prospect *indirectly*, by influencing the policies of large organizations, be they governments, political parties, corporations, environmental action groups, or the World Bank.

This indirect action is crucial, since the proportion of people who are motivated and able to take individual action is still too small to have much influence on such things as deforestation and society's use of fossil fuels—and those on the wrong side of the battle have both economic clout and social inertia on their side. What you can do to influence policy will naturally depend on your personal preferences and who you are. One of us (guess which) likes to work quietly to shape the policies of environmental organizations; the other prefers to cajole decision makers publicly to mend their ways. Some of our friends and colleagues do things as diverse as work within government or corporations as "moles" to effect change, picket the World Bank to speed its conversion from a world-wrecking to a world-saving organization, write letters to public officials telling them to shape up, or simply support good causes financially. Many of the actions that are available to all citizens are detailed in our earlier book, *The Population Explosion*.[2] There is no "right way" to work for a better future—*chacun à son goût*. Each action is a contribution to a growing global movement.

Who You Are

Obviously, who you are to a large degree shapes your opportunities. Not everyone can work within the government; an economist can potentially have more influence on government policy than a biologist can, if only because politicians are more accustomed to listening to the advice of economists, even though they may not always follow it. Politicians themselves and corporate executives are often in especially powerful positions. But the good news is that *anyone* can have an important influence—concern, effort, and readiness to use an opportunity are all that are required.

Our discussion is divided somewhat according to opportunity—starting at home with a relatively small but potentially influential group, scientists. Scientists have a special responsibility to contribute to getting humanity out of its predicament. First of all, it was science that provided the mechanisms for the lowering of the death rate that was largely responsible for the post–World War II population explosion. High birth rates in poor countries would not have led to enormous population increases if antibiotics and the control of malarial mosquitoes with synthetic pesticides had not dramatically reduced mortality among the young. Even earlier, science provided improved agricultural techniques and basic public-health measures that accelerated population growth in what are now the rich countries. And science provided the other tools that made humanity a global force. Second, *Homo sapiens* is now far too abundant to be able to support itself with prescientific technologies. Neither hunting and gathering nor subsistence farming could supply enough food for even today's 5.4 billion people, let alone the additional 4 to 9 billion increase in population size to which humanity is committed. Great efforts by the scientific community will be required to provide an adequate (let alone abundant) living for 9 to 14 billion people. Since achieving even that is problematic, the scientific community as a whole should be mobilized to alert society to its peril.

So far, the performance of scientists in addressing the predicament has been far from sterling; and many, if not most, scientists remain essentially unengaged. In part, the lack of involvement probably arises from the narrow reductionist approach of most science, an approach that has been both a blessing and a curse of the entire scientific enterprise. Scientists are career-oriented, and their principal rewards come from focusing closely on interesting but limited questions: How do chemical reactions on surfaces affect the chemistry of air pollutants? How can a superhot plasma be contained in a magnetic bottle? How does genetic information interact with the environment to produce an individual organism? Indeed, many scientists were trained to bore in on the "truth," ignore the possible implications of their work for society, and stay out of politics.

This situation has been changing, inspired originally by physicists' moral qualms generated by the use of nuclear energy in weapons. Ecologists and evolutionists increasingly have been matching their research with action. The Ecological Society of America and the American Institute of Biological Sciences are both involved in trying to save biodiversity. Academic biologists at the Rocky Mountain Biological Laboratory raised $20,000 to help buy a critical tract of rainforest in Argentina that was in jeopardy and add it to a preserve. But the change is still not sufficiently fast and widespread. Curiously, even most molecular biologists seem, at the moment, relatively uninterested in doing their share to stem the tide of biotic destruction.

That the crucial problem of the decline of organic diversity is receiving so little attention from scientists is doubtless due to the low status of the sciences that deal wtih biodiversity and its functions: taxonomy (the science of classification), evolutionary biology, and ecology. The central importance of these disciplines, not just in biology but in all science,[3] has been eclipsed by rapid progress in more reductionist disciplines and in a gentlemanly consent to the current misallocation of resources devoted to research. For instance, funding for biomedical research, aimed largely at curing heart disease and cancer (and, most recently, AIDS), has been roughly 100 times that available to support ecosystem health. While substantial success in the direct battles with the diseases of adults could extend life expectancy a few years in rich nations, failure to protect ecosystem health could lead to a shortening of life expectancy by decades—to hundreds of millions or more dead children.[4]

So, if you are a scientist, one of the first things you might wish to do personally is to start lobbying your colleagues and the federal government for more support of the disciplines that deal most directly with environmental issues; and if at all possible, direct your own efforts toward them, even if you are not working in an area normally thought of as environmental. The problems are so broad and pervasive that almost any scientific discipline can be applied to them.[5]

Basically, the scientific community must organize itself to deal with the most pressing set of issues it has ever faced. Especially important are people trained in environmental sciences, and the changes in their professional lives are, or soon will be, more profound than those of other disciplines. Ecologists will find the demand for their expertise increasing rapidly, even though they now struggle to find funds to allow them to carry on research or find jobs where they can employ their knowledge and pass it on to students. Systematists (those who study the classification and evolutionary relationships of organisms) will have to change their entire approach to their discipline.[6] In large degree, they must switch away from attempting to describe organic diversity gradually and thoroughly, and working out the patterns in which it has evolved, toward finding ways instead to evaluate quickly the geography of diversity and to help conservation biologists save the objects of systematists' study. A second important task is to make extensive sample collections for preservation by methods (such as very-low-temperature freezers) that will make them available for detailed biochemical examination in the future.[7] Techniques of molecular systematics are shedding new light on the history of life,[8] and the creation of an artificial fossil record (in the sample collections) could clearly be a boon to future scientists working to salvage ecosystem services on a biotically impoverished planet.

Evolutionists also have their work cut out for them putting the human predicament into an appropriate biological and historical context for laypeople. They can help explain the legacy of genetic and cultural evolution that makes it so difficult for people to recognize the predicament and take action to solve it.[9] They can explain why repeated widespread use of synthetic pesticides and antibiotics is a losing tactic in the war against the organisms that eat or compete with our crops, carry disease to us, or cause disease. Evolutionists can make people aware that dreaded diseases such as AIDS are caused by organisms that themselves are products of evolution and will be subject to further evolution—perhaps to the great detriment of our species. Above all, evolutionists can explain how *Homo sapiens* fits into nearly 4 billion years of evolution. They can put people in touch with their own humanity and their place in the universe, and in the process stimulate thought on the purposes of the entire human enterprise and its relationship to our planet, to our only known companions in the universe, and to our descendants.

Great challenges also face individuals in management sciences. For example, much of the ecological destruction in the world's forests, especially in rich nations such as the United States, Canada, Australia, and Sweden, can be laid at the feet of forestry scientists who have been narrowly trained and generally believe that the only good forests consist of even-aged stands of identical trees, preferably exotic pines or eucalyptus, planted in straight rows and harvested regularly as a crop. They are not taught that such stands are, by comparison to mixed stands of native trees, biological deserts that support little biodiversity and, as a result, may be difficult to maintain in the long run. Many apparently are even unaware of the benefits of old-growth stands, not just for the preservation of other organisms and recreational value, but as reservoirs of genetic diversity that may be essential to the long-term health of the local timbering industry. Things are gradually changing in the better forestry schools, but we cannot afford to wait until they are all modernized and an entire new generation of foresters is trained. Enlightened working foresters themselves must initiate many changes in policy, even at the risk of strong opposition from their conservative colleagues and the timber industries that in many nations have strong influence over forestry practices. Some are already doing it.

Entomologists (scientists who study insects, spiders, and mites) have a special double role to play in the human predicament. Since their domain covers the vast majority of biodiversity, their first task is to educate the public about the importance of the "little things that run the world."[10] The second role derives from the still much-too-prevalent misuse of powerful poisons in insect-control programs. This misuse is due both to a long history of deficient training in ecology and evolutionary biology in many entomology departments, and pressure from the petrochemical industry to maximize insecticide use (and long acquiescence from the United States Department of Agriculture and land-grant universities).[11] The situation is changing, though. Entomologists have, happily, been at the forefront of designing systems of integrated pest management (IPM) that work with the natural pest-control functions of ecosystems. These systems have been enormously successful, even in poor nations.[12] Much greater efforts are needed, however, for a universal conversion to systems that employ insecticides sparingly when needed, as a scalpel rather than as a sledgehammer at all times. Individual entomologists need to speak out strongly on these issues within their institutions and professional organizations, even at the risk of some professional penalties such as have plagued others who attempted to rationalize the pest-control system, from Rachel Carson to Robert van den Bosch.

No challenge is likely to be greater in coming decades than producing and distributing adequate food supplies for the fast-growing human population. Not only entomologists, but all agricultural scientists will be deeply interested in increasing production. Here again ecological-evolutionary knowledge must be applied to avoid a substantial lowering of Earth's carrying capacity for humanity by long-term damage from an inappropriate intensification of agriculture. Climate change, loss of biodiversity (including genetically distinct varieties of crops and livestock), evolution of pesticide-resistant pests and disease vectors are among the consequences of pushing to maximize production without adequate attention to ecological side effects. The loss of carrying capacity, combined with the increased need for resources generated by population growth alone, eventually may cause more misery than increased food supplies can alleviate.

The future of increased food production will lie in new directions taken by agricultural scientists. While there still is room for increasing yields of many crops—the traditional strategy—the most fruitful next steps are likely to emphasize an ecological approach. Agricultural scientists, long largely isolated from ecology and evolutionary biology, need to establish more collaborative partnerships with scientists in those fields and begin developing new cultivation systems for different soil-climate regimes.

Biologists from many disciplines will necessarily become involved in a global effort to rehabilitate degraded natural ecosystems. So much of the planet is occupied by ecosystems that are mere shadows of their former selves that restoration (as opposed to simple preservation) is bound to become increasingly important. Efforts will range from large-scale operations such as Dan Janzen's attempt to restore the Guanacaste dry forest in Costa Rica to restoring small habitats in urban areas to make them once again capable of supporting certain butterflies—signs of returning environmental health.[13]

The world has long since passed the point where even 5.4 billion people could be supported at a reasonable standard of living, even for a few decades, without the support of new, environmentally benign technologies—and those technologies must rest on a foundation of sound basic science. The needs are endless, and society's support of science should increase once they are understood. Physicists, for instance, are needed to seek ecologically safer ways of mobilizing energy, help develop better public understanding of the dynamics of the atmosphere and oceans, and explore such potential bonanzas as high-temperature superconductivity. Earth scientists are needed to help with that chore, to find better ways of dealing with changes in sea level, and to continue uncovering the ways in which valuable substances are concentrated in Earth's crust.

The struggle to understand the sources, sinks, and interactions of substances in the atmosphere that contribute to global warming as well as cause other dangerous problems such as ozone depletion and acid deposition, will require collaboration among atmospheric chemists, climatologists, geologists, and biologists, among others. Chemists are needed to design new industrial and agricultural compounds that are less toxic, shorter-lived, or both. Molecular biologists, properly supported, can contribute by producing crops that need less water or fertilizer, tolerate high-salt or low-nutrient conditions, and are more resistant to pests. Perhaps more important, they may be able to develop mechanisms to protect humanity from AIDS and its inevitable successors, and they can contribute to the search for safer, more dependable and effective contraceptives.

While we have often been critical of the allocation of resources in science and the performance of scientists in dealing with the human predicament,[14] the baby must not be thrown out with the bathwater. Scientific expertise will be needed in the coming decades as it has never been needed before, and the scientific community will need unprecedented support for its research. Some of that research will necessarily be targeted to solve immediate social problems, as has been the research support going into the battle against AIDS. But basic scientific advances, which usually flow from the creative insights of individual minds, will be essential to lay groundwork for meeting the challenges of a difficult future and must be strongly supported.

The science-society interface is bound to remain fraught with misunderstanding until the majority of scientists recognize the need to explain their work to the general public. Even those working on the most esoteric "pure" research problems should be able to make the case that understanding the complexities of the universe has intrinsic benefits for humanity. At the same time, the public must accept a responsibility to see that everyone gets an appreciation of science as part of his or her basic education. A scientifically illiterate public cannot even sensibly exercise broad control over the scientific community it supports, let alone understand the scientific aspects of everyday life and society's problems. Science, mathematics, and technology now compose perhaps half of our culture—that is, of the body of nongenetic information that *Homo sapiens* builds, stores, and transmits. That those subjects are often neglected or poorly taught is a central problem of education, especially in the United States.

Engineers, a group usually counted on to apply the advances of the scientific community for the good of society, could be at the cutting edge of solving the human predicament. Some already are, working on problems as diverse as designing solar energy systems or better constructed, more durable and energy-efficient buildings, light bulbs, and appliances, to finding ways to remove toxic substances from aquifers.

However, like many scientists, engineers often have a tendency to focus narrowly that has served them well professionally, but is no longer adaptive for themselves or for society. All too often they practice "suboptimization"—doing in the best way possible something that should never be done at all.

Examples of suboptimization are myriad. One is designing new roads to carry an expected increase in automobiles in places like England and California where motor vehicles are already strangling societies, or designing better "muscle cars" to run on those roads, or, worse yet, off-road vehicles to run off them. Others are improving the design of throwaway plastic products, whether "biodegradable" or not, or designing ever more clever and destructive weapons.

On the other hand, many engineers are already helping to heal Earth. They design sewage-treatment plants and other waste-disposal facilities, design earthquake-resistant and energy-efficient buildings, and help biologists with the physical aspects of environmental restoration projects.[15] The scope for engineering contributions to the global restoration effort is enormous. One example is the development of soil "imprinters," motorized devices that force angular teeth into the soil surface of desertified areas to change the soil structure, forming "funnels" that allow rainwater to infiltrate rapidly during intense storms rather than to simply run off the hardened surface.[16] The funnels also collect seeds, litter, and topsoil, and provide protected sites for seedlings, aiding revegetation.[17] Considering the scale of the desertification problem, such engineering innovations clearly could be vastly beneficial. Restoration engineering is a profession whose time is coming.[18]

Like scientists, all engineers must consider the ecological and social consequences of their own acts and those of their profession. Some may be willing to help make their fields obsolete or even to change fields. Petroleum engineers could proclaim the temporary nature of society's dependence on oil, emphasize the need for energy efficiency and for exploring solar options, and even switch to solar engineering themselves. There will be plenty of jobs for engineers in putting together a sustainable world; they would have everything to gain by helping to push society to move in that direction.

One of the most cheering trends today is the growing interest of the medical community in ecosystem health. In 1988 the World Health Organization (WHO) acknowledged for the first time the problem of a burgeoning human population destroying its own life-support systems.[19] A British physician, Maurice King, writing in *The Lancet*, pointed out that the WHO report "neglects . . . the contribution to planetary ill health made by the industrial one-fifth of the world, which makes greater demands on the global ecosystem than do the remaining four-fifths."[20] King discussed the thorny moral problems confronting those trying to develop public-health policies for poor nations where attempts to reduce infant mortality without complementary efforts toward family planning and developing sustainable agriculture "increase the man years of human misery." We hope King's provocative views, and similar ones voiced by R. Gordon Booth ("It all comes back to one plain fact; *there are too many people*"),[21] will stimulate new dialogue within the medical community and help that community lead the way in tying together in the public mind human and ecosystem health.

That the medical community has special responsibilities in this area has been made explicit by two American internists, Michael McCally and Christine Cassel: "Concern for the health of the general public has been an important responsibility of the profession since the industrial revolution."[22] They point out that "global environmental change will cause illness on a massive scale," and urge physicians to inject their health expertise into the debate on global change. They urge that more physicians join environmental organizations to become informed on the issues and then support appropriate environmental legislation.

Physicians hold a special place in American culture, and are thus in an ideal position to educate people on the risks of environmental degradation and to counsel them on how to protect their own environmental health (including counseling them on how to avoid having more than two children). The organization Physicians for Social Responsibility, which was founded in response to the threat of nuclear war, can be expected to press for programs in environmental health. Even the rather conservative American Medical Association has urged that the role of physicians in environmental education be expanded.[23] So the groundwork has already been laid for the full-scale participation of a key group of professionals in healing Earth.

It is the social scientists, though, whose personal efforts are going to be most essential if civilization is to endure. At center stage will be the economists, some of whom are beginning to grasp both the depth of the crisis facing humanity and the crucial role that their discipline and expertise must play in solving it. Economists are burdened with a set of models derived from an era when the scale of the human enterprise was small enough relative to the biosphere that relegating environmental impacts to the realm of "externalities" was not a major problem. The problem is that in the context of present-day reality, these externalities represent, as one economist put it, "the capacity of the earth to support life."[24] Yet because they are external to the models (and because appreciation of their severity requires some knowledge of other sciences) these externalities are rarely given the weight they deserve.

At the same time, economists know more than do people in other disciplines about how to build models of the world that fit the realities of the late twentieth century. As the discussion of a carbon tax and tradable pollution permits demonstrated, economists can come up with good ideas for internalizing at least some of the environmental externalities that are amenable to inclusion in the market system. They have also explored difficult issues such as how to deal with irreversibility,[25] the depletion of natural capital,[26] the optimal use of renewable resources,[27] pollution control in the presence of uncertainty,[28] and limiting the total throughput of the economy.[29]

Recent work by Robert Repetto of the World Resources Institute on how national accounts deal with natural-resource depletion is especially interesting. Repetto recalculated the growth of the Indonesian economy from 1970 to 1984, taking into account loss of forests, erosion of soils, and pumping of oil. He found that instead of growing 7.1 percent annually, as suggested by standard economic indicators, net growth was only 4.0 percent. In addition, if depreciation of other natural capital such as fossil fuels, metals, and fisheries had been taken into consideration, the real growth rate would be even lower. Repetto concluded that nations like Indonesia that are deeply dependent on the exploitation of nonrenewable natural resources (as are most poor nations) must diversify their economies if they are to develop successfully. In his words, "important measures of economic performance such as income and productivity growth, capital formation, and savings, can be badly distorted by not recognizing natural resource assets as a form of capital."

Much more new thinking is required, however. So much of current macroeconomic thought is based implicitly on notions of perpetual growth and infinite substitution that vast opportunities exist for economists to rework their discipline into one that could make enormous contributions to the design of a sustainable society. No other social science has so great an opportunity, is as well prepared with analytic tools to take advantage of it, and is as likely to be listened to by decision makers.

Economists must also begin to go public on issues that they understand very well but that politicians and the general public often ignore or fail to grasp. A good example is the utter inadequacy of GNP as an indicator of the state of society. Economists should start holding public forums on the complexities of defining growth, what actually is growing, and issues relative to an end of growth (as normally defined). When distinguished economists start saying in public that growth in GNP (or in the physical aspects of the economy) is not necessarily a desirable goal in rich countries, and when they start pointing out repeatedly that a sound economy is utterly dependent on sound ecosystems to support it, they will help legitimate the changes that politicians need to start making.

Economists must also do more to inform the public about the value-neutrality that pervades their discipline (and other sciences). They are specialists at analyzing what *will* happen under given circumstances, but generally eschew voicing opinions on what *should* happen. They must carefully explain that economics generally does not speak to that.

A mission for psychologists would be educating the public on how the human nervous system evolved and how the outcome of that process shapes everyone's perceptions in ways that make it difficult to come to grips with the human predicament.[30] Being aware of our natural handicaps should make it easier to get around them and deal with the real problems. The variety of contributions that psychologists can make will be obvious to them, from helping to negotiate some of the necessary social dislocations to finding methods to ease the psychological traumas that may be caused by them—for example, by designing retraining programs for people phased out of coal mining, logging, or automobile manufacturing.

Other social scientists have major contributions to make as well, since the better society is understood, the better are the chances that it can be changed so as to make it sustainable over the long haul. The breakup of the communist empire and the dismal failure of the Soviet regime to satisfy human wants *or* to protect the environment suggest that one type of political/

social organization can be dismissed as a path to a sustainable world. Uncritical faith in the market to solve all problems, however, in light of the global externalities that must be dealt with, is alarming to environmentalists—even those who recognize that market mechanisms accomplish more social good in many, if not most, situations than central control. A new discussion is very much in order about how human societies can reasonably be organized to preserve democracy, provide opportunities for individual expression and achieving success (but perhaps not with conspicuous consumption as the ultimate criterion), while maintaining sustainability as the central goal of society. It would be very interesting to conduct this discussion with Soviet and Eastern European social scientists participating. Such a discussion has yet to be broached, though, let alone reach the spotlight.

A critical role must be played by journalists, especially those connected with the electronic media. It would be nice if the world had the luxury of waiting for a generation to be educated in school systems to grasp the changes that are needed. But we do not; unless society changes course sharply in the next decade, we see very little chance of a satisfactory denouement to the human predicament. A great deal of public education must be done through newspapers, magazines, radio, and television. A start has been made, from *Time* magazine's recently improved coverage of environmental issues to the introduction of these issues into the story lines of TV shows. Increasing numbers of people in the entertainment business are following Robert Redford's lead and making working for environmental sanity a major commitment.[31]

Business executives, government bureaucrats, judges, and politicians are, more than most, on the front lines of environmental issues—whether they want to be or not. Business executives who inform themselves (and those in their employ) of the dimensions of the human predicament will be doing themselves and their firms a great service.[32] Leaders are needed for such difficult tasks as designing global institutions that can assure that market economies operate so as to internalize major externalities and provide for some reasonable level of equity in the distribution of goods. Discussion of the nature of such institutions has occurred sporadically for a long time, but has begun to take on new urgency as the realities of global change start to sink in.[33]

In our view, in the near future, global institutions (such as the UN and many of its agencies) could usefully provide a forum for discussing how to modify the market system to deal properly with externalities and how to adjust the terms of international trade. Both of these are central issues to growth management and sustainable development in poor nations. These discussions most likely will occur against a backdrop of global political change, of which the recent ethnic struggles in newly liberated eastern Europe are only the latest prominent manifestation. There is a worldwide trend toward separatism and local control by relatively small ethnic and religious (or "tribal" or "peasant") groups.[34] Whether they be Native Americans, Soviet Lithuanians or Georgians, Sikhs in India, Kurds in Iraq, Catholics in Northern Ireland, French Canadians in Quebec, the Basques in Spain, or the Flemish in Belgium, people are demanding more local control over their lives and resources. This could be a useful trend if politicians can steer it into a system in which environmental costs and benefits are more equitably distributed; it could be an utter disaster if it led to isolationism and trade barriers that prevented the cooperation among groups that will be necessary for solving global environmental and economic problems.

We could continue the litany of tasks for those with special expertise or position, but obviously almost anyone with motivation and virtually any kind of expertise can do something, even if the connection between what they know and the human predicament may seem tenuous at first. Attorneys can educate themselves on population-environment issues and then donate some of their time to efforts such as protecting the access of women to abortions or finding new ways of using the law to protect biodiversity.[35] English teachers can assign readings by Garrett Hardin, Aldo Leopold, Ed McGaa, Wallace Stegner, Peter Steinhart, or other writers knowledgeable on environmental issues to classes for analysis (many already are using examples from environmental literature). Social studies teachers can point out why rich nations must be considered overpopulated and emphasize the importance of the poor and minorities getting involved in environmental issues—since they will suffer most from rising food prices and generally are forced to live in the most polluted areas. Physicians can join Physicians for Social Responsibility, become more deeply involved in issues of the allocation of health care and the protection of society against novel epidemics, and follow the lead of pioneering colleagues in using their unique position in society to encourage people to tackle the problems of ecosystem health. Airline pilots can tell their passengers about the overdraft on the Ogallala aquifer as they fly their passengers over the characteristic circular irrigation patterns on the Great Plains. TV weatherpersons could explain that regional record hot spells or cold snaps do not speak for or against the question of global warming—which is an average phenomenon of worldwide climate, not an attribute of local weather.

Parents can work through their local PTA and school board to get more coverage in curricula of the human predicament. And everyone can contribute mightily by becoming determined letter writers. Politicians are much more susceptible to the pressures generated by thoughtful letters, and by persistent demands for replies and action, than most citizens realize.

When writing to a legislator or other government official, be brief and specific, and if possible use professional or business letterhead. Don't say something like "Senator Sloe, I want you to work harder to protect biodiversity." Instead, make a specific request such as "I want you to become a cosponsor and active supporter of Senator Smart's bill S. 10XX, which greatly strengthens the Endangered Ecosystems Act, and to let me know what initiatives you are promoting to deal with the erosion of biodiversity outside the United States." If Senator Sloe's staff writes you a letter for him to sign that basically says, "Thank you for your views," write back and say, "Senator, I'm not satisfied with your vacuous response—I want to know what you are doing about one of the most critical elements of our national security."

If you don't get a satisfactory reply after a few letters, try calling Senator Sloe's office and asking to speak to the member of his staff who deals with environmental issues. If you want to be a real wiseass, ask for the staffer who deals with security issues and then act surprised when he or she claims that biodiversity isn't in his or her bailiwick. As a last resort, write Sloe and explain the steps you and your friends are going to take to return him to his private law practice if he doesn't get moving fast on the biodiversity issue—or whatever critical environmental issue you are writing about. Generally, such tactics will be most effective if you address your questions to legislators who are on the committee appropriate to solving the problems involved.

If possible, get to know your representative and senators personally. They all hold regular meetings in their districts, and they usually are willing to meet with you at least briefly if you travel to Washington. Write or phone in advance for an appointment and have a specific agenda to discuss; go with a few other constituents if possible. Don't expect a great deal of time; congresspeople are busy.

Finally, everyone can do those things that we've largely ignored in this book because they are so well covered in others.[36] Those include recycling, eating less meat, biking or walking instead of driving, conserving water and electricity, and all the other things that help each of us tread more lightly on the planet. And, of course, all parents should see to it that their one or two children are thoroughly educated on environmental issues, if not in school, then at home.

Whatever your role in society, your contributions are needed. Having read this far in the book, you obviously are willing to spend part of your time becoming informed about the human predicament. Once you are informed, the pervasiveness of the predicament puts you in a position to do something about it. We think everyone should donate at least 10 percent of his or her time to learning about the world and acting on the knowledge—as we said in the Preface, to "tithe" their time to society. Politicians and other "leaders" aren't going to get the job done for us—especially if we don't communicate clear instructions. We have to take responsibility ourselves for a far-reaching transformation of our society. We wish you luck; all of us will need it.

Epilogue

All of us want our grandchildren to grow up in a world that is not teetering on the edge of natural disaster or war. We'd like their world to be environmentally secure—where they are assured of freedom, education, plenty of food, and the other good things and opportunities that most of our generation has enjoyed in the rich countries. And we think their best chance of living that kind of life depends heavily on providing some approximation of it for virtually everybody in the world. We believe it will be difficult for people to be happy if more and more other human beings are suffering, even if they can somehow insulate themselves physically from the suffering. We also believe that it will be virtually impossible to insulate ourselves, since serious deterioration in developing regions or the old Soviet bloc could well lead to economic and ecological disasters—or even more major wars. The conflict in the Persian Gulf could all too easily be a harbinger of many bloody global battles over dwindling resources and deteriorating environments. We hope that our grandchildren and yours would rather try to help others, directly or indirectly, than add fuel to struggles that can only wound humanity and Earth further.

To survive the next half-century with some kind of civilization intact, the worst outcomes of the now-unfolding human predicament must be prevented, and that isn't going to be easy. But neither is it hopeless. The rich will have to make life-style changes, but that doesn't mean joining the homeless under bridges and in parks. There are, in fact, many rays of hope and opportunity for rich and poor. If humanity can change course in time, the outlook could be decidedly bright. Yes, a social transformation is required, but we know that societies can be transformed in remarkably short periods. When the time is right, a dominant social paradigm can change with surprising speed.

If you had told us in 1947 that by 1990 many public officials in the South, including dozens of mayors, congresspeople, and the governor of Virginia, as well as many of the highest paid performers and sports stars in the United States, would be black, we would have thought the prediction crazy. In 1947, schools over much of the South were segregated, blacks were often *de facto* denied the right to vote, lynchings still occurred (and would persist into the 1960s), and there were great arguments over whether or not Jackie Robinson could possibly survive as the first major-league black baseball player.

While the United States today still has very far to go in improving the situation of blacks and other minorities, a tremendous revolution was wrought in race relations between about 1957 and 1967. In a single decade, attitudes that had persisted for centuries changed dramatically in a large segment of the population. Similarly, the startling decline in American birth rates in the early 1970s to below replacement level confounded the predictions of social scientists. Most recently, and most dramatically, the dissolution of the Soviet empire in Eastern Europe caught virtually everyone by surprise.

The recent admission of communism's abject failure by the Soviet Union and its former satellites in Eastern Europe makes it clear that societies can face up to their mistakes. Now we must admit that the communist and capitalist systems have jointly brought the world to the brink of ecological disaster. That imminent disaster, and the related problems of widespread poverty, widening gaps between rich and poor, racism, sexism, religious prejudice, xenophobia, and the use of war to maintain access to resources now stand as challenges to the long-term success of democracies with market-based economies—so far the most successful political and economic systems human beings have ever devised. They are challenges that soon, we fear, may be perceived as being far more dangerous than that mounted by the Soviet bloc at the height of its powers, a challenge that may be exacerbated if the newly freed peoples of that bloc move rapidly to emulate the mistakes made in the West.

Humanity faces the task of evolving a democratic market

system, with accompanying national and international regulatory mechanisms, in which negative externalities are minor and under whose influence a worldwide social safety net and stringent environmental safeguards can be created. Above all, billions of poor people must be given the opportunity for a decent life. If the market system fails to accomplish that, capitalist democracies will collapse as surely as the bureacracy-ridden, dictatorial centrally planned economies of the East did. Their failure does not automatically assure the longevity of free societies.[1]

Indeed, the only path we can see that might assure that longevity lies toward the Holdren scenario and beyond. Population growth must be halted and a population decline initiated as soon as possible. Success there will make success in the other endeavors much easier. Everyone will need to become much more careful and efficient in using Earth's bounty, and maintaining the integrity of many extensive natural ecosystems should become a top priority of all governmental bodies. As the scale of the human enterprise is reduced, humanity must try to develop a symbiotic relationship with Earth as it heals. Each of us must recognize that we are part of nature and our survival depends on survival of the biosphere. Any attempt to carry on as a conqueror will lead straight to defeat—and on a very grand scale.

It is an exciting time to be a human being, a member of a species playing the ultimate game of "chicken." By the time today's children are parents, it should be reasonably clear whether humanity will turn aside in time or continue headlong toward civilization's final collision. You can help make that decision.

NOTES

PREFACE

[1] For a discussion of the lack of connection between population size and military capability, see P. Ehrlich and A. Ehrlich, 1990, *The Population Explosion*, Simon & Schuster, New York, p. 170.

[2] A. Lovins and L. Lovins, 1990, "Make Fuel Efficiency Our Gulf Strategy," *New York Times*, 3 December.

[3] B. Barber, 1991, "The Gulf Conflict Is Also Deadly for the Environment," *Washington Post Weekly*, 7–13 January.

[4] A. Leopold, 1953, *Round River*, Oxford University Press, New York (quoted from *A Sand Country Almanac with Essays on Conservation from Round River*, Ballantine Books, 1970, p. 197).

INTRODUCTION

[1] Some of the material in this introduction was first published in *Buzzworm*, March, 1990, under the title "Earth Day: Are We Saving the Planet?"

[2] The EarthWorks Group, 1990, Earthworks Press, Berkeley, Calif.

[3] Although most Earth Day publications did not give adequate attention to the population dimension, there were exceptions. These included B. Anderson, 1990, *Ecologue*, Prentice Hall, New York; D. MacEachern, 1990, *Save Our Planet*, Dell, New York; and W. Steger and J. Bowermaster, 1990, *Saving the Earth: A Citizens' Guide to Environmental Action*, Knopf, New York.

[4] S. Olshansky, B. Carnes, C. Cassel, 1990, "In Search of Methuselah:

Estimating the Upper Limits to Human Longevity," *Science* vol. 250, pp. 634–640. Eliminating ischemic heart disease would have a similar result. Cancer causes a little over a fifth of all deaths in the United States, and heart disease a quarter. Eliminating both would extend life expectancy at birth by only seven years. The reason for these seemingly small gains, of course, is that most of the mortality caused by cancer and heart disease occurs late in life.

[5] See the discussion in chapter 7 of P. Ehrlich and A. Ehrlich, 1990, *The Population Explosion*, Simon & Schuster, New York. Malnutrition depresses immune responses, and disease is easily spread in dense populations, so people in many poor countries are very vulnerable.

[6] References to statistics on population here and elsewhere in this book can be found in Ehrlich and Ehrlich, 1990, *The Population Explosion*. Details on the demographic situation can be found there; we don't discuss the population situation in depth in this book.

[7] One was demographer Dudley Kirk (now retired) of the Hoover Institute at Stanford University.

[8] Oak Ridge National Laboratory, 1968, *Nuclear Energy Centers, Industrial and Agro-industrial Complexes*, Summary Report, ORNL-4291, July. For a general discussion of the projected technological advances of the time, see P. Ehrlich and J. Holdren, 1969, "Population and Panaceas: A Technological Perspective," *BioScience*, vol. 19, pp. 1065–1070.

[9] Norman Myers (ed.), *Gaia: An Atlas of Planet Management* (Doubleday, New York, 1984). See also World Bank, 1990, *World Development Report 1990: Poverty*, Oxford University Press, New York; and World Resources Institute (WRI) and International Institute for Environment and Development (IIED), 1989, *World Resources 1988–89*, Basic Books, New York.

[10] UN report, *Children and the Environment*, quoted in *San Francisco Chronicle*, July 20, 1990. See also United Nations Children's Fund (UNICEF), 1990, *State of the World's Children*, U.N., New York; and L. Timberlake and L. Thomas, 1990, *When the Bough Breaks*, Earthscan, London; World Bank, 1990.

[11] *Washington Post*, 19 November 1988, p. C-15.

[12] B. Wattenberg, 1987, *The Birth Dearth*, Pharos Books, New York; J. Simon and H. Kahn (ed.), 1984, *The Resourceful Earth*, Basil Blackwell, Oxford and New York.

[13] The original formulation can be found in P. Ehrlich and J. Holdren, 1971, "The Impact of Population Growth," *Science*, vol. 171, pp. 1212–1217. A more recent discussion can be found in Ehrlich and Ehrlich, 1990.

[14] AT is, of course, simply per-capita environmental impact, which reduces

the total impact equation to: impact equals population size times per-capita impact.

[15] In terms of commercial (or industrial) energy use per capita (the statistic that is readily available), the figure is 150 times; but probably well over half the actual energy use in Bangladesh is from traditional sources: locally gathered wood, crop residues, or dung. Thus a more realistic comparison is probably about 50 times. The question of how to relate commercial and traditional energy use to environmental degradation is examined in more detail in Chapter Two.

[16] Again, these are very rough estimates corrected from commercial energy statistics that overstate the actual differences between very rich and very poor countries. Commercial energy-use statistics give ratios of roughly 280 (Uganda and Laos), 40 (India), and 14 (China). For the comparisons of industrial countries that follow, commercial energy statistics are relatively accurate reflections of differences.

[17] These limits are based in physical theory as well as practicality. See J. Fremlin, 1964, "How Many People Can the World Support?" *New Scientist*, 21 October.

[18] J. Houghton, G. Jenkins, and J. Ephraums (eds.), 1990, *Climate Change; the IPCC Scientific Assessment* (Report of the Intergovernmental Panel on Climate Change [IPCC], World Meteorological Organization [WMO] and United Nations Environmental Programme [UNEP], Geneva), Cambridge University Press, Cambridge; J. Leggett (ed.), 1990, *Global Warming: The Greenpeace Report*, Oxford University Press, New York.

[19] L. Brown et al., 1990, *State of the World 1990*, W.W. Norton, New York. See L. Brown et al., 1991, *State of the World 1991*, W.W. Norton, New York, for an update.

[20] P. Ehrlich and A. Ehrlich, 1989, "How the Rich Can Save the Poor and Themselves: Lessons from the Global Warming," in S. Gupta and R. Pachauri (eds.), *Proceedings of the International Conference on Global Warming and Climate Change: Perspectives from Developing Countries*, Tata Energy Research Institute, New Delhi, 21–23 February, pp. 287–294.

CHAPTER ONE

[1] E. Wilson, 1989, "Threats to Biodiversity," *Scientific American*, September, p. 108.

[2] For details about the atmosphere, see P.R. Ehrlich, A.H. Ehrlich, and J.P. Holdren, 1977, *Ecoscience: Population, Resources, Environment*, W.H. Freeman, San Francisco, chapter 2.

[3] We are supporters of conversion to the Celsius (sometimes called centigrade) scale, but unfortunately even many educated readers in the United States are unfamiliar with it. We therefore use the Fahrenheit equivalents from here on. We also prefer the metric system, but to make magnitudes easily understandable to today's lay readers, we give them primarily in English units.

[4] The idea is that the glass in a greenhouse similarly reabsorbs long-wavelength radiation from the interior of the greenhouse and reradiates some of it inward. But the name is something of a misnomer since most of the warming in greenhouses is due to the prevention of breezes from cooling the interior.

[5] The system probably is also "chaotic," making it even more difficult to analyze. For a fascinating layperson's view of the role of chaos in science, see J. Gleick, 1987, *Chaos: Making a New Science*, Viking, New York.

[6] Technically, solar radiation between the wavelengths of 0.23 and 0.32 microns.

[7] R. Worrest and L. Grant, 1989, "Effects of Ultraviolet-B Radiation on Terrestrial Plants and Marine Organisms," pp. 197–206 in R. Jones and T. Wigley, *Ozone Depletion*, Wiley, New York.

[8] See P. Ehrlich and A. Ehrlich, 1981, *Extinction: The Causes and Consequences of the Disappearance of Species*, Random House, New York; and P. Ehrlich and H. Mooney, 1983, "Extinction, Substitution, and Ecosystem Services," *BioScience*, vol. 33, pp. 248–254.

[9] J. Lovelock, 1988, *The Ages of Gaia*, Norton, New York, p. 8.

[10] S. Schneider, 1990. "Debating Gaia," *Environment*, vol. 32, no. 4, pp. 5–32; S. Schneider [ed], 1991, *Science of Gaia*, M.I.T. Press, Cambridge, Mass., in press; J. Kirchner, 1991, "The Gaia Hypotheses: Are They Testable? Are They Useful?" in Schneider, *Science of Gaia*.

[11] This is shorthand for a lot of evolutionary theory. For a summary, see P. Ehrlich, 1986, *The Machinery of Nature*, Simon & Schuster, New York. If you are interested in details on how the process works, see D. Futuyma, 1986, *Evolutionary Biology*, 2nd ed., Sinauer Associates, Sunderland, Mass.

[12] S. Schneider and R. Londer, 1984, *The Coevolution of Climate and Life*, Sierra Club Books, San Francisco.

[13] For technical references on recycling of water in the Amazon basin and for an excellent general overview of the role of tropical forests in human affairs, see N. Myers, 1984, *The Primary Source: Tropical Forests and Our Future*, Norton, New York.

[14] T. Lovejoy, 1985, "Amazonia, People and Today," pp. 328–336, in G. Prance and T. Lovejoy, 1985, *Key Environments: Amazonia*, Pergamon, Oxford.

[15] See J. Hughes, 1975, *Ecology in Ancient Civilizations*, University of New Mexico Press, Albuquerque. The thesis in this book, that the ancient world destroyed its own environment and caused its own downfall, is controversial. It is difficult, with the available information, to sort out the roles of political factors and climatic change from deforestation, overgrazing, erosion, and the like. But the latter factors undoubtedly played a role in the decline of the civilizations. See also the excellent D. Hillel, 1991, *Out of the Earth: Civilization and the Life of the Soil*, Free Press, New York.

[16] For more details on soils and technical citations on what follows, see Ehrlich, Ehrlich, and Holdren, 1977, *Ecoscience*.

[17] Ehrlich, 1986, *The Machinery of Nature*.

[18] This discussion of the carbon and nitrogen cycles is necessarily greatly simplified. For more details, see Ehrlich, Ehrlich, and Holdren, 1977.

[19] P. DeBach, 1974, *Biological Control of Natural Enemies*, Cambridge University Press, London.

[20] For a simple explanation of the reasons and of coevolution (discussed below), see Ehrlich, 1986.

[21] R. Roush and B. Tabashnik (eds.), 1990, *Pesticide Resistance in Arthropods*, Chapman and Hall, New York.

[22] R. van den Bosch and P. Messinger, 1973, *Biological Control*, Intext Press, New York.

[23] In 1957, P.R.E. wrote to Secretary of Agriculture Ezra Taft Benson, at the behest of our colleague E.O. Wilson, protesting the first of the Department of Agriculture's ill-conceived plans to "eradicate" the fire ant. See P. Ehrlich and A. Ehrlich, 1972, *Population, Resources, Environment*, 2nd ed., W.H. Freeman, San Francisco, pp. 470–473, for the text of that letter.

[24] P. Lewis, 1990, "Mighty Fire Ants March Out of the South," *New York Times*, 24 July.

[25] For details on the benefits humanity has received and could obtain in the future from the genetic library, see Ehrlich and Ehrlich, 1981, *Extinction;* N. Myers, 1979, *The Sinking Ark*, Pergamon, New York, and 1983, *A Wealth of Wild Species*, Westview Press, Boulder.

[26] P. Waring and A. Müllbacher, 1990, "Fungal Warfare in the Medicine Chest," *New Scientist*, 27 October, pp. 41–44.

[27] Of course, if no more were being produced, photosynthesis would have ceased and we would all promptly starve to death.

[28] Ehrlich and Mooney, 1983.

[29] National Research Council, Committee on the Role of Alternative Farming Methods in Modern Production Agriculture, Board on Agriculture (J. Pesek,

Chairman), 1989, *Alternative Agriculture*, National Academy Press, Washington, D.C.; C.A. Francis, C.B. Flora, and L.D. King (eds.), 1990, *Sustainable Agriculture in Temperate Zones*, John Wiley & Sons, New York.

[30] F. Bormann, 1976, "An Inseparable Linkage: Conservation of Natural Ecosystems and the Conservation of Fossil Energy," *BioScience*, vol. 26, pp. 754–760.

[31] The percentage depends on the diversity of small arthropods in tropical forests, which at the moment has only been *very* roughly estimated. See R. May, 1989, "How Many Species Are There on Earth?" *Science*, vol. 241 (16 September), pp. 1441–1449.

[32] Chlorination, although clearly an important barrier against water-borne disease where supplies are polluted, does not kill some disease organisms— and chlorine compounds (such as chloroform) formed in the water may be carcinogenic or damaging in other ways. See R. Wilson and E. Crouch, 1987, "Risk Assessment and Comparisons: An Introduction," *Science*, vol. 236 (17 April), pp. 267–270.

[33] F. Talbot, 1990, "Earth, Humankind, and Our Responsibility," Plenary Address to the American Association of Museums, June (mimeo).

[34] Much of the material in this section is adapted from P. Ehrlich and E. Wilson, 1991, "Biodiversity Studies: Science and Policy," *Science*, in press.

[35] N. Myers, 1989, *Deforestation Rates in Tropical Forests and Their Climatic Implications*, Friends of the Earth, London.

[36] See P. Raven, "The Scope of the Plant Conservation Problem Worldwide," in *Botanic Gardens and the World Conservation Strategy*, Academic Press, London.

[37] For a fine overview, see E. Wilson, 1989, "Threats to Biodiversity," *Scientific American*, September, pp. 108–116. See also May, 1989.

[38] This very rough estimate depends heavily on how one defines both population and "genetically distinct." Based on our research group's experience with herbivorous insects (which may themselves number millions of species) and a survey of the literature, this number seems to be in the ballpark.

[39] A fine book putting the extinction epidemic in a context of human evolution is J. Diamond, 1991, *The Rise and Fall of the Third Chimpanzee*, Harper Collins, New York.

[40] Ehrlich, Ehrlich, and Holdren, 1977, *Ecoscience*, pp. 142–143; Ehrlich and Ehrlich, 1981.

[41] Assuming T. Erwin's estimates ("The Tropical Forest Canopy," in E. Wilson [ed.], 1988, *Biodiversity*, National Academy Press, Washington, D.C., pp. 123–129) of tropical rainforest diversity are correct, the base of the estimate is a nonconservative total of 30 million species instead of

Wilson's "very conservative" 2 million. See also P. Ehrlich and E. Wilson, 1991.

[42] For an overview of the differentiation of populations (which leads to speciation), see Ehrlich, 1986, *Machinery of Nature*. A more technical treatment can be found in D. Futuyma, 1986.

[43] P. Vitousek, P. Ehrlich, A. Ehrlich, and P. Matson, 1986, "Human Appropriation of the Products of Photosynthesis," *BioScience*, vol. 36, pp. 368–373.

[44] Technically, NPP is the energy remaining after subtracting the respiration of the primary producers (mostly green plants, algae, and bacteria) from the total amount of energy fixed biologically (virtually all solar). NPP is the energy that supports all organisms—animals, fungi, parasitic plants, and other consumers and decomposers—except primary producers.

[45] D. Wright, 1990, "Human Impacts on Energy Flow through Natural Ecosystems, and Implications for Species Endangerment," *Ambio*, vol. 19, July, pp. 189–194. Under the higher estimates listed above, one might assume that roughly one or two million species could be extinct by 2000. That would be 5 percent of 20 to 40 million species, a not unreasonable estimate of organic diversity. But all such estimates, obviously, are only the crudest of approximations.

[46] World Commission on Environment and Development, 1987, *Our Common Future*, Oxford University Press, New York.

[47] W. Clark, 1989, "Managing Planet Earth," *Scientific American*, vol. 261, September, pp. 47–54.

[48] Most consider the founding event to be a symposium organized by Michael Soulé and Bruce Wilcox at the San Diego Wild Animal Park in September 1978, which resulted in the 1980 volume they edited, *Conservation Biology: An Evolutionary-Ecological Perspective*, Sinauer, Sunderland, Mass.

[49] A flavor of the field can be obtained from the journal, from the volume Soulé edited in 1986, *Conservation Biology*; and from the 1988 volume, edited by E. O. Wilson, *Biodiversity* (National Academy Press, Washington, D.C.), that resulted from the 1986 National Forum on Biodiversity sponsored by the National Academy of Sciences and the Smithsonian Institution.

[50] 1985, "A Discipline with a Time Limit," *Nature*, vol. 317, pp. 111–112.

[51] Two excellent recent documents giving overviews are W. Reid and K. Miller, 1989, *Keeping Options Alive: The Scientific Basis for Conserving Biodiversity*, World Resources Institute (WRI), Washington, D.C.; and J. McNeely, K. Miller, W. Reid, R. Mittermeier, and T. Werner, 1990, *Conserving the World's Biological Diversity*, IUCN, Gland, Switzerland; WRI, Conservation International (CI), World Wildlife Fund-U.S. (WWF-US), and

the World Bank, Washington, D.C. One of the best examples of a strategy to preserve biodiversity is the concept of debt-for-nature swaps developed by conservation biologist Thomas Lovejoy, described in Chapter 6.

[52] A. Leopold, 1953, *Round River*, Oxford University Press, Oxford, p. 190.

CHAPTER TWO

[1] Two kinds of energy statistics are commonly encountered. "Commercial" energy refers to energy sources that are sold in the market: oil, gas, coal; electricity generated from those fuels or from hydroelectric or nuclear sources; and commercially sold fuelwood and charcoal. "Traditional" energy refers to fuelwood, crop wastes, and dung used by those who gather it.

[2] Of some 13.1 terawatts (TW) of energy used worldwide in 1990, about 1.5 TW were traditional. (See the next note for definitions of energy units.) Our source is J. Holdren, 1991, "Population and the Energy Problem," *Population and Environment*, vol. 12, no. 3, Spring. Much of this chapter is informed by the work of John Holdren on energy-environment-security issues over the last twenty years. Readers are referred to his classic work, *Energy* (J. Holdren and P. Herrera, Sierra Club Books, San Francisco, 1971), which laid much of the groundwork for today's thinking on issues such as energy efficiency and energy's environmental impacts. We have also leaned heavily on Holdren's 1990 analysis, "Energy in Transition," *Scientific American*, September, pp. 156–163.

[3] A watt is a measure of the flow of energy; one joule per second (10^7 ergs per second). A kilowatt is a flow of 1,000 watts. Watt-seconds and kilowatt hours (kWh) are quantities of energy; one joule and 3.6×10^6 joules respectively. A supply of 24 kWh will run a 1,000-watt electric heater for 24 hours. A kilowatt year is about the amount of energy in a ton of bituminous coal.

[4] Here we are using 1987 per-capita GNP figures, 1989 commercial and traditional energy-use figures (the latter are estimated for categories of nations, although not available country by country); and 1990 population figures. All, of course, are estimates, and in combination they give a revealing picture of the situation around 1990.

[5] Individual country estimates include commercial energy only.

[6] Remember that 1990 population figure for the developing nations is increasing by roughly 90 million each year.

[7] A conservative estimate (basically assuming no great change in the global energy mix) for the period 1986–2010 suggests a tripling of energy use for developing nations: D. Dreyfus and A. Ashby, 1990, "Fueling Our Global Future," *Environment*, vol. 32, no. 4 pp. 17–41.

[8] 11 billion people [P] $\times$ 6 kW [AT] = 66 TW [I]; in 1990 [I] = 13.1 TW, and 66/13.1 = 5.

[9] See especially J. Holdren, 1991, "The Transition to Costlier Energy," in L. Schipper and S. Meyers (eds.), *Energy Transitions*, Stockholm Environmental Institute, Stockholm (in press), a version of which appeared as "Energy in Transition," in *Scientific American*, September, 1990, somewhat marred by that journal's incompetent editing. Our treatment is based on his.

[10] For instance, J. Goldemberg, T. Johansson, A. Reddy, and R. Williams, 1987, *Energy for a Sustainable World*, World Resources Institute, Washington, D.C.

[11] The ratio of CO_2 production per joule of energy yielded is coal, 100: oil, 80: natural gas, 60.

[12] The World Bank, 1990, *World Development Report 1990*, Oxford University Press, New York.

[13] These numbers are based on estimated ultimately recoverable resources, not "proven reserves." Reserves include only material of known location that can be recovered profitably at today's prices with today's technology. The estimates of ultimately recoverable resources attempt to account for effects of future discoveries, future prices, and future technologies.

[14] Economists speak of private and social costs. Private costs are those borne by the business entities involved in, say, producing and marketing a gallon of gasoline. Social costs are private costs plus "externalities" (such as the contribution of that gallon of gas to traffic congestion, air pollution, and global warming) borne by society at large. If there are no externalities (which will be encountered repeatedly in this book), private costs are equal to social costs.

[15] L. Schipper, R. Howarth, and H. Geller, 1990. "United States Energy Use from 1973 to 1987: The Impacts of Improved Efficiency," *Annual Review of Energy*, vol. 15, pp. 455-500.

[16] *Newsweek*, "Fire on the Other Side," 21 January, 1991.

[17] This is in terms of standard costs with no attempt to evaluate the external costs.

[18] See P. Ehrlich, A. Ehrlich, and J. Holdren, 1977, *Ecoscience; Population, Resources, Environment*, Freeman, San Francisco; J. Holdren, 1987, "Global Environmental Issues Related to Energy Supply," *Energy*, vol. 12, pp. 975–992. A less well known possible risk is that from long-lived krypton-85, released from nuclear fuel reprocessing plants, which changes the electrical conductivity of the atmosphere slightly. Some people have been concerned that this change in conductivity could influence the weather, but the likelihood of its being a serious problem seems very low.

[19] The Price-Anderson Act of 1957, in its latest extension, limits the liability of a utility in case of catastrophic accident to $7 billion; above that, the government pays. The total liability in a worst-case accident, killing many thousands of people and making uninhabitable an area the size of Pennsylvania, could easily be $50–100 billion. In a lesser accident, the industry could pay as much as half of the costs.

[20] Those costs could run between between 10 and 150 percent of constructions costs; they are not yet known since no large commercial plants have yet been decommissioned. Those costs, properly discounted (since they occur at the end of the plant's life), should be included in the costs calculated for any future deployment of a new nuclear technology.

[21] J. Holdren, 1986, "Too Much Energy, Too Soon," *Technology Review*, January, p. 118.

[22] Holdren and Herrera, 1971.

[23] L. Schipper, 1991, "Energy Saving in the U.S. and Other Wealthy Countries: Can the Momentum Be Maintained?" in press in D. Wood (ed.), *Energy and the Environment in the 21st Century*, M.I.T. Press, Cambridge, Mass.; L. Schipper, 1987, "Energy Saving Policies of OECD Countries: Did They Make a Difference?" *Energy Policy*, vol. 15, pp. 538–548; L. Schipper, 1990, "Energy Demand and Efficiency in Japan: Myths and Realities," Presentation to U.S. Department of Energy, 1 October.

[24] A. Fickett, C. Gellings, and A. Lovins, 1990, "Efficient Use of Electricity," *Scientific American*, September, pp. 65–74.

[25] This position has been championed for two decades by John Holdren, and we are following his lead when we say "efficiency" rather than "conservation," because "conservation" too often carries a connotation of belt-tightening: uncomfortably cold homes in winter, hot offices in summer, not enough light to read by, underpowered automobiles, and so on. "Efficiency" means using less energy to obtain the same benefits.

[26] A. Lovins, 1990, "The Role of Energy Efficiency," pp. 193–223 in J. Leggett (ed.), *Global Warming: The Greenpeace Report*, Oxford University Press, New York.

[27] R. Bevington and A. Rosenfeld, 1990, "Energy for Buildings and Homes," *Scientific American*, September, pp. 77–86.

[28] Fickett et al., 1990.

[29] R. Carlsmith, W. Chandler, J. McMahon, and D. Santini, 1990, *Energy Efficiency: How Far Can We Go?*, Report ORNL/TM-11441, Oak Ridge National Laboratory, Oak Ridge, Tennessee.

[30] Carlsmith et al., 1990; R. Williams, E. Larson, M. Ross, 1987, "Materials, Affluence, and Industrial Energy Use," *Annual Review of Energy*, vol. 12, pp. 99–144.

[31] Many of the issues in reducing dependence on automobiles are dealt with in M. Lowe, 1990, *Alternatives to the Automobile: Transport for Livable Cities*, Worldwatch Paper 98, Worldwatch Institute, Washington, D.C.; and M. Wachs, 1989, "U.S. Transit Subsidy Policy: In Need of Reform," *Science*, vol. 244 (30 June), pp. 1545–1549.

[32] D. Bleviss and P. Walzer, 1990, "Energy for Motor Vehicles," *Scientific American*, September, pp. 103–109.

[33] M. Wald, 1990, "Improving Gasoline Mileage Seen as Inadequate," *New York Times*, 14 December.

[34] The latter was designed to be especially safe, protecting passengers' lives in head-on crashes at up to 35 mph.

[35] Wald, 1990; J. MacKenzie and M. Walsh, 1990, *Driving Forces: Motor Vehicle Trends and Their Implications for Global Warming, Energy Strategies, and Transportation Planning*, World Resources Institute (December), Washington, D.C.

[36] M. Gladwell, 1990, "Consumers' Choices about Money Consistently Defy Common Sense," *The Washington Post*, 15 February, 1990. In addition to the extremely high positive discount rates implied by what is reported here, these same consumers would rather receive a $40,000 salary over 12 months than as a lump sum at the beginning of the year. (In technical terms, this implies a *negative* discount rate). Overall, it appears that people like even flows of income; they prefer not to receive or spend large lump sums.

[37] N. Lenssen, 1990, "Ray of Hope for the Third World," *World Watch*, vol. 3, September–October, pp. 37–38.

[38] The one impact that cannot be mitigated technologically is the inevitable generation of heat when energy is used. If all other environmental problems connected with energy mobilization were solved, heat pollution would still put a cap on the human enterprise on Earth. See the classic article by J. Fremlin, 1964, "How Many People Can the World Support?" *New Scientist*, 29 October.

[39] C. LaPorta, 1990, "Renewable Energy: Recent Commercial Performance in the U.S.A. as an Index of Future Prospects," pp. 224–259 in Leggett (ed.).

[40] E. Goldsmith and N. Hildyard, 1988, *The Earth Report*, Price Stern Sloan, Los Angeles, p. 127.

[41] Sources of these estimates and many others in this chapter are data from the World Bank, British Petroleum, United Nations, U.S. Department of Energy, and U.S. Environmental Protection Agency (EPA), summarized and supplemented by J. Holdren, 1991, "The Transition to Costlier Energy," and "Population and the Energy Problem." Note that the 13.1 TW of today

and the 30 TW projection figures in Holdren's scenario both refer to primary energy in which the contributions of electricity are calculated in terms of the amount of fossil fuel that would have to be burned to generate the electricity. This accounting convention makes a certain amount of sense in today's world, where most electricity is generated by fossil fuels, but will cause increasing confusion and distortions in a world that gets more of its electricity from nonthermal sources such as hydro, wind, and photovoltaics.

[42] Humanity already has substantial experience with wind power. In the 1920s, before comprehensive rural electrification from central-plant generating facilities was developed, thousands of farms in the United States and other parts of the world depended on electricity supplied by Jacobs wind turbines. The system, consisting of a fifty-foot tower, a generator, and a 21 kWh (kilowatt-hour) battery, powered a line of appliances—refrigerators, radios, electric drills, and so on—especially designed to operate on the 32-volt DC power supplied by the Jacobs (J. Naar, 1990, *Design for a Livable Planet*, Harper & Row, New York).

[43] C. Shea, 1988, "Renewable Energy: Today's Contribution, Tomorrow's Promise," *Worldwatch Paper 81*, Worldwatch Institute, Washington, D.C.

[44] Shea, 1988.

[45] K. Smith (ed.), 1987, *Biofuels, Air Pollution, and Health*, Plenum, New York.

[46] Forestry can usefully be viewed as a form of crop agriculture in which the harvesting cycle is dozens of times longer than that of most annual crops. Problems such as soil depletion still occur, but at a slower pace.

[47] C. Flavin, 1990, "Slowing Global Warming," in L. Brown et al., *State of the World 1990*, Norton, New York.

[48] For more details on fusion, see Ehrlich, Ehrlich, and Holdren, 1977.

[49] Fusion is the process that fuels the sun; it occurs at the opposite end of the weight spectrum of elements from nuclear fission. In fission, the process used in conventional nuclear power plants, very heavy nuclei, such as those of atoms of uranium-235, are split, producing lighter nuclei, some energetic subatomic particles called neutrons, and a great deal of energy. Fission power reactors regulate the rate at which nuclei are split in a controlled "chain reaction," whereby neutrons coming from splitting nuclei induce other nuclei to split. Some of the energy produced by the chain reaction is captured as heat and used to produce steam. The steam in turn is employed to spin turbines and produce electricity just as it is in fossil-fuel-fired power plants.

Nuclear fusion reactions involve the fusing together of very light nuclei (such as those of hydrogen) to make heavier ones (helium or weightier forms of hydrogen). Because of the way energy binds together the subatomic particles of nuclei, a great deal of surplus energy is created when very light nuclei are fused, just as it is when very heavy ones are split (middle-weight nuclei do not have this property). But getting light nuclei to fuse is vastly more difficult than getting heavy ones to split. In order to do so, conditions not unlike those inside the sun must be created. Fusion bombs ("hydrogen" bombs) achieve those conditions by using fission ("atomic") bombs as triggers.

[50] For an excellent, detailed, and optimistic analysis, see J. Ogden and R. Williams, 1989, *Solar Hydrogen: Moving beyond Fossil Fuels*, World Resources Institute, Washington, D.C. The numbers on silicon versus nuclear fuel that follow come from this source.

[51] Ogden and Williams, 1989.

[52] Ehrlich, Ehrlich, and Holdren, 1977.

[53] Those not yet persuaded of this should consult Holdren's *Scientific American* article (September 1990).

[54] Nuclear fission "burners" simply consume uranium fuel. Breeder reactors are often thought of as devices that produce more fuel than they consume. Actually, in the process of "burning," they convert a fuel that is in a form they cannot use to one they can use. Nuclei of uranium-238 and thorium-232, when struck by a neutron going at the right speed, go through a series of nuclear transformations that produce plutonium-239 and uranium-233 respectively. The latter isotopes can sustain chain reactions and are called "fissile"; the former cannot, but because of their potential for yielding fissile materials are called "fertile." Breeder reactors are designed to maximize the rate at which neutrons issuing from their chain reactions come into contact with fertile materials and "breed" fissile materials.

[55] Soviet officials recently stated that the Chernobyl cleanup could take more than 200 years and that some 2 million more people in Byelorussia must be moved from areas contaminated by radiation. The estimated cost of the evacuation was $70 billion (*Denver Post*, "Aid Sought for Victims of Chernobyl," 20 June 1990).

[56] Others believe that simply phasing out the present reactors as they reach the end of their economic lives would be sufficiently safe. We personally prefer the conservative course.

[57] This, of course, raises the question of "satisfactory to whom?" The basic answer is the general public and decision makers, but their attitudes will be strongly influenced by what might be called the "skeptical" element of the scientific community. This element includes technically knowledgeable people such as Holdren, Nobel Laureate physicist Henry Kendall of M.I.T., Princeton physicist Bob Williams, and energy analyst Bill Keepin, who are not opposed to nuclear power on principle, but have been highly critical of important features of today's fission technology. See, for instance, B. Keepin,

1990, "Nuclear Power and Global Warming," pp. 295–316 in Leggett (ed.).

[58] B. O'Neill, 1989, "Nuclear Safety after Chernobyl," *New Scientist*, 24 June, pp. 59–65.

[59] Such as "once-through" light water reactors (LWRs).

[60] J. Holdren, 1989, "Civilian Technologies and Nuclear Weapons Proliferation," pp. 161–198 in C. Schaerf, B. Holden-Reid, and D. Carlton (eds.), *New Technologies and the Arms Race*, Macmillan, London. Much of this section is based on assessments by John Holdren, whose views are summarized in the 1990 *Scientific American* article.

[61] A. Lovins and J. Price, 1975, *Non-nuclear Futures: the Case for an Ethical Energy Strategy*, Ballinger, Cambridge, Mass.

[62] W. Hafele, 1990, "Energy from Nuclear Power," *Scientific American*, September, pp. 136–144.

[63] C. Flavin and N. Lenssen, 1990, *Beyond the Petroleum Age: Designing a Solar Economy*, Worldwatch Paper 100, Worldwatch Institute, Washington, D.C.; C. Weinberg and R. Williams, 1990, "Energy from the Sun," *Scientific American*, September, pp. 147–155.

[64] Flavin, 1990. See also D. Somerville, 1989, "Whatever Happened to Solar Energy?" *American Scientist*, vol. 77, July–August, pp. 328–329.

[65] For instance, Ogden and Williams, 1989. See also L. R. Brown et al., 1991, *State of the World 1991*, Norton, New York, chapter 2.

[66] C. Weinberg and R. Williams, 1990, "Energy from the Sun," *Scientific American*, September, pp. 146–155.

[67] When oil prices dropped and extraction costs soared in the mid-1980s, the Soviet Union found itself faced with an "energy crisis" of the sort their propagandists had claimed only affected capitalist economies. Russia had squandered the hard currency it garnered during the boom on vast amounts of aid to client states and nonaligned nations it wished to influence: modern jet fighters went to Iraq, steel mills went to India, the Cuban adventure in Angola was supported, and Soviet armed forces invaded Afghanistan. Meanwhile, at home the incredibly inefficient Soviet economy was shored up by the financial injections from oil exports, rather than being reformed. The disappearance of those injections left such an economic shambles that the door was opened to Gorbachev's reforms; indeed, many would claim that the decline of oil revenues mandated them. (M. Dobbs, 1990, "Black Gold, Red Tape, and Hard Times," *Washington Post Weekly*, 4–10 June, pp. 7–8.)

[68] For a detailed discussion of China's energy situation, see R. Gränzer, 1989, "The Energy Impediment to China's Growth," *OECD Observer*, no. 157, April–May, pp. 9–14.

[69] China's coal resources, between 700 and 1,000 billion tons, are the third biggest in the world behind those of the Soviet Union (about 3,000 billion tons) and the United States (1,500 billion tons). ("Reserves" of coal are the amounts whose location is known and that can be extracted at costs similar to today's, with today's technologies. "Resources" include reserves plus estimated amounts that are less well located and delimited geologically or that cannot be extracted with today's technologies at today's costs. For long-term projections, it is often most useful to compare resources.) About three quarters of China's energy now comes from coal; most important, over 70 percent of the electrical energy generated by steam-driven turbines in China (as opposed to hydroelectric and wind power) is coal-based. Some 40 percent of the nation's total railroad capacity is involved in energy transport, largely coal. This coal-dependence is reflected by a series of recent investments in new large-scale mining projects financed by the World Bank, Japanese banks, and a U.S. energy firm; and by China's progress in developing an industry capable of producing giant mining equipment. At the other end of the spectrum, China has encouraged the development of small local coal mines, which now number some 65,000, account for about a third of coal production, and provide jobs for 10 million people.

[70] P. Ehrlich and A. Ehrlich, 1989, "How the Rich Can Save the Poor and Themselves: Lessons from the Global Warming," in S. Gupta and R. Pachauri (eds.), *Proceedings of the International Conference on Global Warming and Climate Change: Perspectives from Developing Countries*, Tata Energy Research Institute, New Delhi, 21–23 February, pp. 287–294.

[71] Information on Chinese energy planning is from S. Meyer (ed.), 1988, *Proceedings of the Chinese-American Symposium on Energy Markets and the Future of Energy Demand, Nanjing, China, June 22–24, 1988*, published by Lawrence Berkeley Laboratory (available from NTIS, Springfield, Va. 22161).

[72] Of course, we are making the extremely optimistic assumption that other factors will not raise death rates and bring Indian population growth to a halt long before such numbers are reached.

[73] Lenssen, 1990, "Ray of Hope for the Third World," pp. 37–38.

[74] Goldemberg, Johansson, Reddy, and Williams, 1987.

[75] This paragraph and part of the next is based on Holdren, 1990, "Energy in Transition."

CHAPTER THREE

[1] This chapter owes much to discussions with climatologist Stephen H. Schneider of the National Center for Atmospheric Research. His 1989 book,

Global Warming (Sierra Club Books, San Francisco), is the best discussion of "greenhouse science" written for the general public. An earlier volume, S. Schneider and R. Londer, *The Coevolution of Climate and Life* (Sierra Club Books, San Francisco, 1984), is the best available for nonscientists on how the climate works. Other good popular books on global warming are M. Oppenheimer and R. Boyle, 1990, *Dead Heat*, Basic Books, New York; and J. Leggett (ed.), 1990, *Global Warming: The Greenpeace Report*, Oxford University Press, New York. A fine, more technical volume is D. Abrahamson (ed.), 1989, *The Challenge of Global Warming*, Island Press, Washington, D.C.

[2] R. Revelle and H. Suess, 1957, "Carbon Dioxide Exchange between Atmosphere and Ocean and the Question of an Increase in Atmospheric CO_2 during the Past Decades," *Tellus*, vol. 9, pp. 18–27.

[3] We also follow the convention, which was established by scientists studying the carbon cycle, of measuring CO_2 emissions in tons of carbon. The atmospheric pool of carbon is part of the carbon cycle, and carbon combines with many different other elements in various parts of the cycle. Since the atomic weight of carbon is 12 and that of oxygen is 16, the molecular weight of CO_2 is $12 + (2 \times 16) = 44$. To convert tons of carbon into tons of carbon dioxide, simply multiply by 44/12 or 3.67.

[4] J. Leggett, 1990, "The Nature of the Greenhouse Threat," in Leggett (ed.), *Global Warming*, pp. 14–43; R. Watson, H. Rodhe, H. Oeschger, U. Seigenthaler, 1990, "Greenhouse Gases and Aerosols," in J. Houghton, G. Jenkins, and J. Ephraums (eds.), *Climate Change: The IPCC Scientific Assessment* (Intergovernmental Panel on Climate Change, World Meteorological Organization [WMO] and United Nations Environmental Programme [UNEP]), Cambridge University Press, Cambridge.

[5] The most recent estimates of CO_2 emissions from deforestation are higher than earlier ones, due in part to the acceleration of tropical deforestation since 1980. Sources include: R. Detwiler and C. Hall, 1988, "Tropical Forests and the Global Carbon Cycle," *Science*, vol. 239, pp. 42–47; P. Ciborowski, 1989, "Sources, Sinks, Trends, and Opportunities," pp. 213–230, and G. Woodwell, 1989, "Biotic Causes and Effects of the Disruption of the Global Carbon Cycle," pp. 71–81, both in Abrahamson (ed.), *The Challenge of Global Warming*; and R. Houghton, 1989, "Emissions of Greenhouse Gases," pp. 53–62 in N. Myers, *Deforestation Rates in Tropical Forests and Their Climatic Implications*, Friends of the Earth, London; N. Myers, 1990, "Tropical Forests," in Leggett, *Global Warming*, pp. 372–399; "Policymakers Summary," p. xxxii in Houghton, Jenkins, and Ephraums, 1990; and P. Crutzen and M. Andreae, 1990, "Biomass Burning in the Tropics: Impact on Atmospheric Chemistry and Biogeochemical Cycles," *Science*, vol. 250, pp. 1669–1678. The CO_2 sources are broken down by Crutzen and Andreae into burning associated with shifting agriculture and permanent

deforestation, savanna fires, and burning of firewood and agricultural wastes.

[6] "Policymakers Summary," in Houghton, Jenkins, and Ephraums, 1990, p. xxxii.

[7] The nature of the exchange of CO_2 between atmosphere and oceans is poorly understood, however, and remains one of the major mysteries in the global carbon cycle and thus in dealing with the CO_2 component of global warming.

[8] Sources vary somewhat, owing in part to the uncertainties of oceanic uptake, but recent measurements and higher estimates of the contribution from deforestation put the rate of rise at about 0.5 percent per year (Watson et al., 1990).

[9] We wrote: "The greenhouse effect is being enhanced now by the greatly increased level of carbon dioxide in the atmosphere. In the last century our burning of fossil fuels raised the level some 15%. The greenhouse effect today is being countered by low-level clouds generated by contrails, dust, and other contaminants that tend to keep the energy of the sun from warming the Earth in the first place.

"At the moment we cannot predict what the overall climatic results will be of our using the atmosphere as a garbage dump. We do know that very small changes in either direction in the average temperature of the Earth could be very serious. . . . In short, when we pollute, we tamper with the energy balance of the Earth. The results in terms of global climate and in terms of local weather could be catastrophic. Do we want to keep it up and find out what will happen? What do we gain by playing 'environmental roulette'?" (P. Ehrlich, 1968, *The Population Bomb*, Ballantine, New York, pp. 60–61).

[10] See T. Wigley, 1989, "Possible Climate Change Due to SO_2 Derived Cloud Condensation Nuclei," *Nature*, vol. 339, pp. 365–367; and the discussion in Schneider, 1990, *Global Warming*, Vintage Books, New York (paperback), n. 17, pp. 313–314.

[11] A. Lacis, J. Hansen, P. Lee, T. Mitchell, and S. Lebedeff, 1981, "Greenhouse Effect of Trace Gases, 1970–1980," *Geophysical Research Letters* vol. 8, pp. 1035–1038; V. Ramanathan, 1975, "Greenhouse Effect Due to Chlorofluorocarbons: Climate Implications," *Science*, vol. 190, pp. 50–51.

[12] V. Ramanathan, R. Cicerone, H. Singh, and J. Kiehl, 1985, "Trace Gas Trends and Their Potential Role in Climate Change," *Journal of Geophysical Research*, vol. 90, pp. 5547–5566; R. Dickinson and R. Cicerone, 1986, "Future Global Warming from Atmospheric Trace Gases," *Nature*, vol. 319, pp. 109–114. A recent summary of the various greenhouse factors can be found in T. Graedel and P. Crutzen, 1989, "The Changing Atmosphere," *Scientific American*, September, pp. 58–68.

[13] Watson et al., 1990; H. Mooney, P. Vitousek, and P. Matson, 1987,

"Exchange of Materials between Terrestrial Ecosystems and the Atmosphere," *Science*, vol. 238, pp. 926–932; D. Lashof and D. Tirpak (eds.), 1989. *Policy Options for Stabilizing Global Climate* (Draft report to Congress), EPA, Washington, D.C.

[14] Crutzen and Andreae, 1990.

[15] R. Cicerone, 1989, "Global Warming, Acid Rain, and Ozone Depletion," pp. 231–238, and P. Ciborowski, 1989, "Sources, Sinks, Trends, and Opportunities," pp. 213–230, both in D. Abrahamson (ed.), *The Challenge of Global Warming*.

[16] J. Chappellaz, J. Barnola, D. Reynaud, Y. Korotkevich, and C. Lorius, 1990, "Ice-Core Record of Atmospheric Methane over the Past 160,000 Years," *Nature*, vol. 345, pp. 127–131.

[17] Calculation of effective residence times is especially tricky for CO_2, since rather than being destroyed, CO_2 molecules can move to nonatmospheric pools (the oceans or biomass), from which it can return to the atmosphere. Because of this mobility, a perturbation of the CO_2 concentration is likely to last 50 to 200 years, which is the meaning of "residence time" as used here and differs from that used for methane. See D. Lashof and D. Ahuja, 1990, "Relative Contributions of Greenhouse Gas Emissions to Global Warming," *Nature*, vol. 344, pp. 529–531.

[18] "Hydroxyl" is, more technically, free hydroxyl radical (OH), which is sometimes called "nature's cleanser." See K. Shine, R. Derwent, D. Wuebbles, and J.-J. Morcrette, 1990, "Radiative Forcing of Climate," in Houghton, Jenkins, and Ephraums, 1990.

[19] R. Weiss, 1981, "The Temporal and Spatial Distribution of Tropospheric Nitrous Oxide," *Journal of Geophysical Research*, vol. 86, pp. 7185–7195; R. Rasmussen and M. Khalil, 1986, "Atmospheric Trace Gases: Trends and Distributions over the Last Decade," *Science*, vol. 232, pp. 1623–1624.

[20] Watson et al., 1990; Cicerone, 1989; Lashof and Tirpik, 1989.

[21] J. Leggett, 1990, "Playing Roulette with the Atmosphere," *New Scientist*, 7 July, p. 16.

[22] Watson et al., 1990.

[23] This is a simplification, since the warming influence of each gas is also a function of its concentration and that of all the other greenhouse gases present. Consideration of warming potential also may include a "discount rate"—that is, present contributions to warming may be considered more important than those made in the distant future. This might be because we care little about the fates of our remote descendants, or because we believe science will produce a "cure" for global warming in the intervening time. In general, the higher the discount rate, the less importance CO_2 has, as

well as other gases with long residence times (such as nitrous oxide or CFC-12), since the high discount rate reduces the effective residence time. For details, see Lashof and Ahuja, 1990, from which these numbers are taken. A similar calculation, but less heavily discounted, is that of H. Rodhe, 1990, "A Comparison of Various Gases to the Greenhouse Effect," *Science*, vol. 248, pp. 1217–1219. A more conservative analysis can be found in Shine, Derwent, Wuebbles, and Morcrette, 1990.

[24] Watson et al., 1990. The relative potency of CFC-12 is 5750 times that of CO_2 (Leggett, 1990, "Playing Roulette . . .").

[25] Measured this way, carbon dioxide's share of global warming potential through 1985 was estimated to be 71.5 percent, and that of CFCs was 9.5 percent, methane 9.2 percent, nitrous oxide 3.1 percent, and carbon monoxide (CO) 6.6 percent. CO contributes to the greenhouse effect when it is converted to CO_2 in the atmosphere. In the atmosphere, methane is broken down by hydroxyl (OH) radicals into carbon dioxide and water; methane combustion also results in a release of CO_2. Adding this to CO_2's account raises its warming potential to about 80 percent (Lashof and Ahuja, 1990).

[26] Schneider, 1989.

[27] W. Broecker, 1989, "Greenhouse Surprises," in Abrahamson, pp. 196–209.

[28] J. Barnola, D. Raynaud, Y. Korotkevich, and C. Lorius, 1987, "Vostok Ice Core Provides 160,000-Year Record of Atmospheric CO_2," *Nature*, vol. 329, pp. 408–414.

[29] Oxygen, like many other elements, exists in different forms called "isotopes," which have the same chemical properties but different masses (weights) because they have different numbers of neutrons in their nuclei. Atmospheric temperatures associated with the air in the core bubbles can be estimated by looking at the concentration of heavy oxygen (O^{18}), which is scarcer in cold conditions.

[30] See S. Schneider, 1989, "Global Warming: Causes, Effects, and Implications," pp. 33–51 in J. Cairns and P. Zweifel (eds.), *On Global Warming*, Virginia Polytechnic Institute and State University, Blacksburg, Virginia; and S. Schneider, 1989, for details on the theories of glacial cycles. For a recent analysis reinforcing this speculation, see Chappellaz, Barnola, Raynaud, Korotkevich, and Lorius, 1990.

[31] S. Kuo, C. Lindberg, and D. Thomson, 1990, "Coherence Established between Atmospheric Carbon Dioxide and Global Temperature," *Nature*, vol. 343, pp. 709–714.

[32] "Policymaker's Summary," in Houghton, Jenkins, and Ephraums, 1990; P. Jones, P. Groisman, M. Coughlan, N. Plummer, W.-C. Wang, and T. Karl, 1990, "Assessment of Urbanization Effects in Time Series of Surface

Air Temperature over Land," *Nature*, vol. 347, pp. 169–172. Further evidence is reported in J. Bethoux, B. Gentili, J. Raunet, and D. Tailliez, 1990, "Warming Trend in Western Mediterranean Deep Water," *Nature*, vol. 347, pp. 660–662.

[33] In descending order, the seven warmest years on record through 1990 were 1990, 1988, 1983, 1987, 1944, 1989, and 1981. W. Booth, 1991, "Hot Time in the Old Town," *Washington Post Weekly*, 21–27 January, p. 31.

[34] W. Stevens, 1991, "2 Studies Rank 1990 as the Warmest Year," *New York Times*, 10 January.

[35] T. Wigley and S. Raper, 1990, "Natural Variability of the Climate System and Detection of the Greenhouse Effect," *Nature*, vol. 344, pp. 324–327. There has been an attempt by a small group of conservative scientists to tell the Bush administration what it wants to hear—that there is no need to take action to attempt to slow the buildup of greenhouse gases. Their science has been as bad as their policy judgments. See J. Gribbin, 1990, "An Assault on the Climate Consensus," *New Scientist*, 15 December, pp. 26–31; L. Roberts, 1989, "Global Warming: Blaming the Sun," *Science*, vol. 246 (24 November), pp. 992–993; and R. Kerr, 1990, "New Greenhouse Report Puts Down Dissenters," *Science*, vol. 249, pp. 481–482.

[36] V. Ramanathan, 1989, "Observed Increases in Greenhouse Gases and Predicted Climatic Changes," pp. 239–247 in Abrahamson.

[37] Cicerone, 1989 (in Abrahamson), p. 233; D. Lashof, 1989, "The Dynamic Greenhouse: Feedback Processes That May Influence Future Concentrations of Atmospheric Gases and Climatic Changes," *Climatic Change*, vol. 14, pp. 213–214. See also F. Pearce, 1989, "Methane: The Hidden Greenhouse Gas," *New Scientist*, 6 May; Watson et al., 1990; J. Leggett, 1990, "Critique of the Policymakers Summary of the Scientific Assessment of Climate Change Report to IPCC from Working Group 1, Third Draft, 2 May 1990," Greenpeace International submission for the WG1 Plenary, 23–25 May, Windsor, United Kingdom; and D. Schimel, 1990, "Biogeochemical Feedbacks in the Earth System," in Leggett, *Global Warming*, pp. 68–82.

[38] S. Whalen and W. Reeburgh, 1990, "Consumption of Atmospheric Methane by Tundra Soils," *Nature*, vol. 346, pp. 160–162.

[39] G. MacDonald, 1990, "Role of Methane Clathrates in Past and Future Climates," *Climate Change*, vol. 16, pp. 247–281.

[40] Pearce, 1989.

[41] MacDonald, 1990.

[42] G. Woodwell, 1989, "Biotic Causes and Effects of Disruption of the Global Carbon Cycle," pp. 71–81 in Abrahamson; G. Woodwell, 1990, "The Effects of Global Warming," in Leggett, *Global Warming*.

[43] Leggett, 1990, "Nature of the Greenhouse Threat"; "Policymakers Summary," in Houghton, Jenkins, and Ephraums, 1990; and R. Houghton, 1989, in Myers.

[43A] See W.K. Stevens, 1991, "An Oceanic Indication That Earth's Climate Might Regulate Itself," *New York Times*, May 7. V. Ramanathan and W. Collins calculated that huge clouds over the tropical oceans could eventually put a cap on global warming by reflecting away sunlight, but that, despite this negative feedback mechanism, global warming "could still cause vast climatic disruptions."

[44] For an excellent discussion of the relative uncertainties in global warming, see S. Schneider, 1989, "The Greenhouse Effect: Science and Policy," *Science*, vol. 243, pp. 771–781.

[45] There is substantial debate today on how much soil moisture might be reduced, the impacts of reduced rainfall on runoff (and reservoir storage), and related issues. Much of the uncertainty is centered on the effects of increased CO_2 on the use of water by plants. See F. Pearce, 1990, "High and Dry in the Global Greenhouse," *New Scientist*, 10 November, pp. 34–37.

[46] R. Adams, B. McCarl, and D. Dudek, 1988, "Implications of Global Climate Change for Western Agriculture," *Western Journal of Agricultural Economics*, vol. 13, 348–356; M. Parry, 1990, *Climate Change and World Agriculture*, Earthscan, London.

[47] The Canadian shield is the mass of Precambrian rock, centered on Hudson Bay, that forms the great stable mass of the North American continent. It underlies eastern Canada and a small part of the north central United States, south and west of Lake Superior.

[48] J. Harte, D. Jensen, and M. Torn, 1991, "The Nature and Consequences of Indirect Linkages between Climate Change and Biological Diversity," in R. Peters and T. Lovejoy (eds.), *Global Warming and Biological Diversity*, Yale University Press, New Haven.

[49] Of course, climates do help produce characteristic soils, and in a few thousand years those poor ones might be transformed into quite good agricultural soils.

[50] Pearce, 1990, "High and Dry . . ."; P. Gleick, 1988, "Climate Change and California: Past, Present, and Future Vulnerabilities," in M. Glantz, *Societal Responses to Climate Change: Forecasting by Analogy*, Westview Press, Boulder, Colo.; and P. Waggoner (ed.), 1990, *Climate Change and U.S. Water Resources*, Wiley-Interscience, New York.

[51] A. Pittock, 1989, "The Greenhouse Effect, Regional Climate Change and Australian Agriculture," Proceedings of the 5th Australian Agronomy Conference, Perth (September 24-29), Australian Society of Agronomy.

[52] C. Blumenthal, S. Barlow, and T. Wrigley, 1990, "Global Warming and Wheat," *Nature*, vol. 347 (20 September), p. 235.

[53] J. Gribbin, 1990, "Britain Must Learn to Farm the Greenhouse," *New Scientist*, 24 February, p. 28.

[54] D. Lincoln, D. Couvet, and N. Sionit, 1986, "Response of an Insect Herbivore to Host Plants Grown in Carbon Dioxide Enriched Atmosphere," *Oecologia* (Berlin), vol. 69, pp. 556–560.

[55] D. Lincoln and D. Couvet, 1989, "The Effect of Carbon Supply on Allocation to Allelochemicals and Caterpillar Consumption of Peppermint," *Oecologia* (Berlin), vol. 78, pp. 112–114. Interestingly, the plants allocated enough additional energy to producing defensive chemicals that their concentration in the relatively nitrogen-poor leaves did not drop.

[56] E. Fajer, M. Bowers, and F. Bazzaz, 1989, "The Effects of Enriched Carbon Dioxide Atmospheres on Plant-Insect Herbivore Interactions," *Science*, vol. 243, pp. 1198–1200.

[57] For instance: R Adams et al., "Global Climate Change and U.S. Agriculture," *Nature*, vol. 345, pp. 219–224; T. Maugh, 1990, "The Flip Side of Global Warming," *Los Angeles Times*, 17 May, 1990; Parry, 1990.

[58] P. Matson and P. Vitousek, 1990, "Ecosystem Approach to a Global Nitrous Oxide Budget," *BioScience*, vol. 40, pp. 667–672.

[59] A. Ehrlich, 1990, "Agricultural Contributions to Global Warming," in Leggett (ed.), *Global Warming*.

[60] For a comprehensive review of CO_2 effects, see F. Bazzaz, 1990, "The Response of Natural Ecosystems to the Rising Global CO_2 Levels," *Annual Review of Ecology and Systematics*, vol. 21, pp. 167–196. For a general evaluation of ecosystem change, see J. Clark, 1991, "Ecosystem Sensitivity to Climate Change and Complex Responses," in R. Wyman (ed.), *Global Climate Change and Life on Earth*, Chapman and Hall, New York, pp. 65–98.

[61] R. Peters, "Consequences of Global Warming for Biological Diversity," pp. 99–118; R. Lester and J. Myers, "Double Jeopardy for Migrating Wildlife," pp. 119–133; and R. Wyman, "Multiple Threats to Wildlife: Climate Change, Acid Precipitation, and Habitat Fragmentation," pp. 134–155, all in R. Wyman (ed.), 1991, *Global Climate Change and Life on Earth*.

[62] D. Botkin, D. Woodby, and R. Nisbet, 1991, "Kirtland's Warbler Habitats: A Possible Early Indicator of Climatic Warming," *Biological Conservation*, vol. 56, no. 1, pp. 63–78.

[63] D. Schindler, K. Beaty, E. Fee, D. Cruikshank, E. DeBruyn, D. Findlay, G. Linsey, J. Shearer, M. Stainton, and M.A. Turner, 1990, "Effects of Climatic Warming on Lakes of the Central Boreal Forest," *Science*, vol. 250, pp. 967–970.

[64] L. Roberts, 1990, "Warm Waters, Bleached Corals," *Science*, vol. 249 (12 October), p. 213; R. Langreth, 1990, "Bleached Reefs," *Science News*, vol. 138 (8 December), pp. 364–365.

[65] P. Parsons, 1989, "Conservation and Global Warming; A Problem in Biological Adaptation to Stress," *Ambio*, vol. 18, no. 6, pp. 322–325; R. Graham, M. Turner, and V. Dale, 1990, "How Increasing CO_2 and Climate Change Affect Forests," *BioScience*, vol. 40, no. 8, pp. 575–587.

[66] The West Antarctic ice sheet, which is anchored at some points on the ocean floor, could be detached as warmer water flowed beneath it or air above it. That would allow a huge volume of glacial ice to slip into the sea. Eventually, West Antarctica could be "deglaciated," with all the ice disappearing in a matter of 500 years or so. See J. Titus, 1989, "The Causes and Effects of Sea Level Rise," pp. 161–195 in Abrahamson.

[67] The uncertainty has been exaggerated by two papers in *Science* claiming that global warming was already causing the southern Greenland ice dome to grow, and assuming that this growth would help slow the sea-level rise. Unfortunately for that argument, the wider region around Greenland was *cooling* during the time period in question. See S. Schneider, 1991, "Is the Recent Downward Revision of Global Sea Level Rise Projections Premature?" *Nature*, in press.

[68] If sea level rose 15 to 30 feet, the East and Gulf coasts of the United States would lose much land along their shorelines, and further large areas would be at risk of flooding during storms. Parts of the lower Mississippi valley and the Central valley of California would be inundated, and neighborhoods in many major cities would be drowned. One would be able to tie one's boat up to the Washington Monument. The same would happen to much of the fringes of northwestern Europe and the northern Soviet Union.

In Africa, Egypt's Nile Delta would be the most dramatic loss, but parts of coastal West Africa are also vulnerable. A big chunk of Bangladesh would disappear, along with fertile lower river valleys in India and China. Part of the Amazon basin would be an inland sea, and many of the world's great cities would face the fate of Venice. We trust that humanity will never have to face that dire prospect, but what we are likely to experience will be difficult enough.

[69] H. Gavaghan, 1989, "Effect of Global Warming on Sea Levels 'Overestimated,' " *New Scientist*, 16 December, p. 11; "Policymakers Summary," Houghton, Jenkins, and Ephraums, 1990.

[70] U. Mikolajewicz, B. Santer, and E. Maier-Reimer, 1990, "Ocean Response to Greenhouse Warming," *Nature*, vol. 345, pp. 589–593.

[71] Titus, 1989.

[72] S. Leatherman, 1991, "Impact of Climate-induced Sea-Level Rise on

Coastal Areas," in Wyman (ed.), *Global Climate Change and Life on Earth,* pp. 170–179.

[73] See for instance, J. Hecht, 1990, "The Incredible Shrinking Mississippi Delta," *New Scientist,* 14 April, pp. 36–41.

[74] Titus, 1989.

[75] A. Wijkman and L. Timberlake, 1984, *Natural Disasters: Acts of God or Acts of Man?* Earthscan, Washington, D.C.

[76] Pearce, 1990, "High and Dry." The most recent disaster was in May 1991.

[77] "Policymakers Summary," in Houghton, Jenkins, and Ephraums (eds.), 1990.

[78] Schneider, 1989, *Global Warming.*

[79] Including killing many endangered red-cockaded woodpeckers (K. Gruson, 1989, "Healing Comes Slowly to Woods and Streams Where Hurricane Left Death," *New York Times,* 11 November).

[80] P. Gleick, 1987, "Regional Hydrological Consequences of Increases in Atmospheric CO_2 and Other Trace Gases," *Climatic Change,* vol. 10, pp. 137–160; Schneider, 1989, *Global Warming,* p. 138.

[81] M. Reisner, 1986, *Cadillac Desert: The American West and Its Disappearing Water,* Viking, New York. See also Chapter 6 here.

[82] J. Maurits La Rivière, 1989, "Threats to the World's Water," *Scientific American,* September, pp. 80–94.

[83] A. Haines, 1990, "The Implications for Health," in Leggett (ed.), *Global Warming,* pp. 149–162.

[84] D. Abrahamson, 1989, "Global Warming: The Issue, Impacts, Responses," in Abrahamson (ed.), pp. 3–34.

[85] Mikolajewicz, Santer, and Maier-Reimer, 1990; Harte et al., 1990.

[86] W. Brown, 1990, "Flipping Oceans Could Turn up the Heat," *New Scientist,* 25 August, p. 27.

[87] P. Ehrlich, 1968, *The Population Bomb,* Ballantine, New York, p. 61.

[88] S. Schneider, 1990, "Prudent Planning for a Warmer Planet," *New Scientist,* 17 November, pp. 49–51.

[89] We will return frequently to the issue of externalities, which can be either positive (that is, benefits) or negative, and which are a central factor in the human predicament. Most externalities that we discuss are negative, as economists say, making social costs often far higher than private costs.

[90] Feasibility here obviously involves complex economic considerations that

are beyond our discussion here but are dealt with in principle later in this chapter and in Chapter 8.

91 "Policymakers Summary," in Houghton, Jenkins, and Ephraums, 1990, p. xi.

92 Of course, care must be taken to minimize methane leakage from natural gas distribution systems.

93 W. Nordhaus, 1982, "How Fast Should We Graze the Global Commons?" *American Economic Review*, vol. 72, pp. 242–246. See also R. Stavins, 1990, "Innovative Policies for Sustainable Development: The Role of Economic Incentives for Environmental Protection," *Harvard Public Policy Review*, Spring, pp. 13–25; and Congressional Budget Office, 1990, *Carbon Charges as a Response to Global Warming*, U.S. Government Printing office, Washington, D.C., August.

94 Congressional Budget Office, 1986, *The Budgetary and Economic Effects of Oil Taxes*, U.S. Government Printing Office, Washington, D.C., April.

95 One of the objections to a gasoline tax is that, like a carbon tax, it would be regressive, but it probably would be so only for very low income groups. Generally, the more affluent families own more cars and more gas guzzlers. The regressiveness at the bottom could be cured by many mechanisms, perhaps the best of which would be a compensatory reduction in the regressive FICA tax. It might be a good idea to supplement carbon taxes with higher gasoline taxes to encourage even more efficiency and more use of alternative transport and to emphasize the problems that automobile-dependence cause for society.

96 S. Boyle, 1990, "Lessons from the Past: What People Do When Energy Costs More," *New Scientist*, 3 November, p. 38.

97 *Science*, vol. 249 (10 August), p. 622.

98 A prime example of one-sided analysis is A. Manne and R. Richels, 1990, "CO_2 Emission Limits: An Economic Cost Analysis for the USA," *The Energy Journal*, April, pp. 51–74. An article on an early draft of this paper appeared in the *New York Times*: P. Passell, 1989, "Cure for Greenhouse Effect: The Costs Will Be Staggering," November 19. See the commentaries on pp. 302–303 in the 1990 Vintage edition of Schneider's *Global Warming*, and pp. 188–190 of his chapter 9 in Leggett (ed.), 1990, *Global Warming*.

99 R. Houghton, 1989, "Emissions of Greenhouse Gases," in Myers. We say "as much as" since Houghton's estimate, based on the high deforestation rate of 1989, is near the high end of the 1 to 1.5 billion-ton range (p. 59). The IPCC cites an estimate of 1.6 billion tons per year for tropical deforestation ("Policymakers Summary," in Houghton, Jenkins, and Ephraums, 1990).

[100] N. Myers, 1988, "Tropical Deforestation and Climatic Change," *Environmental Conservation*, vol. 15, pp. 293–298.

[101] R. Houghton and G. Woodwell, 1989, "Global Climatic Change," *Scientific American*, April, pp. 36–44. See also W. Booth, 1988, "Johnny Appleseed and the Greenhouse," *Science*, vol. 242, pp. 19–20.

[102] A. Ehrlich, 1990; Lashof and Tirpik, 1989; Watson et al., 1990.

[103] O. Tickell, 1990, "Up in the Air," *New Scientist*, 20 October, pp. 41–43; *New Scientist*, 1990, "Total CFC Ban Needed to Halt Global Warming," 8 September, p. 37.

[104] R. Benedick, 1991, *Ozone Diplomacy; New Directions for Safeguarding the Planet*, Harvard University Press, Cambridge, Mass.; W. Nitze, 1990, "A Proposed Structure for an International Convention on Climate Change," *Science*, vol. 249, pp. 607–608.

[105] P. Gleick, 1989, "Climate Change and International Politics: Problems Facing Developing Countries," *Ambio*, vol. 18, no. 6, pp. 333–339.

[106] S. Schneider, 1990, "Cooling It," *World Monitor*, July, pp. 30–38.

[107] D. Helm, 1990, "Who Should Pay for Global Warming?" *New Scientist*, 3 November, pp. 36–39.

[108] F. Pearce, 1990, "Bids for the Greenhouse Auction," *New Scientist*, 4 August. Apparently, the Bush administration flirted briefly with the idea of tradable permits in 1990, but then abandoned it as too complex (M. Sun, 1990, "Emissions Trading Goes Global," *Science*, vol. 247 (2 February), pp. 520–521). See also "Warm World, Cool Heads," in *The Economist*, 27 October, 1990, pp. 13–14, an editorial that overstated the costs of abating global warming, but warmly approved this approach.

[109] J. Swisher and G. Masters, 1991, "Buying Environmental Insurance: Prospects for Trading of Global Climate Protection Services," in press, *Climatic Change*.

[110] W. Stevens, 1991, "EPA Fans Debate over Warming," *New York Times*, 13 January.

[111] The exchange was described in Schneider, 1990, "Cooling It."

[112] "Policymakers Summary," Houghton, Jenkins, and Ephraums, 1990; Kerr, 1990; Leggett, 1990, in Leggett, *Global Warming*.

[113] J. Gribbin, 1990, "Why Caution Is Wrong on Global Warming," *New Scientist*, 28 July. See also Chapter 8 of this book.

[114] Such as Manne and Richels, 1990.

[115] Based on the government's "Economic Report to the President" of February 1990.

[116] W. Nordhaus, 1990, "Greenhouse Economics: Count Before You Leap," *The Economist*, 7 July, pp. 21–24; and 1990, "To Slow or Not to Slow: The Economics of the Greenhouse Effect," mimeo, 5 February; and 1990, "Economic Policy in the Face of Global Warming," mimeo, 9 March.

[117] Of course, we're not going to be completely without food under any reasonable warming scenario, but obviously the 3-percent figure grossly understates the fundamental importance of the agricultural and forestry sectors of our economy. One imagines that, if confronted with this point, Nordhaus would simply claim that the United States could import food from areas that gained in the greenhouse lottery. But the availability of supplies to import would be *far* from certain, especially considering that expected patterns of precipitation would bring drier climates and reduced harvests to the world's main breadbaskets and that the prospective change will be continuous. Nordhaus also expressed much faith in the fertilizing effects of CO_2 enrichment. The potential disruption of rainfall patterns he considers merely an inconvenience for some industries (no concern about farming). Because of air conditioning and other technological compensations for uncomfortable weather, the impacts on most people or businesses would, he predicted, be negligible.

[118] Nordhaus, 1990, "To Slow . . .", p. 23.

[119] The amount recommended by IPCC Scientists Working Group (Houghton, Jenkins, and Ephraums, 1990); C. Whitney, "Scientists Warn of Dangers in Warming Earth," *New York Times*, May 26, 1990.

[120] For a summary, see R. Borchelt, 1990/91, "Enlisting Marine Algae to Ease Global Warming," *NRC News Report* (National Research Council), December–January, pp. 5–7; and G. Chiu, 1991, "Scientists Doubt Plan to Sow Iron in Ocean," *San Jose Mercury News*, 17 January.

[121.] Borchelt, 1990/91.

[122] In early 1990, Sununu said on television: "[There's] a little tendency by some of the faceless bureaucrats on the environmental side to try and create a policy in this country that cuts off our use of coal, oil, and natural gas. . . . I don't think America wants not to be able to use their automobiles" (quoted in Schneider, 1990, "Cooling It," p. 38).

Of course, no responsible analyst has ever advocated that Americans give up their cars—although many have recommended heavily taxing fuel-inefficient models to internalize the costs they impose on the rest of us, and designing cities and mass transit systems so the need for their use is reduced. More rational, less energy-intensive transport systems in which autos played a smaller role both would be possible and would carry many benefits, including more convenient and efficient people movement.

[123] *CBS News* (radio), 10 July, 1990.

[124] *Newsweek*, 1991, "Will Bush Be Bold?" 7 January.

[125] J. Jager (ed.), 1990, *Responding to Climate Change: Tools for Policy Development*, Stockholm Environment Institute, Stockholm; see also *New Scientist*, 1990, "Off-the-Shelf Technologies to Cool the Planet," 20 October, p. 16.

CHAPTER FOUR

[1] Told brilliantly for the layperson in Sharon Roan's 1989 book *Ozone Crisis: The 15 Year Evolution of a Sudden Global Emergency*, Wiley, New York.

[2] The coolant in automobile air conditioners is CFC-12 (CCl_2F_2), while that in home air conditioners is not in today's terminology a CFC, but HCFC-22 ($CHClF_2$).

[3] Cl, not Cl_2.

[4] M. Molina and F.S. Rowland, 1974, "Stratospheric Sink for Chlorofluoromethanes: Chlorine Atom Catalysed Destruction of Ozone," *Nature*, vol. 249, pp. 810–814.

[5] For details of the affair, see Roan, 1989. We participated in the debate on the side of Rowland and Molina: P. Ehrlich, A. Ehrlich, and J. Holdren, 1977, *Ecoscience: Population, Resources, Environment*, Freeman, San Francisco, pp. 674–675.

[6] Some of the assaults on Molina and Rowland came from people who should have known better. James Lovelock had earlier invented an instrument capable of detecting tiny amounts of gases in the atmosphere. With that instrument, he discovered that CFCs hung around in the atmosphere for a long time, and thus he laid the groundwork for Molina's and Rowland's work. Lovelock, however, was critical of Rowland for attempting to alert the public to the possible danger—one that he himself had entirely overlooked. Another British scientist critical of the CFC-ozone depletion connection, Richard Scorer, toured the United States at the CFC industry's behest. He attacked Molina and Rowland as "doomsayers" and, like Lovelock, was eventually shown to be dead wrong. Doom was really in the cards.

[7] A. Burford, 1986, *Are You Tough Enough?* McGraw-Hill, New York, p. 134.

[8] At this writing, there is a scientific dispute over whether ozone depletion has had an impact on Antarctic life. See L. Roberts, 1989, "Does the Ozone Hole Threaten Antarctic Life?" *Science*, vol. 244, pp. 288–289.

[9] For recent work on the Antarctic ozone hole and its connection to CFCs, see J. Anderson, D. Toohey, and W. Brune, 1991, "Free Radicals within the Antarctic Vortex: The Role of CFCs in Antarctic Ozone Loss," *Science*, vol. 251 (4 January), pp. 39–45; and M. Shoeberl and D. Hartmann, 1991,

"The Dynamics of the Stratospheric Polar Vortex and Its Relation to Spring-time Ozone Depletions," *Science*, vol. 251, pp. 46–52.

[10] I. Isaksen, 1989, "The Beginnings of a Problem," in R. Jones and T. Wigley (eds.), *Ozone Depletion: Health and Environmental Consequences*, Wiley, New York.

[11] Quoted in Roan, 1989, p. 2; confirmed by Rowland, personal communication, 1990. The quote was originally published in an article by Michael Drosnin in the now defunct *New Times*.

[12] J. van der Leun, 1989, "Experimental Photocarcinogenesis," pp. 161–168 in Jones and Wigley. Nonmelanoma skin cancers are basal-cell and squamous-cell carcinomas.

[13] R. Jones, 1989, "Consequences for Human Health of Stratospheric Ozone Depletion," in Jones and Wigley, pp. 207–227.

[14] J. Elwood, 1989, "Epidemiology of Melanoma: Its Relationship to Ultraviolet Radiation and Ozone Depletion," pp. 169–189 in Jones and Wigley.

[15] Technically between 2030 and 2074; U.S. Environmental Protection Agency (EPA), 1987, *Assessing the Risks of Trace Gases That Can Modify the Atmosphere*, USEPA, Washington, D.C.

[16] S. Gianinni, 1990, "Effects of UV-B on Infectious Diseases," in *Proceedings of the Conference on Global Atmospheric Change and Public Health*, Center for Environmental Information, New York; P. Hersey, G. Haran, E. Hasic, and A. Edwards, 1983, "Alterations of T Cell Subsets and Induction of Suppressor T Cell Activity in Normal Subjects after Exposure to Sunlight," *Journal of Immunology*, vol. 131, pp. 171–174.

[17] M. McCally and C. Cassel, 1990, "Medical Responsibility and the Global Environment," *Annals of Internal Medicine*, vol. 113, 15 September, pp. 467–473.

[18] R. Worrest and L. Grant, 1989, "Effects of Ultraviolet-B Radiation on Terrestrial Plants and Marine Organisms," pp. 197–206 in Jones and Wigley.

[19] A. Termura and N. Murali, 1986, "Intraspecific Differences in Growth and Yield of Soybeans Exposed to Ultraviolet-B Radiation under Greenhouse and Field Conditions," *Environmental and Experimental Botany*, vol. 26, pp. 89–95.

[20] R. Worrest and D. Häder, 1989, "Effects of Stratospheric Ozone Depletion on Marine Organisms," *Environmental Conservation*, Autumn, pp. 261–263.

[21] J. Sinclair, 1990, "Ozone Loss Will Hit Health and Food, Says UN Study," *New Scientist*, 3 February, p. 27.

[22] G. Daily and P. Ehrlich, 1990, "An Exploratory Model of the Impact of Rapid Climate Change on the World Food Situation," *Proceedings of the*

Royal Society (B), vol. 241, pp. 232–244; L. Brown et al., 1990, *State of the World 1990*, Norton, New York. For details, see Chapter 7.

[23] J. Sinclair, 1990.

[24] Worrest and Grant, 1989.

[25] Worrest and Häder, 1989.

[26] Jones, 1989, pp. 207–227.

[27] R. Benedick, 1991, *Ozone Diplomacy: New Directions for Safeguarding the Planet*, Harvard University Press, Cambridge, Mass.

[28] S. Davis, "The European Dimension," pp. 109–113, and P. Usher, "The Montreal Protocol on Substances That Deplete the Ozone Layer: Its Development and Likely Impact," pp. 115–140, both in Jones and Wigley, 1989.

[29] Usher, 1989; Benedick, 1991.

[30] R. Jones and T. Wigley, 1989, "A Fact Sheet about Ozone," in Jones and Wigley.

[31] J. Gribbin, 1990, "Why Arctic Ozone Has Survived—So Far," *New Scientist*, 14 April, p. 23.

[32] J. Porritt, 1989, "Environmental Imperatives," pp. 243–249 in Jones and Wigley.

[33] The story of Du Pont's performance is well told in Roan, 1989, from which the quote is taken.

[34] Standard measurement of stratospheric ozone consists of comparing the ground-level flux of one wavelength of UV-B versus one wavelength of UV-A. "Decreased ozone" is reported when there is an increase in the received ratio of UV-B/UV-A, since there is no indication that the UV-A flux is changing. The standard measurement is made by focusing an instrument directly at the sun; wide-angle measurements may underestimate UV-B flux because tropospheric ozone pollution may filter the indirect rays of UV-B, effectively canceling an increase in direct UV-B (UV-A is not so affected).

[35] P. Shabecoff, 1990, "Scientists Report More Deterioration of Earth's Ozone Layer," *New York Times*, 24 June.

[36] M. Blumthaler and W. Ambach, 1990, "Indication of Increasing Solar Ultraviolet-B Radiation Flux in Alpine Regions," *Science*, vol. 248, pp. 206–207.

[37] P. Shabecoff, 1990, "U.S., in Switch, Will Back Fund Shielding Ozone," *New York Times*, 16 June. See also T. Wicker, 1990, "Cheering a Flip-Flop," *New York Times*, 21 June, which properly takes the Bush administration to task for its backward stance on the issue of global warming.

[38] Partially halogenated CFCs—that is, ones without halogens substituted for all the hydrogen atoms on the carbon backbone.

[39] In HFC compounds, all halogen substitutions are fluorine atoms; they contain no chlorine, but do contain hydrogen.

[40] A. Makhijani, A. Bickel, and A. Makhijani, 1990, "Still Working on the Ozone Hole," *Technology Review*, Spring, pp. 53–59.

[41] J. Holusha, 1990, "Du Pont to Construct Plants for Ozone-safe Refrigerant," *New York Times*, 23 June.

[42] D. Fisher et al., 1990, "Model Calculations of the Relative Effects of CFCs and Their Replacements on Stratospheric Ozone," *Nature*, vol. 344, pp. 508–512. See also J. Houghton, G. Jenkins, and J. Ephraums (eds.), 1990, *Climate Change: The IPCC Assessment*, Cambridge University Press, Cambridge.

[43] J. Leggett, 1990, "Playing Roulette with the Atmosphere," *New Scientist*, 7 July, p. 16.

[44] In Jones and Wigley, 1989, p. 224.

[45] See, for example, the calculations in M. Prather and R. Watson, 1990, "Stratospheric Ozone Depletion and Future Levels of Atmospheric Chlorine and Bromine," *Nature*, vol. 344, pp. 729–734.

[46] *New Scientist*, 1990, "Ozone Depletion Quickens at Both Poles," 4 August, p. 32. The 1991 discovery made front page news: *New York Times*, 5 April 1991.

[47] Indeed, there are signs that, even as late as 1988, the chemical industry had not yet learned its lesson. For instance, Britain's Imperial Chemical Industries (ICI), while calling for a phase-out of CFC use in England, would not rule out *exporting* quantities at least as great as those to be forgone in the United Kingdom (Porritt, 1989, p. 248). ICI's response to the announcement by the United States Environmental Protection Agency of the ozone-depleting potential of methyl chloroform was that the report was "somewhat alarmist" and contained "wholly unrealistic projections." This is the same type of irresponsible behavior threatening the common good pioneered (and persisted in) by the tobacco industry for more than a quarter-century.

[48] See R. Benedick, 1989, "Ozone Diplomacy," *Issues in Science and Technology*, Fall, pp. 43–50.

CHAPTER FIVE

[1] G. Speth, 1988, "Environmental Pollution," *Earth '88: Changing Geographic Perspectives*, National Geographic Society, Washington, D.C.

[2] EPA statistics reported in *New York Times*, 17 August 1990.

[3] A. Nero, 1988, "Controlling Indoor Air Pollution," *Scientific American*,

May, pp. 42–48; K. Smith, 1988, "Air Pollution: Assessing Total Exposure in the United States," *Environment*, vol. 30, no. 8 (part I; October).

⁴ Urban air pollution is composed of dozens of gases and other chemical materials mixed in varying proportions. The different materials also often interact or chemically react with each other to form new compounds, such as ozone formed when hydrocarbons react with nitrogen oxides under the influence of sunlight. See J. Seinfeld, 1989, "Urban Air Pollution: State of the Science," *Science*, vol. 243 (10 February), pp. 745–752.

⁵ R. Watson, 1989, "Report on Reports," *Environment*, May, pp. 25–27. A good summary of air pollution's health effects can be found in H. French, 1990, "Clearing the Air," *Worldwatch Paper 94* (January), Worldwatch Institute, Washington, D.C.

⁶ Mexico City has been named the world's smoggiest city: W. Branigan, 1988, "Mexico City's Smoggy Nightmare," *Washington Post Weekly*, 12–18 December. For a global overview, see United Nations Environment Programme (UNEP) and World Health Organization (WHO), 1989, "An Assessment of Urban Air Quality," *Environment*, vol. 31, no. 8 (October).

⁷ K. Smith, 1989, "Dialectics of Improved Stoves," *Economic and Political Weekly* (New Delhi, 11 March); K. Smith, 1987, *Biofuels, Air Pollution, and Health: A Global Review*, Plenum, New York; K. Smith, 1988, "Air Pollution: Assessing Total Exposure in Developing Countries," *Environment*, vol. 30, no. 10 (part II; December).

⁸ For an overview, see J. MacKenzie and M. El-Ashry (eds.), 1989, *Air Pollution's Toll on Forests and Crops*, Yale University Press, New Haven.

⁹ R. Smith, 1852, "On the Air and Rain in Manchester," *Memoirs of the Literary and Philosophic Society of Manchester*, ser. 2, vol. 10, pp. 207–217.

¹⁰ See the chronology of ecosystem acidification in E. Gorham, 1989, "Scientific Understanding of Ecosystem Acidification: A Historical Review," *Ambio*, vol. 18, no. 3, pp. 150–154.

¹¹ G. Likens, F. Bormann, and N. Johnson, 1972, "Acid Rain," *Environment*, vol. 14, pp. 33–40.

¹² H. Rodhe, 1989, "Acidification in a Global Perspective," *Ambio*, vol. 18, no. 3, pp. 155–160.

¹³ S. Woodin and U. Skiba, 1990, "Liming Fails Acid Test," *New Scientist*, 10 March, pp. 50–53.

¹⁴ A. Blaustein and D. Wake, 1990, "Declining Amphibian Populations: A Global Phenomenon," *Trends in Ecology and Evolution*, vol. 5, pp. 203–204.

¹⁵ P. Ehrlich, D. Dobkin, D. Wheye, and S. Pimm, 1992, *A Field Guide*

to the Natural History of the Birds of Europe, Oxford University Press, Oxford, in press; J. McIntyre, 1988, *The Common Loon*, University of Minnesota Press, Minneapolis.

[16] In contrast, no thinning was seen in pied flycatchers, which migrate and start egg-laying soon after arrival. Decreases in eggshell quality, caused by insufficient calcium deposition, are related to soil types in breeding areas. Thinning has been most severe in woods growing on poor sandy soils but not detectable in woods on loam or clay soils. Uptake of calcium by trees is impaired by acidification of the poorer soils, and caterpillars (a major food source for the breeding birds) feeding on those trees also contain less calcium.

[17] E.-D. Schulze, 1989, "Air Pollution and Forest Decline in a Spruce (*Picea abies*) Forest," *Science*, vol. 244, (19 May), pp. 776–783; F. Pearce, 1990, "Whatever Happened to Acid Raid?" *New Scientist*, 15 September, pp. 57–60. See also World Resources Institute (WRI), 1986, *World Resources 1986*, Basic Books, New York, chapter 12, for a general discussion.

[18] International Institute for Applied Systems Analysis (IIASA), 1990, "Acid Facts," *Options*, September, p. 5.

[19] Pearce, 1990.

[20] J. Aber, K. Nadelhoffer, P. Steudler, and J. Melillo, 1989, "Nitrogen Saturation in Northern Forest Ecosystems," *BioScience*, vol. 39, no. 6, pp. 378–385.

[21] E. Cowling, 1989, "Recent Changes in Chemical Climate and Related Effects on Forests in North America and Europe," *Ambio*, vol. 18, no. 3, pp. 167–171.

[22] For instance, M. Haus and B. Ulrich, 1989, "Decline of European Forests," and reply by L. Blank, R. Sheffington, and T. Roberts, *Nature*, vol. 339, pp. 265–266. A good summary can be found in Pearce, 1990.

[23] See discussion in G. Likens, 1989, "Some Aspects of Air Pollution Effects on Terrestrial Ecosystems and Prospects for the Future," *Ambio*, vol. 18, no. 3, pp. 172–178.

[24] H. Rodhe and R. Herrera, 1988, *Acidification in Tropical Countries*. SCOPE Report 36, Wiley, Chichester, England.

[25] J. Kirchner and J. Harte, 1991, "Validation of a Predictive Geochemical Model of Lake Acidification," submitted to *Nature*.

[26] Cowling, 1989.

[27] W. Winner, H. Mooney, and R. Goldstein, 1985, *Sulphur Dioxide and Vegetation: Physiology, Ecology, and Political Issues*, Stanford University Press, Stanford, California.

[28] D. Charles, 1990, "East German Environment Comes into Light," *Science*,

vol. 247 (19 January), pp. 274–276; M. Simons, 1990, "Central Europe's Grimy Coal Belt: Progress, Yes, but at What Cost?" (1 April) and "Rising Iron Curtain Exposes Haunting Veil of Polluted Air" (8 April), *New York Times;* and P. Wallach, 1990, "Dark Days," *Scientific American,* August, pp. 16–20.

[29] M. Simons, 1990, "Pollution's Toll in Eastern Europe: Stumps Where Great Trees Used to Grow," *New York Times,* 19 March.

[30] H. Dovland, 1987, "Monitoring European Transboundary Air Pollution," *Environment,* vol. 29, no. 10, pp. 10–29.

[31] J. Caroll, 1989, "The Acid Challenge to Security," *The Bulletin of the Atomic Scientists,* October, pp. 32–34.

[32] W. Bown, 1990, "Europe's Forests Fall to Acid Rain," *New Scientist,* 11 August, p. 17; IIASA, 1990, "The Price of Pollution," *Options,* September, pp. 4–8.

[33] IIASA, 1990, "The Price of Pollution."

[34] Summarized in Pearce, 1990.

[35] Aber et al., 1989; H. Mooney, P. Vitousek, and P. Matson, 1987, "Exchange of Materials between Terrestrial Ecosystems and the Atmosphere," *Science,* vol. 238, pp. 926–932.

[36] W. Booth, 1990, "The Ugly Truth about the Chemical that Causes Acid Rain," *Washington Post Weekly,* 24–30 September, p. 38; Pearce, 1990.

[37] A. Weisman, 1989, "L.A. Fights for Breath," *New York Times Magazine,* 30 July.

[38] World Commission on Environment and Development, 1987, *Our Common Future,* (often called the "Brundtland Report"), Oxford University Press, New York.

[39] Sierra Club, 1990, "Clean Air Act Signing Begins New Era of Environmental Protection," *National News Report,* vol. 22, no. 14 (3 December), and "Congress and President Heed Cries for Clear Air: Clean Air Act of 1990 Enacted!" *Pollution Activist,* 20 November; *New York Times,* 1990, "Bush Signs First Major Revision of Anti-Pollution Law since 1977," 16 November.

[40] K. Schneider, 1990, "Lawmakers Reach an Accord on Reduction of Air Pollution," *New York Times,* 23 October; Trout Unlimited, "Clean Air Act of 1990," *Action Line,* Fall.

[41] This is a hot topic in the economic literature; a recent survey is R. Hahn, 1989, "Economic Prescriptions for Environmental Problems: How the Patient Followed the Doctor's Orders," *Journal of Economic Perspectives,* vol. 3, no. 2 (Spring), pp. 95–114.

[42] French, 1990; S.Boehmer-Christianson, 1990, "Curbing Auto Emissions in Europe," *Environment*, vol. 32, no. 6; R. Gould, 1989, "The Exhausting Options of Modern Vehicles," *New Scientist*, 13 May, pp. 42–47.

[43] This dilemma is well illustrated in M. Wald, 1990, "How Dreams of Clean Air Get Stuck in Traffic," *New York Times*, 11 March (Week in Review section, p. 1); M. Wachs, 1989, "U.S. Transit Subsidy Policy: In Need of Reform," *Science*, vol. 244 (30 June), pp. 1545–1549.

[44] "Beware of the Catch," *Ann Arbor News*, 7 January 1989; "Great Lakes Bird Count Is Increasing; Contamination of Animals Now Rising, Study Says," *Detroit News*, 8 June 1989; "Mercury in Everglades fish worries experts," *New York Times*, 14 March 1989; "Clean Water Won't Be Cheap as State Aims to Keep It Safe," *Star-Ledger*, 16 May 1989; "2 EPA Studies Confirm Threat to Fish of Dioxin from Paper Plants," *New York Times*, 14 March 1989; "Mining, Growth, Farming Weigh upon the Suwannee," *Miami Herald*, 21 August 1989.

[45] P. Abelson, 1990, "Volatile Contaminants of Drinking Water," *Science*, vol. 247 (12 January), p. 141; "EPA Finds Pesticides in Water of 38 States," *New York Times*, 14 December 1988. Aluminum mobilized in soil by acidification is washed into rivers and aquifers; there may be a connection to Alzheimer's disease: "Alzheimer's Linked to Aluminum in Water," *Wall Street Journal*, 13 January 1989; G. Ferry, 1990, "Alzheimer's and Aluminum—The Guesswork Goes On," *New Scientist*, 18 February, p. 27.

[46] S. Armstrong, 1988, "Marooned in a Mountain of Manure," *New Scientist*, 26 November, pp. 51–55; N. Harle, 1990, "The Ecological Impact of Over-Development: A Case Study from the Limbourg Borderlands," *The Ecologist*, vol. 20, no. 5, (September/October), pp. 182–189.

[47] T. Colburn et al., 1990, *Great Lakes, Great Legacy?* The Conservation Foundation (Washington, D.C.) and the Institute for Research on Public Policy (Ottawa, Canada). A summary of the study can be found in R. Liroff, 1990, "The Great Lakes Basin: A Great Resource at Risk," *Conservation Foundation Letter*, no. 5. More on the Great Lakes dilemma can be found in M. Brown, 1987, "Toxic Wind," *Discover*, November, pp. 42–49.

[48] Colburn et al., 1990.

[49] For details of this long-hidden scandal of waste disposal, see A. Ehrlich and J. Birks, 1990, *Hidden Dangers: The Environmental Consequences of Preparing for War*, Sierra Club Books, San Francisco. Much of this dangerous material has seeped down into aquifers, contaminating citizens' well water.

[50] J. Carpi, 1990, "Metal Illness," *E Magazine*, November/December, pp. 34–39.

[51] D. Laurence and B. Wynne, 1989, "Transporting Waste in the European Community: A Free Market?" *Environment*, vol. 31, no. 6, pp. 12–35.

[52] *Newsweek*, 1988, "The Global Poison Trade," 7 November, pp. 66–68; Laurence and Wynne, 1989; M. Uva and J. Bloom, 1989, "Exporting Pollution: The International Waste Trade," *Environment*, vol. 31, no. 5, pp. 4–43.

[53] Office of Technology Assessment, 1989, *Facing America's Trash: What Next for Municipal Solid Waste?* OTA-O-424, U.S. Government Printing Office, Washington, D.C.

[54] One must be cautious in drawing conclusions about percentages, because to the uninitiated it might at first glance seem that if population went up 34 percent and waste by 80 percent, the AT factor must be the major one. By doing the calculation in decimals, one sees that population is actually the dominant of the three factors, since the product of A × T just equals it.

[55] *Newsweek*, 1989, "Buried Alive," 27 November, pp. 67–78.

[56] H. Levenson, 1990, "Wasting Away: Policies to Reduce Trash Toxicity and Quantity," *Environment*, vol. 32, no. 2, pp. 11–36; J. Harte, C. Holdren, R. Schneider, and C. Shirley, 1991, *Toxics from A to Z*, University of California Press, Berkeley.

[57] H. Daly, 1977, *Steady-State Economics: The Economics of Biophysical Equilibrium and Moral Growth*, W.H. Freeman, San Francisco.

[58] Office of Technology Assessment, 1989.

[59] F. Trippett, 1989, "Summer of the Spills," *Time*, 13 July, p. 18; A. Davidson, 1990, *In the Wake of the Exxon Valdez*, Sierra Club, San Francisco.

[60] But it has since been greatly exceeded by the apparently deliberate spill in the Persian Gulf during the Gulf War in January 1991.

[61] M. Turner, 1990, "Oil Spill: Legal Strategies Block Ecology Communications," *BioScience*, vol. 40, no. 4, pp. 238–242; L. Dayton, 1990, "Alaska's Silent Spring," *New Scientist*, 24 March, pp. 25–26.

[62] S. Pain, 1989, "Alaska Has Its Fill of Oil," *New Scientist*, 12 August, pp. 34–40.

[63] L. Dayton, 1989, "Exxon Valdez's Human Toll Is Still Unknown," *New Scientist*, 12 August, p. 23; M. Barinaga, 1989, "Alaskan Oil Spill: Health Risks Uncovered," *Science*, vol. 245 (4 August), p. 463.

[64] Including impacts on their edible quality: L. Dayton, 1989, "Oil Contamination Makes Fish Tough and Rancid," *New Scientist*, 19 August, p. 27.

[65] For a broad view of the environmental risks of society's mobilization of vast amounts of oil, see *Wildlife Conservation*, 1990, "The Ills of Oil" (special section), November/December, pp. 35–81.

[66] J. Jackson et al., 1989, "Ecological Effects of a Major Oil Spill on Panamanian Coastal Marine Communities," *Science*, vol. 243, pp. 37–44 (6 January).

[67] E. Paté-Cornell, 1990, "Organizational Aspects of Engineering System Safety: The Case of Offshore Platforms," *Science*, vol. 250 (30 November), pp. 1210–1217.

[68] P. Byrnes, 1990, "The Price of Addiction," *Wilderness*, Summer, pp. 3–4; *Discover* listed the top several hundred for January through October 1990, and included seven million-gallon-plus spills.

[69] A. Toufexis, 1988, "The Dirty Seas," *Time*, 1 August, pp. 44–50.

[70] M. MacQuitty, 1988, "Pollution beneath the Golden Gate," *New Scientist*, 30 June, pp. 62–66; Toufexis, 1988.

[71] Personal observations by P.R.E. in 1989.

[72] Personal communication, 1989.

[73] *New Scientist*, 1990, "Drift-Net Ban Snagged by Britain," 27 October, p. 13.

[74] *New Scientist*, "Deformities Found in North Sea Fish Embryos," 27 January, p. 24; Toufexis, 1988.

[75] K. Forestier, "The Degreening of China," *New Scientist*, 1 July, pp. 52–58.

[76] Toufexis, 1988.

[77] J. Stansell, "A Mediterranean Holiday from Pollution," *New Scientist*, 5 May, pp. 28–29; M Batisse, 1990, "Probing the Future of the Mediterranean," *Environment*, vol. 32, June, pp. 4–9, 28–34.

[78] For a brief discussion, see P. Ehrlich and A. Ehrlich, 1990, *The Population Explosion*, Simon & Schuster, New York, pp. 85–87.

[79] J. Cairns, 1978, "Waterway Recovery," *Water Spectrum*, Fall, p. 28.

[80] E. Odum, 1989, "Input Management of Production Systems," *Science*, vol. 243 (13 January), pp. 177–182.

[81] H. Needleman, 1989, "The Quiet Epidemic: Low Dose Lead Toxicity in Children," speech delivered on receiving the Charles A. Dana Award for Pioneering Achievement in Health; C. Joyce, 1990, "Lead Poisoning 'Lasts Beyond Childhood,' " *New Scientist*, 13 January, p. 26; S. Young, 1990, "An Insidious Legacy Is Dulling America's Mind," *Washington Post Weekly*, 16–22 April, p. 25.

[82] H. Daly, 1973, "The Steady-State Economy: Toward a Political Economy of Biophysical Equilibrium and Moral Growth," in H. Daly (ed.), *Toward a Steady-State Economy*, W. H. Freeman, San Francisco, pp. 149–174. This

source gives details on the quotas, including suggestions on how to handle imports and an economic justification for the superiority of quotas over pollution taxes. See also Daly's 1977 book, *Steady-State Economics*.

CHAPTER SIX

[1] Considerably more than that is grazed occasionally, mainly in arid regions. Land-use data are from United Nations Food and Agriculture Organization (FAO), 1988, *FAO Production Yearbook, 1987*, FAO, Rome; summarized also in World Resources Institute (WRI), United Nations Environment Programme (UNEP), and United Nations Development Programme (UNDP), 1990, *World Resources 1990–91*, Basic Books, New York. FAO uses a more conservative definition of forest that accounts only for "closed" forest, not open woodland or small plots. Their land-use data thus show only 30 percent of the land in forests, while other estimates are closer to 40 percent.

[2] P. Vitousek, P. Ehrlich, A. Ehrlich, and P. Matson, 1986, "Human Appropriation of the Products of Photosynthesis," *BioScience*, vol. 36, pp. 368–373. The 60 percent or so of NPP on land not used or diverted by people is mostly produced in forests (especially tropical forests), wetlands, natural grasslands, and desert vegetation. In the study, only the biomass of trees actually killed each year in natural forests and of the grass consumed by livestock in natural grasslands was ascribed to human use. Wetlands are highly productive and do supply some food for human beings, but they were not included in the study except where their destruction has contributed to NPP losses. Much desert vegetation also is grazed, but only the fraction consumed and losses due to desertification were included.

[3] L. Brown et al., 1990, *State of the World 1990*, Norton, New York; L. Brown, 1988, "The Changing World Food Prospect: The Nineties and Beyond," *Worldwatch Paper 85*, Worldwatch Institute, Washington, D.C. See also previous editions of *State of the World*.

[4] L. Brown, 1989, "Reexamining the World Food Prospect," in L. Brown et al., *State of the World 1989*, Norton, New York.

[5] E. Eisenberg, 1989, "Back to Eden," *Atlantic Monthly*, November, pp. 57–89.

[6] Brown, 1988.

[7] U.S. Department of Agriculture, Economic Research Service, 1989, *World Agriculture Situation and Outlook Report*, Washington, D.C. Dregne estimated that slight deterioration had occurred on percentages of agricultural land ranging from 38 percent in Australia to 70 percent in North America; moderately damaged land (enough to reduce yield potential by 10 to 50 percent) ranged from 17 percent (South America) to 55 percent (Australia);

and severe damage (more than a 50 percent reduction in yield potential) on 6 percent (Europe) to 17 percent (Africa).

[8] Brown, 1988.

[9] B. Forse, 1989, "The Myth of the Marching Desert," *New Scientist,* 4 February, pp. 31–32.

[10] See discussion and references in Ehrlich and Ehrlich, 1990, *The Population Explosion,* Simon & Schuster, New York; N. Myers (ed.), 1984, *Gaia: An Atlas of Planet Management,* Doubleday, New York.

[11] S. Postel, 1989, "Halting Land Degradation," in L. Brown et al., *State of the World 1989,* Norton, New York; United Nations Environmental Programme (UNEP), 1984, *General Assessment of Progress in the Implementation of the Plan of Action to Combat Desertification 1978–1984,* Nairobi.

[12] Forse, 1989; *IUCN Bulletin,* "The Sahel: Facts and Fallacies," vol. 20, no. 7–9 (July/September), pp. 16–29.

[13] For details on cattle as engines of destruction in the American West, see D. Ferguson and N. Ferguson, 1983, *Sacred Cows at the Public Trough,* Maverick Books, Bend, Ore. Much of what follows is from this source and from P.R.E's 30 years of fieldwork in every state of the West. See also G. Wuerthner, 1990, "The Price Is Wrong," and R. Strickland, "Taking the Bull by the Horns," *Sierra,* September/October, pp. 38–49; and G. Wuerthner, 1991, "How the West Was Eaten," *Wilderness,* Spring, pp. 27–37.

[14] T. Egan, 1990, "In West, a Showdown over Rules on Grazing: Ranchers Defy Grazing Laws," *New York Times,* 19 August. The outcome is described in E. Marston, 1991, "Rocks and Hard Places," *Wilderness,* Spring, pp. 38–45.

[15] Council on Environmental Quality, 1981, *Desertification of the United States,* U.S. Government Printing Office, Washington, D.C.

[16] T. Egan, 1990. There are rumblings of change at the BLM too: R. Coniff, 1990, "Once the Secret Domain of Miners and Ranchers, the BLM Is Going Public," *Smithsonian,* September, pp. 30–47.

[17] In high-rainfall areas such as the East River drainage in Gunnison County, Colorado, cattle do relatively little ecological damage when stocked at reasonable rates.

[18] P. Ehrlich, 1986, *The Machinery of Nature,* Simon & Schuster, New York, pp. 264–268.

[19] See, for instance, E. Royte, 1990, "Showdown in Cattle Country," *New York Times Magazine,* 16 December, pp. 60–70.

[20] H. Dregne, 1984, "Combating Desertification: Evaluation of Progress," *Environmental Conservation,* vol. 11, no. 2, pp. 115–121.

[21] World Resources Institute (WRI), International Institute for Environment and Development (IIED), and United Nations Environmental Programme (UNEP), 1988, *World Resources 1988–89*, Basic Books, New York, chap. 13, "Rehabilitating and Restoring Degraded Lands"; Postel, 1989.

[22] F. El-Baz, 1990, "Do People Make Deserts?" *New Scientist*, 13 October, pp. 41–44.

[23] Postel, 1989; WRI, 1988.

[24] O. Sattaur, 1990, "The Green Solution for India's Poor," *New Scientist*, 15 September, pp. 28–29; A. Agarwal and S. Narain, 1990, *Towards Green Villages: A Strategy for Environmentally Sound and Participatory Rural Development*, Center for Science and Environment, New Delhi.

[25] Sattaur, 1990.

[26] M. Tobias, 1988, "Desert Survival by the Book," *New Scientist*, 17 December, pp. 29–31.

[27] Postel, 1989. Loess is wind-deposited soil, which can be many feet deep and is usually very fertile. The Loess Plateau was blessed with such soil, but has been heavily cultivated for centuries.

[28] This section has benefited from discussions with Imhoff. The basic reference is M. Imhoff et al., 1986, "Monsoon Flood Boundary Delineation and Damage Assessment Using Space-borne Imaging Radar and Landsat Data," *Photogrammetric Engineering and Remote Sensing*, vol. 53, pp. 405–413.

[29] WRI, 1988, p. 42; P.R.E, personal observation, 1989.

[30] Council on Environmental Quality, 1979, *Environmental Quality—1979*, Government Printing Office, Washington, D.C., p. 396.

[31] L. Brown, 1988, "The Changing World Food Prospect: The Nineties and Beyond," *Worldwatch Paper 85* (October), Worldwatch Institute, Washington, D.C.

[32] FAO figures cited in L. Brown et al., 1990, *State of the World 1990*, Worldwatch Institute, Washington, D.C.

[33] *Newsweek*, 1989, "Buried Alive," 27 November, pp. 67–78.

[34] N. Myers, 1988, "The Future of Forests," in L. Friday and R. Laskey (eds.), *The Fragile Environment*, Cambridge University Press, Cambridge, pp. 22–40.

[35] S. Postel, 1988, "Reforesting the Earth," in L. Brown et al., *State of the World 1988*, Norton, New York.

[36] S. Hamburg and C. Cogbill, 1988, "Historical Decline of Red Spruce Populations and Climatic Warming," *Nature*, vol. 331, pp. 428–430.

[37] N. Myers, 1989, *Deforestation Rates in Tropical Forests and Their Climatic Implications*, Friends of the Earth, U.K., London. Estimates vary, depending on sources of information and how deforestation is measured; for instance, the FAO does not count replacement of natural forest with palm or coffee plantations as deforestation. Myers used data from satellites as well as both governmental and nongovernmental sources within the tropical nations at risk.

[38] Myers, 1989.

[39] J. Brooke, 1990, "Trying to Reclaim the Rain Forest," *New York Times*, 9 October, p. B7.

[40] T. Lovejoy, 1985, "Amazonia, People, and Today," in G. Prance and T. Lovejoy (eds.), *Key Environments: Amazonia*, Pergamon, Oxford, pp. 328–338.

[41] For a fuller explanation of the irreversibility of tropical deforestation, see P. and A. Ehrlich, 1981, *Extinction: The Causes and Consequences of the Disappearance of Species*, Random House, New York. An overview of the tropical forest dilemma can be found in S. Head and R. Heinzman (eds.), 1990, *Lessons of the Rainforest*, Sierra Club Books, San Francisco.

[42] N. Myers, 1984, *Gaia: An Atlas of Planet Management*, Doubleday, New York.

[43] R. Repetto, 1988, *The Forest for the Trees? Government Policies and Misuse of Forest Resources*, World Resources Institute, Washington, D.C.

[44] Brooke, 1990.

[45] D. Janzen, 1988, "Guanacaste Park: Tropical Ecological and Biocultural Restoration," in J. Cairns, Jr. (ed.), *Rehabilitating Damaged Ecosystems*, CRC Press, Boca Raton, Fla.

[46] Janzen, 1988; see also WRI, IIED, and UNEP, 1988, p. 220.

[47] Myers, 1984.

[48] M. Daria, quoted by the Associated Press, 27 October, 1987. Similar sentiments were expressed by the late Prime Minister Rajiv Gandhi the previous year: Government of India, 1986, "Strategies, Structures, Policies: National Wastelands Development Board," 6 February (mimeo), New Delhi.

[49] B. Barber, 1991, "The Gulf Conflict Is Also Deadly for the Environment," *Washington Post Weekly*, 7–13 January, 1991. Barber cites the World Bank as his source for the energy figures provided.

[50] S. Postel and L. Heise, 1988, "Reforesting the Earth," in L. Brown et al., *State of the World 1988*, Norton, New York.

[51] E. Eckholm, G. Foley, G. Bernard, and L. Timberlake, 1984, *Fuelwood: The Energy Crisis That Won't Go Away*, Earthscan, Washington, D.C.

⁵² R. Hutchinson, 1988, "A Tree Hugger Stirs Villagers in India to Save Their Forests," *Smithsonian*, February, pp. 185–195.

⁵³ R. Hanbury-Tenison, 1990, "No Surrender in Sarawak," and M. Cross, "Logging Agreement Fails to Protect Sarawak," *New Scientist*, 1 December, pp. 23, 28–29; *Environesia*, "Who Is Violating Whose Laws?" Friends of the Earth Indonesia, Jakarta, April/August.

⁵⁴ At the least, large-scale deforestation may change the regional climate enough to make reestablishment of the forest problematic: J. Lean and D. Warrilow, 1989, "Simulation of the Regional Climatic Impact of Amazon Deforestation," *Nature*, vol. 342 (23 November), pp. 411–413.

⁵⁵ Repetto, 1988.

⁵⁶ L. Dayton, 1990, "New Life for Old Forest," *New Scientist*, 13 October, pp. 25–29; E. Norse, 1990, *Ancient Forests of the Pacific Northwest*, Island Press, Washington, D.C.

⁵⁷ Dayton, 1990; Norse, 1990.

⁵⁸ M. Lipske, 1988, "Who Runs America's Fore$t$?" *National Wildlife*, October/November, pp. 24–28.

⁵⁹ E. Norse, 1990, See also R. Rice, 1990 "Old-Growth Logging Myths: The Ecological Impact of the U.S. Forest Service's Management Policies," *The Ecologist*, vol. 20, no. 4 (July), pp. 141–146.

⁶⁰ For instance, L. Kysar, 1990, "My Turn: A Logger's Lament," *Newsweek*, 22 October, p. 10.

⁶¹ T. Egan, 1990, "U.S. Declares Owl to Be Threatened by Heavy Logging," *New York Times*, 23 June.

⁶² P. Raven, in Foreword to Norse, 1990, p. xx (our emphasis).

⁶³ A. Gillis, 1990, "The New Forestry," *BioScience*, vol. 40, no. 8, pp. 558–562; Dayton, 1990; J. Luoma, 1990, "New Logging Approach Tries to Mimic Nature, *New York Times*, 12 June; W. Booth, 1990, "A New Leaf from the Forest Service," *Washington Post Weekly*, 13–19 August.

⁶⁴ J. Lancaster, 1989, "The Amazon Isn't the Only Rain Forest That's Disappearing," *Washington Post Weekly*, 11–17 September.

⁶⁵ G. Laycock, 1987, "Trashing the Tongass," *Audubon*, November, pp. 108–127.

⁶⁶ S. Holmes, 1990, "Senate Passes Measure to Protect Alaskan Forest," *New York Times*, 14 June. The House and Senate bills were signed into law in late 1990.

⁶⁷ Inflation is also a factor in discounting, which is usually calculated by economists in terms of the real interest rate—that is, the interest rate corrected for inflation.

[68] G. Hardin, 1968, "The Tragedy of the Commons," *Science*, vol. 162, pp. 1243–1248.

[69] H. Brough, 1991, "A New Lay of the Land," *World Watch*, January/February, pp. 12–19.

[70] R. Peters, A. Gentry, and R. Mendelsohn, 1989, "Valuation of an Amazonian Rainforest," *Nature*, vol. 339, pp. 655–656.

[71] T. Lovejoy, 1985, "Rehabilitation of Degraded Tropical Forest Lands," Commission on Ecology Occasional Paper Number 5, International Union for Conservation of Nature and Natural Resources; Brooke, 1990.

[72] J. Berger (ed.), 1990, *Environmental Restoration*, Island Press, Washington, D.C. See also Berger's earlier book, focused on restoration work being done in the United States: *Restoring the Earth*, Knopf, New York (1985); and Cairns (ed.), 1988, *Rehabilitating Damaged Ecosystems* (2 volumes).

[73] J. Terborgh, 1989, *Where Have All the Birds Gone?*, Princeton University Press, Princeton.

[74] N. Collar and P. Andrew, 1988, *Birds to Watch: The ICBP World Check-List of Threatened Birds*, International Council for Bird Preservation, Technical Publication no. 8, ICBP, Cambridge.

[75] See Ehrlich and Ehrlich, 1981; N. Myers, 1979, *The Sinking Ark*, Pergamon, New York; and N. Myers, 1981, *The Primary Source*, Norton, New York.

[76] E. Wilson, 1989, "Threats to Biodiversity," *Scientific American*, September.

[77] A. Gentry, 1990, paper presented at Crafoord Prize Symposium, Swedish Royal Academy of Sciences, Stockholm, September 28.

[78] M. Soulé, 1990, Crafoord Prize Symposium, Swedish Royal Academy of Sciences, Stockholm, September 28.

[79] P. Raven, 1990, "Botanists in a Fast Moving World," *New Scientist*, 13 October.

[80] J. Ayers, 1989, "Debt-for-Equity Swaps and the Conservation of Tropical Rain Forests," *Trends in Ecology and Evolution*, vol. 4, pp. 331–332.

[81] A. Umana Quesada, 1990, "Banks, Debt, and Development," *International Environmental Affairs*, vol. 2, pp. 140–149.

[82] Lengthy negotiations between donors and host governments are required to put the projects together and settle the details of discounts and methods of payment of local currency. Debt-for-equity swaps in general can have the undesirable side effect of fueling inflation in poor nations, but on any anticipated scale, debt-for-nature swaps are unlikely to make a significant difference.

[83] R. Wagner, 1990, "Doing More with Debt-for-Nature Swaps," *International Environmental Affairs*, vol. 2, pp. 160–165.

[84] Wilson, 1989.

[85] A. Leopold, 1949, *A Sand County Almanac and Sketches from Here and There*, Oxford University Press, New York.

[86] P. Ehrlich and A. Ehrlich, 1970, *Population, Resources, Environment*, W.H. Freeman, San Francisco.

[87] J. Cooley, 1984, "The War over Water." *Foreign Policy*, no. 54, pp. 3–26. See also J. Starr, 1991, "Water Wars," *Foreign Policy*, no. 82, pp. 17–36.

[88] C. Murphy, 1990, "Could a Water Shortage Cause a War?" *Washington Post Weekly*, 19–25 March, p. 15.

[89] Cable News Network Headline News, 14 October 1990.

[90] Wang Jusi, 1989, "Water Pollution and Water Shortage Problems in China," *Journal of Applied Ecology*, vol. 26, pp. 851–857.

[91] See World Resources Institute (WRI), 1988, *World Resources 1988–89*, Basic Books, New York, p. 129, table 8.2. There is a misprint in the table—the heading of area column should be "thousand cubic meters per hectare," not "thousand cubic kilometers per hectare"!

[92] For a more thorough discussion of water resources, see P. Ehrlich, A. Ehrlich, and J. Holdren, 1977, *Ecoscience: Population, Resources, Environment*, Freeman, San Francisco, chap. 6.

[93] See WRI, 1988, p. 128.

[94] M. Reisner, 1986, *Cadillac Desert: The American West and Its Disappearing Water*, Viking, New York.

[95] The story of the "capture" of water to support the Los Angeles basin is brilliantly told in Reisner, 1986.

[96] M. Reisner, 1990, "The Big Thirst," *New York Times Magazine*, 28 October, pp. 36–60.

[97] In particular, going to enormous expense and doing untold damage to support undesirable expansion of car-jammed and smog-strangled cities.

[98] In 1990, the United States Forest Service filed a water claim to maintain minimum flows in western Colorado's streams, to protect watersheds, prevent erosion and floods, and help preserve wildlife and ecological values (R. Suro, 1990, "U.S. Fights Colorado for Rockies Water," *New York Times*, 4 February.)

[99] *Time*, "The Last Drops," subtitled "Population Growth and Development Have Depleted and Polluted the World's Water Supply, Raising the Risk of

Starvation, Epidemics, and Even Wars" (20 August, pp. 58–61). The article described the water problems of agriculture, the impact of water projects on ecosystems, and the effects of ecosystem disruption on rainfall and river flows. The potential for conflict over water was outlined, and the population component was present throughout. The article closed with a discussion of some of the measures that should be taken to deal with the problem.

Time awakened late to a sophisticated recognition of the population-resource crisis, and we have been critical of its coverage, but in this case they did it right. We had only minor quibbles with the article, one being that the authors apparently thought that the water problem "crept up on a world distracted by fears of global warming," although it has long been recognized by environmental scientists, who have issued many warnings. The authors also seemed unaware of the major water component in the 1967 Arab-Israeli War and did not mention the likelihood that water problems in key areas could be exacerbated by global warming.

[100] S. Postel, 1989, "Water for Agriculture: Facing the Limits," *Worldwatch Report 93* (December), Worldwatch Institute, Washington, D.C.

[101] P. Gleick, 1988, "Climate Change and California: Past, Present, and Future Vulnerabilities," in M. Glantz (ed.), *Societal Responses to Regional Climate Change: Forecasting by Analogy*, Westview Press, Boulder.

[102] As we saw for ourselves in November 1990.

[103] P. Fearnside, 1989, "Brazil's Balbina Dam: Environment versus the Legacy of the Pharoahs in Amazonia," *Environmental Management*, vol. 13, no. 4, pp. 401–423.

[104] R. Gribel, 1990, "The Balbina Disaster: The Need to Ask Why?" *The Ecologist*, vol. 20, no. 4, pp. 133–135; M. Simons, 1989, "Brazil Wants Its Dams, but at What Cost?" *New York Times*, 12 March; Fearnside, 1989.

[105] P. Fearnside, 1988, "China's Three Gorges Dam: 'Fatal' Project or Step toward Modernization?" *World Development*, vol. 16, no. 5, pp. 615–630; K. Forestier, 1989, "China Moves towards a Damming Verdict," *New Scientist*, 1 April, p. 23; O. Sattaur, 1989, "Dam Unleashes India's Reservoirs of Anger," *New Scientist*, 11 March, pp. 32–33; M. Fineman, 1990, "A Scheme to Harness India's Sacred Waters Brings Tempers to a Boil," *Smithsonian*, November, pp. 118–133.

[106] S. Postel, 1989.

[107] Ehrlich, Ehrlich, and Holdren, 1977; WRI, IIED, and UNEP, 1988; Postel, 1989.

[108] C. Darwin, 1952, *The Next Million Years*, Hart-Davis, London.

[109] See T. Jacobsen and R. Adams, 1958, "Salt and Silt in Ancient Mesopotamian Agriculture," *Science*, vol. 128, pp. 1251–1258. The exact

degree to which land degradation from irrigation led to the collapse of the Sumerian civilization is in dispute.

[110] See, for an experimental new method for draining usually difficult clay soils, R. Wiseman, 1989, "Quick Purge Puts Salty Land Back into Production," *New Scientist*, 14 October, p. 36.

[111] Postel, 1989.

[112] Reisner, 1986.

[113] Eisenberg, 1989.

[114] Postel, 1989.

[115] B. Keller, 1988, "Developers Turn Aral Sea into a Catastrophe," *New York Times*, 20 December; W. Ellis, 1990, "A Soviet Sea Lies Dying," *National Geographic*, February, pp. 73–93; Postel, 1989; V. Kotlyakov, 1991, "The Aral Sea Basin: A Critical Environmental Zone," *Environment*, vol. 33, no. 1, pp. 4–38.

[116] NBC *Today Show*, 29 October, 1990.

[117] *New Scientist*, 1989, "Soviet Cotton Threatens a Region's Sea—and Its Children," 18 November, p. 22.

[118] The NPP of an acre of wetland ecosystem is exceeded only by that of an acre of tropical forest.

[119] J. Gosselink et al., 1990, "Landscape Conservation in a Forested Wetland Watershed," *BioScience*, vol. 40, no. 8, pp. 588–600.

[120] P. Steinhart, 1990, "No Net Loss," *Audubon*, July, pp. 18–21. The July issue of *Audubon* was dedicated entirely to wetlands.

[121] J. Madison, 1990, "Green Suits, Gray Suits, and White Hats," *Audubon*, July, pp. 108–111.

[122] Steinhart, 1990. Environmental writer Peter Steinhart notes a long cultural tradition of aversion to swamps and marshes, which may well have been based in an association with insect-borne diseases such as malaria.

[123] L. Leopold, 1990, "Ethos, Equity, and the Water Resource," *Environment*, vol. 32, no. 2, pp. 16–41; W. Veissman, Jr., 1990, "A Framework for Reshaping Water Management," *Environment*, vol. 32, no. 4, pp. 11–35.

CHAPTER SEVEN

[1] World Bank, 1990, *World Bank Report*, Oxford University Press, New York. See also World Resources Institute (WRI), International Institute for Environment and Development (IIED), and United Nations Environmental

Programme (UNEP), 1988, *World Resources 1988–89*, Basic Books, New York. The most recent report on nutritional levels focuses on children, the most seriously affected age group: B.A. Carlson and T.M. Wardlaw, 1990, *A Global, Regional and Country Assessment of Child Malnutrition*, UNICEF Staff Working Paper no. 7, UNICEF, New York.

[2] The differences between the quoted numbers reflect three things: true uncertainties because data are difficult to obtain (especially where hunger is most widespread or acute, since governments are not anxious to collect accurate information about their own failures); different assumptions in estimating nutritional needs, food availability, and distribution among groups; and the practice of making "high" and "low" estimates. The lower estimates of hungry perople are mostly generated by the UN Food and Agriculture Organization (FAO), using unrealistic measures of food availability. For instance, per-capita "dietary energy supply" (DES) expressed as kilocalories per day is used as an index of consumption, even though it does not allow for wastage of food between harvest and consumption or for uneven distribution.

[3] United Nations Administrative Committee on Coordination, Subcommittee on Nutrition (ACC/SCN), 1987, *First Report on the World Nutrition Situation*, November.

[4] World Bank figures in WRI, 1988, *World Resources 1988–89*. See also WRI, UNEP, and United Nations Development Programme (UNDP), 1990, *World Resources 1990–91*, Oxford University Press, chap. 6.

[5] In 1990, over 1.5 billion people resided in nations where total food supplies are chronically too meager to supply everyone with adequate nutrition; an additional 500 million or so were in countries where supplies in 1989–1990 were below usual levels of consumption, and 90 million more lived in nations reporting famines. (Extrapolated from R. Chen, 1990, "The State of Hunger in 1990," in R. Chen [ed.], *The Hunger Report: 1990*, The Alan Shawn Feinstein World Hunger Program, Brown University, Providence, and the Executive Summary of the volume.)

[6] United Nations Children's Fund (UNICEF), 1990, *State of the World's Children*, UNICEF, New York. UNICEF estimates that some 14 million children die each year from some combination of malnutrition, unsanitary conditions, and poverty.

[7] FAO, 1987, *The Fifth World Food Survey*, Food and Agriculture Organization of the United Nations, Rome.

[8] For an excellent representation of this general view (as well as a fine summary of the world food situation as seen by professionals in agricultural development), published as the grain gluts reached their recent maximum, see M. Swaminathan and S. Sinha (eds.), 1986, *Global Aspects of Food Production*, Tycooly International, Oxford.

[9] A good summary is E. Messer, 1990, "Food Wars: Hunger as a Weapon of War," in Chen, pp. 27–35.

[10] A good example is F. Lappé and J. Collins, 1977, *Food First*, Houghton Mifflin, Boston. More recent editions of their work do acknowledge that population growth is a factor in causing hunger.

[11] L. Brown et al., 1989, *State of the World 1989*, Norton, New York.

[12] That is, assuming that grains now fed to animals were consumed directly by people.

[13] Numbers from R. Kates et al., 1988, *The Hunger Report: 1988*, The Alan Shawn Feinstein World Hunger Program, Brown University. Similar figures were produced in the 1990 report, based on the global harvest of 1988, which was considerably smaller than that of 1985. With that harvest, the numbers fall to 5.5 billion vegetarians, 3.7 near-vegetarians, and 2.8 billion getting 25 percent of their calories from animal products.

[14] For a recent overview of the difficulties facing agriculture, see D. Pimentel and C. Hall (eds.), 1989, *Food and Natural Resources*, Academic Press, New York.

[15] The average rate of increase in food production for the decade 1980–1990 was calculated from grain production figures of the United Nations Food and Agriculture Organization (FAO), and the 1989–1990 increase was based on preliminary USDA figures.

[16] Food and Agriculture Organization (FAO), 1987, *The Fifth World Nutrition Survey*, FAO, Rome.

[17] For details on Chinese agriculture, see S. Wittwer, Yu Y., Sun H., and Wang L., 1987, *Feeding a Billion*, Michigan State University Press, E. Lansing.

[18] L. Brown et al., 1990, *State of the World 1990*, Norton, New York.

[19] United States farm policy changes also accounted for some of the production drop in 1987, as several million acres of marginal, highly erodable land were taken out of production and set aside in "conservation areas." While this move caused a brief drop in the grain harvest (but a rise in average yields), the productivity of adjacent croplands will be enhanced and preserved over the long term (WRI, UNEP, and IIED, 1988; Brown et al., 1990).

[20] S. Schneider, 1989, *Global Warming*, Sierra Club Books, San Francisco.

[21] According to United States Department of Agriculture figures. The FAO, which starts its "years" with a different month, showed the drop as about 7 percent.

[22] United States Department of Agriculture (USDA), Foreign Research Ser-

vice (1990, August), *World Agricultural Production*. WAP 8-90, USDA, Washington, D.C.

[23] L. Brown, 1988, "The Changing World Food Prospect: The Nineties and Beyond," *Worldwatch Paper 85* (October), Worldwatch Institute, Washington, D.C.; Brown et al., 1990.

[24] Population Reference Bureau, 1990, *1990 World Population Data Sheet*, Population Reference Bureau, Washington, D.C.

[25] Brown et al., 1990.

[26] G. Daily and P. Ehrlich, 1990, "An Exploratory Model of the Impact of Rapid Change on the World Food Situation," *Proceedings of the Royal Society* (B), vol. 241, 22 September, pp. 232–244; also available as Paper no. 0034, Morrison Institute for Population and Resource Studies, Stanford University. See also Schneider, 1989.

[27] Land use figures are from FAO, 1989, *1988 Food Production Yearbook*, FAO, Rome.

[28] H. Dregne, 1982, "Impact of Land Degradation on Future World Food Production," ERS-677, U.S. Department of Agriculture, ERS, Washington, D.C.

[29] Brown et al., 1990.

[30] Modern agriculture is very dependent on fossil fuels, which are used to manufacture, transport, and apply fertilizers and pesticides, and to build irrigation systems and pump water. They are also employed to run farm machinery, dry grains, and power food distribution systems and storage facilities, Agriculture in poor nations is much less dependent on fossil fuels than in rich nations, but dependence is sufficient to cause great difficulties when oil prices rise and increase costs that the poor can ill afford to pay.

[31] Brown et al., 1990. See also K.A. Dahlberg, 1979, *Beyond the Green Revolution*, Plenum, New York, for details of the green revolution's shortcomings in less developed regions.

[32] Dahlberg, 1979.

[33] D. Andow and D. Davis, 1989, "Agricultural Chemicals: Food and Environment," in Pimentel and Hall, pp. 192–235.

[34] P. Ehrlich, A. Ehrlich, and J. Holdren, 1977, *Ecoscience: Population, Resources, Environment*, W.H. Freeman, San Francisco.

[35] A. Ehrlich, 1990, "Agricultural Contributions to Global Warming," in J. Leggett (ed.), *Global Warming: The Greenpeace Report*, Oxford University Press, New York; R. Watson, H. Rodhe, H. Oeschger, and U. Siegenthaler, 1990, "Greenhouse Gases and Aerosols," in J. Houghton, G. Jenkins, and

J. Ephraums, *Climate Change: The IPCC Assessment*, Cambridge University Press, New York.

[36] D. Pimentel and C. Hall (eds.), 1984, *Food and Energy Resources*, Academic Press, New York.

[37] Ehrlich, Ehrlich, and Holdren, 1977.

[38] Swaminathan and Sinha, 1986; Ehrlich, Ehrlich, and Holdren, 1977.

[39] An overview can be found in V. Cervinka, 1984, "Water Use in Agriculture," in Pimentel and Hall (eds.), pp. 142–163.

[40] S. Postel, 1989, "Water for Agriculture: Facing the Limits," *Worldwatch Report 93*, December, Worldwatch Institute, Washington, D.C.

[41] S. Postel, 1989; Brown et al., 1990.

[42] Weeds and plant pathogens are the other major competitors.

[43] National Research Council, 1989, *Alternative Agriculture*, National Academy Press, Washington, D.C.

[44] The cosmetic problem, of course, is one that consumer education could probably solve, but agribusiness has shown little interest in moving in that direction. The world (and consumers) would doubtless be healthier with a few more worms in apples and some blemishes on the skins of peaches.

[45] A. Brown, 1978, *Ecology of Pesticides*, Wiley, New York; J. Kalmakoff and J. Miles, 1980, "Ecological Approaches to the Use of Microbial Pathogens in Insect Control," *BioScience*, vol. 30, pp. 344–347.

[46] R. Roush and B. Tabashnik (eds.), 1990, *Pesticide Resistance in Arthropods*, Chapman & Hall, New York.

[47] D. Pimental et al., 1978. "Benefits and Costs of Pesticide Use in United States Food Production," *BioScience*, vol. 28, p. 778; D. Pimentel et al., 1989, "Environmental and Economic Impacts of Reducing U.S. Agricultural Pesticide Use" (manuscript).

[48] See also R. Lal, 1989, "Land Degradation and Its Impact on Food and Other Resources," in Pimentel and Hall, pp. 86–141.

[49] S. Postel, 1989, "Halting Land Degradation," in L. Brown et al., *State of the World 1989*, Norton, New York; S. Postel, 1989, "Water for Agriculture."

[50] D. Pimentel et al., 1989, "Benefits and Risks of Genetic Engineering in Agriculture," *BioScience*, vol. 39, pp. 606–614.

[51] P. Vitousek, P. Ehrlich, A. Ehrlich, and P. Matson, 1986, "Human Appropriation of the Products of Photosynthesis," *BioScience*, vol. 36, no. 6, pp. 368–373.

[52] P. Ehrlich and A. Ehrlich, 1990, *The Population Explosion*, Simon &

Schuster, New York; and P. Ehrlich and A. Ehrlich, 1981, *Extinction: The Causes and Consequences of the Disappearance of Species*, Random House, New York.

[53] D. Pimentel, W. Dazhong, S. Eigenbrode, H. Lang, D. Emerson, and M. Karasik, 1985, "Deforestation: Interdependency of Fuelwood and Agriculture," *Oikos*, vol. 46, pp. 440–412.

[54] Brown, 1988; Brown et al., 1990.

[55] Schneider, 1989.

[56] Daily and Ehrlich, 1990.

[57] Based on UNICEF's annual reports, *State of the World's Children*, UNICEF, New York; See also Carlson and Wardlaw, 1990.

[58] Using Lester Brown's estimate of future growth in grain production (Brown et al., 1990).

[59] Almost counterintuitively, in model scenarios where the trend in growth of food production was below that of population growth, imposing adverse weather changes caused by global warming on the food system did not have dramatic effects. Production deficits and great increases in mortality occur early in such simulations, reducing the population growth rate. This creates increases in per-capita production (since there are fewer people to eat what is produced) and lowers total consumption, allowing stocks to build up as a buffer against climate-induced reductions in harvests that occur later (this follows from the design of the model, which assumes that the food production trend will continue upward without being affected by the reductions in population size). Thus, climate-induced shortages had relatively little effect as long as the underlying upward trend of production increase (low though it was) could be maintained.

[60] A. Ehrlich, 1988. "Development and Agriculture," in P. Ehrlich and J. Holdren (eds.), *The Cassandra Conference*, Texas A & M University Press, College Station, Texas; Pimentel and Hall, 1984; G. Douglass (ed.), 1984, *Agricultural Sustainability in a Changing World Order*, Westview, Boulder.

[61] Pimentel and Hall, 1984.

[62] J. Power and R. Follett, 1987, "Monoculture," *Scientific American*, March, pp. 79–86.

[63] National Academy of Sciences, 1972, *Genetic Vulnerability of Major Crops*, NAS, Washington, D.C.

[64] E. Eisenberg, 1989, "Back to Eden," *Atlantic Monthly*, November, pp. 57–89.

[65] See National Academy of Sciences, 1972, and discussions in Ehrlich, Ehrlich, and Holdren, 1977, and Ehrlich and Ehrlich, 1981.

[66] J. Williams, 1986, "Germplasm Resources," in Swaminathan and Sinha, *Global Aspects of Food Production*, pp. 117–128. See also for an excellent general presentation, R. Rhoades, 1991, "The World's Food Supply at Risk," *National Geographic*, vol. 179, no. 4 (April), pp. 74–105.

[67] National Research Council, Committee on the Role of Alternative Farming Methods in Modern Production Agriculture, Board on Agriculture, 1989, *Alternative Agriculture*, National Academy Press, Washington, D.C.; C. Francis, C. Flora, and L. King (eds.), 1990, *Sustainable Agriculture in Temperate Zones*, Wiley, New York; C. Edwards, R. Lal, P. Madden, R. Miller, and G. House [eds.], 1990, *Sustainable Agricultural Systems*, Soil and Water Conservation Society, Ankeny, Ia.

[68] W. Jackson, W. Berry, and B. Colman (eds.), 1984, *Meeting the Expectations of the Land*, North Point Press, San Francisco.

[69] Brown et al., 1990; WRI, IIED, and UNEP, 1988, *World Resources 1988–1989*; information on the 1990 Farm Act from Sierra Club, 1990, *National News Report*, 1 November.

[70] P. Weber, 1990, "U.S. Farmers Cut Soil Erosion by One-Third," *World Watch*, July–August, pp. 5–6.

[71] R. Goodland, C. Watson, and G. Ledec, 1984, *Environmental Management in Tropical Agriculture*, Westview, Boulder; P. Fearnside, 1987, "Rethinking Continuous Cultivation in Amazonia," *BioScience*, vol. 37, no. 3, pp. 209–214; A. Ehrlich, 1988; Ehrlich and Ehrlich, 1981.

[72] N. Myers, 1989, *Deforestation Rates in Tropical Forests and their Climatic Implications*, Friends of the Earth, UK, London.

[73] Goodland, Watson, and Ledec, 1984; P. Fearnside, 1987.

[74] P. Fearnside, 1989, "Extractive Reserves in Brazilian Amazonia," *BioScience*, vol. 39, no. 6, pp. 387–393; C. Peters, A. Gentry, and R. Mendelsohn, 1989, "Valuation of an Amazonian Rainforest," *Nature*, vol. 339 (29 June), pp. 655–656; G. Prance, 1990, "Fruits of the Rainforest," *New Scientist*, 13 January, pp. 42–45; D. Dufour, 1990, "Use of Tropical Rainforests by Native Amazonians," *BioScience*, vol. 40, no. 9, pp. 652–659. A careful analysis of the various agricultural activities of local people and the possibilities as dictated by soils and climate, see P. Fearnside, 1985, "Agriculture in Amazonia," in G. Prance and T. Lovejoy, *Key Environments: Amazonia*, Pergamon Press, New York.

[75] *International Agricultural Development*, 1990, "A Better Way for the Forest: How Agroforestry Is Reviving Forest Land in the Philippines," September/October.

[76] G. Marten (ed.), 1986, *Traditional Agriculture in Southeast Asia: A Human Ecology Perspective*, Westview Press, Boulder. A more conventional view

can be found in G. Wrigley, 1982, *Tropical Agriculture: The Development of Production*, Longman, New York; and Swaminathan and Sinha, 1986.

[77] M. Dover and L. Talbot, 1987, *To Feed the Earth: Agro-Ecology for Sustainable Development*, World Resources Institute, Washington, D.C.; G. Douglass (ed.), 1984, *Agricultural Sustainability in a Changing World Order*, Westview Press, Boulder; Marten, 1986.

[78] Eisenberg, 1989; Jackson, Berry, and Colman, 1984.

[79] M. Altieri, 1983, *Agroecology: The Scientific Basis of Alternative Agriculture*, Division of Biological Control, University of California, Berkeley.

[80] Altieri, 1983.

[81] Wilson, 1989, "Threats to Biodiversity," *Scientific American*, September.

[82] R. Prescott-Allen and C. Prescott-Allen, 1990, "How Many Plants Feed the World?" *Conservation Biology*, vol. 4, no. 4 (December), pp. 365–374.

[83] M. Oldfield and J. Alcorn, 1987, "Conservation of Traditional Agroecosystems," *BioScience*, vol. 37, no. 3, pp. 199–208; Dufour, 1990; Fearnside, 1985.

[84] Ehrlich and Ehrlich, 1981, *Extinction*, p. 63.

[85] D. Vaughan and L. Sitch, 1991, "Gene Flow from the Jungle to Farmers," *BioScience*, vol. 41. no. 1, pp. 22–28.

[86] Cassava, unfortunately, is a nutritionally poor crop, but it would still be helpful if yields could be improved, and with new technologies it might also be possible to improve its content of protein, vitamins, and minerals.

[87] Swaminathan and Sinha, 1986; Wrigley, 1982; Marten, 1984.

[88] Pimentel et al., 1979; Pimentel et al., 1989.

[89] R. Sinha and F. Watters, 1985, *Insect Pests of Flour Mills, Grain Elevators, and Feed Mills, and Their Control*, Research Branch, Agriculture, Canada, Publication 1776, Ottawa.

[90] See, for instance, Andow and Davis, 1989.

[91] D. Horn, 1988, *Ecological Approach to Pest Management*, Elsevier, London.

[92] Sierra Club, 1990.

[93] R. Frisbie and P. Adkisson, 1985, *CIPM: Integrated Pest Management on Major Agricultural Systems*, Texas Agricultural Experiment Station MP-1616, Texas A & M University.

[94] P. Adkisson, R. Frisbie, J. Thomas, and G. McWhorter, 1985, "Impact of IPM on Several Major Crops of the United States," in Frisbie and Adkisson, *CIPM*.

⁹⁵ National Research Council, 1990.

⁹⁶ K. Holl, G. Daily, and P. Ehrlich, 1990, "Integrated Pest Management in Latin America," *Environmental Conservation*, vol. 17, no. 4, pp. 341–350.

CHAPTER EIGHT

¹ Perhaps the best general source on risk is R. Kates, C. Hohenemser, and J. Kasperson (eds.), 1985, *Perilous Progress: Managing the Hazards of Technology*, Westview Press, Boulder. See also W. Lowrance, 1976, *Of Acceptable Risk*, W. Kaufman, Los Altos, California; E. Crouch and R. Wilson, 1982, *Risk/Benefit Analysis*, Ballinger, Cambridge, Mass.; R. Wilson and E. Crouch, 1987, "Risk Assessment and Comparisons: An Introduction," *Science*, vol. 236, pp. 267–270; P. Slovic, 1987, "Perception of Risk," *Science*, vol. 236, pp. 280–285; M. Russell and M. Gruber, 1987, "Risk Assessment in Environmental Policy-Making," *Science*, vol. 236, pp. 286–290; and L. Lave, 1987, "Health and Safety Risk Analyses: Information for Better Decisions," *Science*, vol. 236, pp. 291–295.

² Basal-cell carcinoma or "rodent ulcer."

³ For a more detailed treatment of pitfalls in risk comparisons, see J. Holdren, 1982, "Energy Hazards: What to Measure, What to Compare," *Technology Review*, vol. 85, no. 3, pp. 32–38, 74–75.

⁴ Risk-benefit analyses generally only make much sense when the nonrisk components of cost are small—as when deciding whether or not to cross a street against the light when one is in a hurry. There are clear risks and benefits, but any monetary costs of the action are negligible. In the real world the problem is much more often ignoring very large environmental and sociopolitical risk components of cost in cost-benefit analyses, because they are difficult to monetize. Leaving them out can be a big mistake.

⁵ Logically, risks are subsets of costs. Other components of costs include capital costs, operating costs, maintenance costs, decommissioning costs, etc. There are also financial risks associated with many activities, and the financial community has well-established ways of accounting for these (e.g., the concept of "risk premium" in financial markets basically means that people want a higher return on a risky investment).

⁶ Economists generally speak of "cost-benefit" analyses, and look at the "expected value" of outcomes (the probability of a particular outcome multiplied by the consequence, either good or bad). In their cost-benefit analyses, the costs of a potential investment are usually balanced against the expected benefits to determine whether the investment is wise. In risk assessments (or risk-benefit analyses), the benefit may simply be avoidance of the risk,

or the analysis may be used to evaluate which of two necessary risks should be chosen (that is, which of two medical procedures to use to solve a heart problem), where economic costs are not considered. Technically, the relationship between probability and consequence in calculating risk is not necessarily strictly multiplicative, but we will not go into that in this book.

[7] Or the "opportunity cost" of the land—that is, the benefits that might be reaped from it in the economically highest alternative use of the land—say, growing a high-priced crop on it or developing it into an office complex. In a perfectly competitive market with no externalities, the price would accurately measure the opportunity cost. The costs of maintaining and policing the reserve should be counted only to the extent they are incremental over those associated with the highest alternative use.

[8] Given the following estimates: the present value of preserving the species in perpetuity $= \$1,000,000$; the probability of survival (P) without the reserve $= 0.1$; and P with the reserve $= 0.9$; then the expected benefit of the reserve $= (0.9 - 0.1) \times \$1,000,000 = \$800,000$. If the costs of creating and maintaining the reserve in perpetuity are more than $\$800,000$, the analysis suggests the reserve shouldn't be created.

[9] Cited in J. Holdren, 1983, "The Risk Assessors," *Bulletin of Atomic Scientists*," June/July, pp. 33–38.

[10] It is associated with what has become known as "physics envy"—basically the idea that only by being highly quantitative can one be "scientific." Physics envy has led to disciplines such as "micropolitical analysis" where political scientists do things like counting manhole covers to determine why the city council votes in a certain way.

[11] H. Daly, 1977, *Steady-State Economics*, Freeman, San Francisco, p. 172. The phenomenon is similar to the attitude toward the hideous slaughter of 100,000 people with an indiscriminate nuclear weapon impersonally delivered by an ICBM, and the personal shooting of that many people in the head by individuals.

[12] Holdren, 1983, p. 33.

[13] U.S. Nuclear Regulatory Commission, 1975, *Reactor Safety Study: An Assessment of Accident Risks in U.S. Commercial Nuclear Power Plants*, WASH-1400, NUREG-75/014, National Technical Information Service, Washington, D.C.

[14] CBS *Sixty Minutes*, December 30, 1990.

[15] G. Likens, 1990, "The Science of Nature, the Nature of Science," *The Scientist*, in press, November; J. Harte, Energy and Resources Group, University of California, Berkeley, personal communication, January 2, 1991.

[16] This is related to how people perceive the world—see R. Ornstein and P. Ehrlich, 1989, *New World/New Mind*, Doubleday, New York.

[17] See J. Holdren, 1976, "The Nuclear Controversy and the Limitations of Decision-Making by Experts," *Bulletin of the Atomic Scientists* vol. 32, pp. 20–22, March; and J. Holdren, 1976, "Zero-Infinity Dilemmas in Nuclear Power," *Reactor Safety Study*, Oversight Hearing before the Subcommittee on Energy and the Environment, Committee on Interior and Insular Affairs, U.S. House of Representatives, Serial 94-61, USGPO, Washington, D.C., pp. 357–364.

[18] P. Slovic, 1987, "Perception of Risk," *Science*, vol. 236 (17 April), pp. 280–286.

[19] One, of course, can argue that advertising and the extremely addictive nature of cigarettes removes some of the choice, especially in the case of teenagers, but this does not affect the basic point. The risk as calculated would be about .25 in a lifetime of 70 years (the annual risk is calculated as averaged over all years of life). The risk of death from smoking one cigarette is very roughly 1/2,000,000 (Wilson and Crouch, p. 173).

[20] For an attempt at technical treatment of risk in which the analysts appear to believe that the morality of risk does not count, see M. Douglas and A. Wildavsky, 1982, *Risk and Culture*, University of California Press, Berkeley. The analysis is largely incompetent, for example confusing stability and reversibility and mixing up the chance of harm with risk, although it contains some scattered excellent insights. For those with a professional interest in risk, it is interesting for what it reveals about the authors' attitudes.

[21] There are still some disagreements about the mechanisms of evolution, but all competent scientists agree that present-day organisms evolved from much simpler organisms in the past and that the process is on-going. In common language, evolution is a fact, although technically it is a theory, just as Earth's circling of the sun rather than vice versa is technically a theory. "Scientific creationism" is the antithesis of science— since it is based on authority, rather than being subject to constant test and revision. Any scientist would be delighted to demonstrate that the notion of evolution was as erroneous as the view that continents were immobile. In that direction lie fame and fortune for a scientist—but being able to do it is about as likely as showing that Earth does not circle the sun.

[22] T. Colburn et al., 1990. *Great Lakes, Great Legacy?* Conservation Foundation, Washington, D.C.

[23] B. Wattenberg, 1987, *The Birth Dearth*, Pharos Books, New York.

[24] See, for instance, P. Kennedy, 1987, *The Rise and Fall of Great Powers*, Random House, New York, and the analysis and other references in P. Ehrlich and A. Ehrlich, 1990, *The Population Explosion*, Simon & Schuster, New York, p. 170.

[25] S. Schneider, 1990, "Cooling It," *World Monitor*, July, pp. 30–38.

[26] See J. Gribbin, 1990, "An Assault on the Climate Consensus," *New Scientist*, 15 December, pp. 26–31.

[27] The ethics of animal testing have been the topic of extensive debate in recent years with, at the extremes, some claiming that all animal testing is immoral and others claiming that virtually all of it is justified. We believe that in some circumstances (such as testing medicines for serious human diseases and some food preservatives), animal testing, if done humanely, is not only justified but essential. For other substances, such as dyes to "improve" the appearance of food, it is not justified—indeed, such products should be banned. But there is a large gray area here, as there is in the use of animal testing in medical research, much of which has led to enormous improvements in human health, and some of which has been cruel and useless.

[28] A preservative, chemical compound X, to be added in very small quantities to food, would be considered unacceptable if it gave cancer to one in 10,000 people who ate normal quantities of the food. But it is impractical to test the equivalent low quantities, normally the same proportion of body weight, of X on tens of thousands of rats. Instead much heavier doses of X are given to many fewer rats, and an estimate of the low-dose risk made from that. This is a perfectly legitimate procedure, but it contains some assumptions that may not be true. It assumes that rats and people will react in the same manner to X, and it assumes a simple (that is, linear) relationship between dose and response—in essence, if X gives cancer to one out of ten rats in a test group at 1,000 times the normal dose, X would have given cancer to one in 10,000 rats at the normal dose.

The standard use of massive doses in testing leads to many jokes on late-night television about overdosed rats, or industry comments of the "you'd have to drink a tank-car full every day" variety, but they all miss the point. There simply is no other practical way to assess experimentally the low-dose risks that concern most of us, although epidemiological research and cross-cultural analyses can sometimes provide useful information. For knowledgeable people, decision making is even more complicated. For instance, the rats might be more susceptible to X than people, as X's manufacturer is likely to claim. Maybe so. Then again, maybe people are more susceptible to X than are rats. Even after X has been on the market for years, epidemiological studies, if carried out, would have a hard time separating a 1/5000 chance of X causing cancer from a 1/10,000 chance. For most substances, it is very difficult to get accurate estimates of people's exposures to a substance and to control for other factors.

For example, X might cause liver cancer, but so might peanut butter and a couple of hundred other foods. For an accurate test, a researcher would need full dietary information on a huge number of people, along with information on other factors that might cause cancer (working in a gas station, smoking, living in a smoggy locality, etc.). The tasks become gargantuan, which is why so few such studies are done and so much reliance is placed on animal testing. Indeed, smoking is one of the relatively few cases in which enough testing of various sorts has been done to provide utterly convincing evidence of carcinogenesis and other harms from a substance widely used in the population.

Things are even more complicated than that. The various tests done on animals examine only the effects of a substance by itself. But the presence of other substances can either reduce its effects or enhance them. Thus, the effect of X may be less in the presence of Z than it would be alone. For example, eating food containing high concentrations of vitamin C might help prevent the induction of liver cancer by aflatoxins, a common fungal contaminant in nut products. People with little vitamin C in their diets might be much more susceptible. In other cases, the effects of X and Y are greater than the sum of their individual effects; that is, they act *synergistically*. Smoking reduces the efficiency of the natural cleansing processes of the lungs and makes smokers more susceptible to carcinogens (such as asbestos) deposited in the lungs by increasing their residence time. Considering the difficulties of evaluating low-dose exposures to a single substance like X, it is easy to see why X's possible two-way interactions with literally tens of thousands of other substances can never be tested, to say nothing of any possible three-way, four-way, five-way, and so on interactions.

The same sorts of problems that plague determination of carcinogenesis of various chemical compounds also make determining the healthiness of various foods relatively difficult. A high-fat diet might increase the chances of heart attack in people who are genetically susceptible to coronary artery disease or who do not exercise, but might not pose a significant health risk to people with other genetic endowments or to athletes. When someone drops dead of a heart attack, no little sign appears on his forehead saying, "Too many eggs killed me," any more than a cancer victim might have "Red dye 23 did me in" written on her body. So arguments continue on the role of diet in various diseases, although most scientists agree that diet does play a significant role in many of them.

In diet the benefit side of the risk-benefit analysis is often more prominent than in questions of safety in food additives, cosmetics, and the like. While some food additives or cosmetic ingredients may provide significant benefits, most, in our view, could be relatively easily forgone or replaced. Foods like beef or eggs are another matter. They provide important dietary benefits, and for many people are important components in what they consider to be an adequate and tasty diet. The perceived benefits of various foods can be very large, and people are usually very hesitant to alter their diets drastically.

[29] See his article by that title in *Society*, vol. 27, no. 1, 1989, and the subsequent responses. For an excellent analysis of Wildavsky's confused thinking on risk and insight into disagreements in the field of risk assessment, see J. Holdren's 1983 joint review of M. Douglas and A. Wildavsky, 1982, *Risk and Culture* (University of California Press, Berkeley) and Wilson and Crouch, *Risk Benefit Analysis, Bulletin of Atomic Scientists*, June/July, pp. 33–37, and the exchange of correspondence in August/September, pp. 59–60.

[30] The assumptions behind Wildavsky's views are heroic: that all consumers are equally well informed and that prices reflect social costs—that is, there are no externalities associated with individual actions.

[31] See W. Catton, Jr., 1989, "Choosing Which Danger to Risk," *Society*, vol. 27, pp. 6–8; P. Ehrlich and A. Ehrlich, 1989, "Intelligent Planning for Safety," *Society*, vol. 27, pp. 15–16.

[32] See F. Dyson, 1975, "The Hidden Cost of Saying No!", and P. Ehrlich, 1975, "The Benefits of Saying Yes!", *Bulletin of Atomic Scientists*, vol. 31, pp. 23–27 and 49–51.

[33] For a nice discussion of applied cost-benefit analysis, see J. Calfee and J. Pappalardo, 1989, *How Should Health Claims for Foods Be Regulated? An Economic Perspective*, Bureau of Economics, Federal Trade Commission, September.

[34] As when they successfully conspired to destroy the streetcar system in Los Angeles.

[35] One hundred thousand human generations is roughly 2.5 million years. It would take *at least* that long for an impoverished biota to be significantly reenriched by the process of speciation. Recovery from the extinctions at the cretaceous-tertiary boundary took tens of millions of years.

[36] For more on the social consequences of human perceptual limitations, see Ornstein and P. Ehrlich, 1989.

[37] Ornstein and Ehrlich, 1989.

[38] Numbers that are reasonable on the basis of the results of G. Daily and P. Ehrlich, 1990, "An Exploratory Model of the Impact of Rapid Climate Change on the World Food Situation," *Proceedings of the Royal Society*, ser. B, vol. 241, pp. 232–244.

[39] Even in poor African nations, the life of a young person represents on the order of $20,000 in future earnings.

[40] As a result, for example, of retraining coal miners to work in less dangerous occupations.

CHAPTER NINE

[1] Population issues are not discussed in detail in this book. For those details, see P. Ehrlich and A. Ehrlich, 1990, *The Population Explosion*, Simon & Schuster, New York.

[2] The United States is a particularly niggardly provider of foreign aid as a fraction of the nation's GNP; Scandinavian countries give four or five times as much. Moreover, most American aid goes to nations it sees as strategically important, such as Israel and Egypt (World Bank, 1990, *World Development Report, 1990*, Oxford University Press, New York).

[3] The population situation and possible policies addressing it are discussed in depth in Ehrlich and Ehrlich, 1990. See also L. Grant (ed.), 1991, *The Elephants in the Volkswagen; Perspectives on Optimum U.S. Population*, Freeman, New York.

[4] World Commission on Environment and Development, 1987, *Our Common Future*, Oxford University Press, Oxford (called the "Brundtland Report" after the chairman of the commission, Gro Harlem Brundtland, Prime Minister of Norway).

[5] The gross national product is the value of all the goods and services produced by a nation in a single year. GNP is for the moment the standard measure of "affluence" or "development."

[6] For example, R. Lipsey and P. Steiner, 1975, *Economics*, Harper & Row, New York.

[7] The following section is adapted from P. Ehrlich, 1989, "The Limits to Substitution: Meta-Resource Depletion and a New Economic-ecological Paradigm," *Ecological Economics*, vol. 1, pp. 9–16. See also Ehrlich and Ehrlich, *The Population Explosion*, chap. 8. We owe a great debt here to pioneering economist Herman Daly, whose writings on sustainable economic systems have greatly influenced our thinking.

[8] A Pareto-optimal distribution of a given set of goods (or resources) is one in which no person can be made better off without making someone else worse off. If Pareto optimality is defined in an intergenerational sense (so that the welfare of future as well as present consumers is considered), then only economies of a sustainable scale would be likely to meet the Pareto optimality criterion.

[9] The global energy economy has grown from about 1 terawatt in 1890 to about 13 terawatts today.

[10] Some farsighted economists have appreciated this and have attempted to swim against the tide of dogma in this area, Herman Daly being an outstanding example today. But Daly's push for consideration of a steady-state or sustainable economy (building on the work of John Stuart Mill, Kenneth

Boulding, and Nicholas Georgescu-Roegen) has thus far been largely ignored by the establishment within the economic community.

[11] C. Perrings, *Economy and Environment*, 1987, Cambridge University Press, London; R. Hahn and R. Stavins, 1991, "Incentive-based Environmental Regulation: A New Era from an Old Idea?" *Ecology Law Quarterly* (in press); R. Stavins, 1990, "Innovative Policies for Sustainable Development: The Role of Economic Incentives for Environmental Protection," *Harvard Public Policy Review*, vol. 7, Spring, pp. 13–25. See also the new journal (started in 1989) of the International Society for Ecological Economics, *Ecological Economics*, edited by Robert Costanza (Center for Environmental and Estuarine Studies, University of Maryland, Solomons, Md.).

[12] Some economists share with Wilfred Beckerman the notion that economic growth has gone on since the time of Pericles and can continue for another 2,500 years into the future (W. Beckerman, 1972, "Economists, Scientists, and Environmental Catastrophe," Inaugural lecture, University College, London, 24 May [unpublished]. A few simple calculations show that idea to be ludicrous (J. Parsons, 1977, *Population Fallacies*, Elek/Pemberton, London.)

[13] D. Colander and A. Klamer, 1987, "The Making of an Economist," *Economic Perspectives*, vol. 1, pp. 95–111.

[14] R. Stavins, 1990.

[15] We are especially grateful to Kenneth Arrow, Herman Daly, Timothy and Lisa Daniel, Partha Dasgupta, Wally Falcon, and Lawrence Goulder for conversations that have helped shape our thinking on economic issues— even though they certainly would not endorse all of our conclusions on economy-ecology relationships.

[16] M. Frome, 1990, "Free Speech and Forestry," *Authors' Guild Bulletin*, Spring, pp. 23–24.

[17] Quoted in Frome, 1990.

[18] Those two brilliant thinkers did have a strong ecological element in their world view. (See H. Parsons [ed.], 1977, *Marx and Engels on Ecology*, Greenwood Press, Westport, Conn. Readers might also be interested in S. Avineri [ed.], 1969, *Karl Marx on Colonialism and Modernization*, Doubleday [Anchor], New York). Marx basically endorsed colonialism because, as Avinera put it: "Just as the horrors of industrialization are dialectically necessary for the triumph of communism, so the horrors of colonialism are dialectically necessary for the world revolution of the proletariat since without them the countries of Asia (and presumably also Africa) will not be able to emancipate themselves from their stagnant backwardness" (p. 13). Marx's attitude toward what Rudyard Kipling called "lesser breeds without the law" is encapsulated in the following statement from a letter Marx wrote to the

New York Daily Tribune (8 August, 1853): "Indian society has no history at all, at least no known history." Marx could not help being a man of his own century; that many of his ideas are wrong or offensive after the passage of 150 years does not diminish his greatness, but it should long ago have alerted Marxists to the dangers of assuming that everything Marx said was of value or pertinence today. Marx and Engels were opposed to Malthus's ideas on population (as was appropriate at the time they were writing). Unhappily, most of their followers seem to believe that Marx and Engels, if alive, would still be opposed—that they would believe today what they believed in the middle of the last century. They were too bright for that.

Modern Marxists too often forget how often Marx changed his views during his life and ignore his statement in the preface to the first volume of *Capital:* "I pre-suppose, of course, a reader who is willing to learn something new and therefore to think for himself" (International Publishers edition, 1967, p. 8, translation from preface of first edition). In fact, the ritual blindness of some left-wing environmentalists (such as B. Commoner, 1971, *The Closing Circle*, Knopf, New York) to the population element in the I = PAT equation has had some divisive effect on the environmental movement. In general, though, people from all parts of the political spectrum realize that both "population and pollution" are important elements of the environmental crisis (S. Fox, 1981, *John Muir and His Legacy: The American Population Movement*, Little, Brown, Boston).

[19] K. Phillips, 1990, *The Politics of Rich and Poor*, Random House, New York.

[20] D. Pirages and P. Ehrlich, 1974, *Ark II: Social Response to Environmental Imperatives*, Viking, New York.

[21] R. Ornstein and P. Ehrlich, 1989, *New World/New Mind*, Doubleday, New York.

[22] Phillips, 1990.

[23] This might not be true if the educational system continues to decay and only the children of the rich gain the training necessary to keep society functioning. Measures to curb reproduction must be coupled with measures to increase equality of education and opportunity. The problem of high birth rates among the poor, especially those too poor to support children adequately, is extremely complex and cannot be solved by tax or other financial penalties, which only harm the children and eventually result in higher costs to society.

[24] The data are not available to quantify this implicit cost-benefit test, but obviously only a small portion of American productivity is used to offset crucial problems created by consumption.

[25] Between 1970 and 1990, the United States population grew by 20 percent and America's vehicle population grew by 40 percent.

²⁶ P. Berg, 1983 (editorial), "Raise the Stakes."

²⁷ Quoted in B. Devall and G. Sessions, 1985, *Deep Ecology*, Gibbs M. Smith, Salt Lake City. A good introduction to this complex topic is available in the combination of this book and in the fine new volume by W. Fox, 1990, *Toward a Transpersonal Ecology: Developing New Foundations for Environmentalism*, Shambhala, Boston.

²⁸ Fox, 1990, deals in detail with most of these issues. See also D. Ehrenfeld, 1978, *The Arrogance of Humanism*, Oxford, New York. For a Native American's view of the relationship between human beings and Earth, see Ed McGaa (Eagle Man), 1990, *Mother Earth Spirituality*, Harper & Row, New York.

²⁹ For an introduction to what we have called "deep economics," see H. Daly and J. Cobb, Jr., 1989, *For the Common Good: Redirecting the Economy toward Community, the Environment, and a Sustainable Future*, Beacon Press, Boston. Daly is an economist who deals directly with issues of value, and we strongly agree with his goals (although sometimes differing with him on points of both economics and philosophy). There are also "deep physicists" groping for a transformed worldview and values; see F. Capra, 1982, *The Turning Point: Science, Society, and the Rising Culture*, Simon & Schuster, New York.

³⁰ The "gap" between human beings and other animals assumed by older philosophies has been shown to be much less wide than previously thought; see Ornstein and Ehrlich, 1986, *New World/New Mind*, Simon & Schuster, New York.

³¹ See Fox, 1990.

³² A. Lovins and L. Lovins, 1990, "Make Energy Efficiency Our Gulf Strategy," *New York Times*, 3 December.

³³ M. Closson, 1990, "Eco-Conversion: Beyond Oil," *Earth Island Journal*, Fall, p. 30.

³⁴ P. Webos, 1990, "Energy and Population: Transitional Issues and Eventual Limits," *The NPG Forum*, August.

³⁵ See the following issues of *The Defense Monitor* (Center for Defense Information, Washington, D.C.): "Wasteful Weapons," 1989 (vol. 18, no. 7); "Preparations for Nuclear War: Still More than $1 Billion a Week," 1990 (vol. 19, no. 7); "Star Wars Reality: The Emperor Has No Clothes," 1988 (vol. 17, no. 1).

³⁶ The other part would be needed to clean up the mess already made by weapons programs (A. Ehrlich and J. Birks, 1990, *Hidden Dangers; The Environmental Consequences of Preparing for War*, Sierra Club, San Francisco). Much of it would almost certainly be diverted to deficit reduction

and reducing the tax burden. The actual economics would be *much* more complex than represented here.

[37] J. Holdren, 1990, "Energy in Transition," *Scientific American*, September, pp. 157–163.

[38] Ehrlich and Birks, 1990.

[39] D. Meadows, 1990, "CIA Doesn't Need Enemies to Do Crucial Intelligence Work," *Valley News* (Hanover, New Hampshire), 1 December.

[40] L. Brown, 1977, *Redefining National Security*, Worldwatch Institute Paper no. 14; N. Myers, 1987, *Not Far Afield: U.S. Interest and the Global Environment*, World Resources Institute, Washington, D.C.; P. Ehrlich and A. Ehrlich, 1989, "The Environmental Dimensions of National Security," in J. Rotblat and V. Goldanskii (eds.), *Global Problems and Common Security: Annals of Pugwash 1988*, Springer, Berlin, pp. 180–190; P. Gleick, 1990, "Environment, Resources, and International Security and Politics," in *Science and International Security: Responding to a Changing World*, American Association for the Advancement of Science, Washington, D.C.

[41] The best estimate is 200 million deaths from hunger and hunger- or poverty-related causes during two decades, while about 16 million people perished in the four decades of wars following World War II (N. Myers, 1984, *Gaia: An Atlas of Planet Management*, Doubleday, New York). The two sets of numbers are not independent, however, since war often causes hunger. Nevertheless, given the above numbers, the estimate of a tenfold difference seems conservative.

[42] For instance, George Will stated that the Gulf war was basically a justification for past large arms buildups, "Star Wars," the military-industrial complex, and so on, implying that American military power would be much needed in the future to defend American interests. (ABC *This Week*, 20 January, 1991).

[43] The number is very approximate. It assumes that about 5 percent of total United States military expenditures (in 1990 almost $1 billion per day) could be assigned to operations and preparations related to the Middle East in connection with Persian Gulf oil flow of roughly 2 million barrels of oil a day to the United States. Thus $.05 \times 1$ billion = 50 million, and 50/2 = $25/barrel.

[44] The cost of the first two months of the military deployment after the invasion of Kuwait was estimated to be about an additional $50 million daily, or about $25 each for the 2 million barrels. But one could argue that the military action should be viewed as an attempt to safeguard the entire 12-million-barrel flow of oil from the Gulf to other nations, in which case the additional cost would only be around $4 per barrel.

[45] At this writing (April 1991), it is not possible to estimate it accurately,

although estimates of war costs by the press ranged around $1 billion per day, or over $40 billion for six weeks of combat. No estimates were available for the aftermath, including care for millions of refugees.

[46] C. Tudge, 1990, "Grasping the Green Nettle," *New Scientist*, 6 October.

[47] Nonlinearities, threshold effects, and interactions generally make the environmental impact of large populations grow disproportionately.

CHAPTER TEN

[1] The Earthworks Group, 1990, Earthworks Press, Berkeley, California.

[2] Simon & Schuster, New York, 1990.

[3] Scientists themselves, after all, are the outcome of an evolutionary process.

[4] Life expectancies, unless otherwise stated, are life expectancies at birth. Most of the variation in life expectancy in the past, and we suspect in the future, will be governed by death rates in the first few years of life; stating that life expectancies will be dramatically lower means that many more infants and young children will be dying prematurely.

[5] Environmental sciences are not just the domain of biologists. Chemists like Sherwood Rowland, Mario Molina, and Susan Solomon, physicists like John Holdren and Steve Schneider, and many other physical scientists have already made extremely important contributions. Prominent scientists of all disciplines can call public attention to various aspects of the dilemma, as biologist Garrett Hardin, botanists Peter Raven and Hugh Iltis, plant physiologist Harold Mooney, earth scientists Preston Cloud and Earl Cook, behaviorists Thomas Eisner and E.O. Wilson, entomologist Richard Southwood, ornithologists Thomas Lovejoy and Pete Myers, ecosystem scientists Gene Likens, George Woodwell, and Peter Vitousek, physicists John Fremlin and John Harte, geophysicist Harrison Brown, climatologists Thomas Malone and Walter Orr Roberts, astronomer Carl Sagan, and innumerable others have done.

[6] See P. Ehrlich, 1964, "Some Axioms of Taxonomy," *Systematic Zoology*, vol. 13, pp. 109–143; M. Soulé, 1990, "The Real Work of Systematics," *Annals of the Missouri Botanical Garden*, vol. 77, pp 4–12.

[7] Such changes were suggested to taxonomists three decades ago (Ehrlich, 1964), but the museums of the world have yet to undertake a systematic sampling program, and the time available is becoming very short.

[8] C. Woese, O. Kandler, and M. Wheelis, "Towards a Natural System of Organisms: Proposal for the Domains Archaea, Bacteria, and Eucarya," 1990, *Proceedings of the National Academy of Sciences*, vol. 87, pp. 4576–4579.

[9] R. Ornstein and P. Ehrlich, 1989, *New World/New Mind*, Doubleday, New York.

[10] E. Wilson, 1987, "The Little Things That Run the World (the Importance and Conservation of Invertebrates)," *Conservation Biology*, vol. 1, pp. 344–346.

[11] R. van den Bosch, 1978, *The Pesticide Conspiracy*, Doubleday, New York.

[12] K. Holl, G. Daily, and P. Ehrlich, 1990, "Integrated Pest Management in Latin America," *Environmental Conservation*, vol. 17, no. 4, pp. 341–350.

[13] S. Weiss and D. Murphy, 1990, "Thermal Microenvironments and the Restoration of Rare Butterfly Habitat," in J. Berger (ed.), *Environmental Restoration: Science and Strategies for Restoring the Earth*, Island Press, Washington, D.C., pp. 50–60.

[14] P. Ehrlich, 1986, *The Machinery of Nature*, Simon & Schuster, New York; P. Ehrlich and E. Wilson, 1991, "Biodiversity Studies: Science and Policy," (in press) *Science*.

[15] Berger, 1990.

[16] R. Dixon, 1979, *Land Imprinter, Vegetative Rehabilitation and Equipment Workshop, 33rd Annual Report*, USDA/Forest Service, Casper, Wyoming.

[17] R. Dixon, 1990, "Land Imprinting for Dryland Revegetation and Restoration," and R. Virginia, 1990, "Desert Restoration: The Role of Woody Legumes," pp. 14–22 and 22–30 in Berger, 1990.

[18] For example, see J. Cairns, 1988, *Rehabilitating Damaged Ecosystems*, vols. I and II, CRC Press, Boca Raton, Fla.

[19] World Health Organization, 1988, *From Alma Ata to the Year 2000: Reflections at the Midpoint*, WHO, Geneva.

[20] M. King, 1990, "Health Is a Sustainable State," *The Lancet* (15 September), vol. 336, pp. 664–667.

[21] R. Booth, 1990, "*Homo sapiens*—a Species Too Successful," *Journal of the Royal Society of Medicine*, vol. 83, pp. 757–759.

[22] M. McCally and C. Cassel, 1990, "Medical Responsibility and the Global Environment," *Annals of Internal Medicine* vol. 113, 15 September, pp. 467–473.

[23] American Medical Association Council on Scientific Affairs, 1989, "Stewardship of the Environment," AMA House of Delegates Report: G (I-89), American Medical Association, Chicago.

[24] H. Daly and J. Cobb, 1989, *For the Common Good*, Beacon Press, Boston, p. 37.

25 K. Arrow and A. Fisher, 1974, "Environmental Preservation, Uncertainty, and Irreversibility," *Quarterly Journal of Economics*, vol. 88, pp. 313–319.

26 R. Repetto, 1989, "Balance-Sheet Erosion: How to Account for the Loss of Natural Resources," *International Environmental Affairs*, vol. 1, pp. 103–137.

27 H. Hotelling, 1931, "The Economics of Renewable Resources," *Journal of Political Economy*, vol. 39, pp. 137–175; R. Stavins, 1990, "Alternative Renewable Resource Strategies: A Simulation of Optimal Use," *Journal of Environmental Economics and Management*, vol. 19, pp. 143–159.

28 P. Dasgupta, P. Hammond, and E. Maskin, 1980, "On Imperfect Information and Optimal Pollution Control," *Review of Economic Studies*, vol. 47, no. 4, pp. 857–860.

29 H. Daly, 1977, *Steady-State Economics*, Freeman, San Francisco.

30 Ornstein and Ehrlich, 1989.

31 A new organization of actors, performers, and production workers in televison and recording, dedicated to "creating environmental awareness," was formed in 1990: Earth Communications Office, or ECO. It gives us particular pleasure that one of the most dedicated environmentalists in Hollywood and an ECO board member is Ed Begley, Jr., who became famous for his fine portrayal of a Dr. Ehrlich on the TV show *St. Elsewhere*.

32 For instance, Esprit Corps of San Francisco sponsors regular seminars for its employees on important public issues, including environmental issues.

33 See J. Harris, 1991, "Global Institutions and Ecological Crisis," *World Development*, 19 (in press).

34 See W. McNeill, 1991, "The Peasantry's Awakening All over the World," *Washington Post National Weekly Edition*, 7–13 January.

35 An example of such an attempt was C. Stone, 1974, *Should Trees Have Standing? Towards Legal Rights for Natural Objects*, Kaufmann, Los Altos, California.

36 Some examples in addition to *50 Simple Things* are B. Anderson (ed.), 1990, *Ecologue*, Prentice Hall, New York; D. MacEachern, 1990, *Save Our Planet: 750 Everyday Ways You Can Help Clean Up the Earth*, Dell, New York; and W. Steger and J. Bowermaster, 1990, *Saving the Earth: A Citizen's Guide to Environmental Action*, Knopf, New York.

EPILOGUE

1 R. Broad, J. Cavanaugh, and W. Bello, 1990, "Development: The Market Is Not Enough," *Foreign Policy*, no. 81, pp. 144–162.

ACKNOWLEDGEMENTS

Innumerable colleagues have helped us over the years to understand the dimensions of the human predicament, and we are very conscious of our debt to them. We are deeply appreciative of those who took the time to read the entire manuscript or major portions of it: Gretchen C. Daily, Harold A. Mooney, and Peter Vitousek, Department of Biological Sciences, Stanford University; Lisa M. Daniel and Timothy Daniel, Bureau of Economic Research, Federal Trade Commission; Pamela Matson, NASA Ames, Lawrence Goulder, Department of Economics, Stanford University; John Harte and John P. Holdren, Energy and Resources Group, University of California, Berkeley; Mary Ellen Harte, Rocky Mountain Biological Laboratory; Sally Mallam and Robert Ornstein, Institute for the Study of Human Knowledge; David Pimentel, Department of Entomology, Cornell University; Sherwood Rowland, Department of Chemistry, University of California, Irvine; Lee Schipper, Energy Analysis Program, Lawrence Berkeley Laboratory; and Stephen H. Schneider, National Center for Atmospheric Research. We are especially grateful to Lisa and Tim Daniel and Gretchen Daily, all of whom struggled over the entire manuscript in several versions, and whose words and thoughts were often adopted without attribution. We, of course, must accept the responsibility for any errors that made it into the final version.

Our agent Ginger Barber (Virginia Barber Agency) and editor Jane Isay (Addison-Wesley) also were very helpful in both criticizing the manuscript and in doing very competently and

355

promptly all of those other things that agents and editors do. And once again the staff of the Falconer Biology Library of Stanford's Department of Biological Sciences, under the able leadership of Joe Wible, was an enormous help in running down obscure references, and Steve Masley and Pat Browne handled many copying chores quickly and accurately. Finally, Jessica and Mara gave us much of the incentive to get the rather onerous task of researching and writing done. *Healing the Planet* is really for their generation.

We are deeply indebted to the W. Alton Jones Foundation for support of this project. And finally, we would like once again to express our love and thanks to LuEsther, whose support meant so much to us over the past two decades, and who is sorely missed.

INDEX

357